Bioinorganic Chemistry

Bioinorganic Chemistry

Maria Jose de Leone
Santiago Salinas

Kruger Brentt
Publishers
2025

Kruger Brentt Publishers UK. LTD.
Company Number 9728962

Regd. Office: 68 St Margarets Road, Edgware, Middlesex HA8 9UU

ISBN: 978-1-78715-445-2

For information on all our publications visit our website at http://krugerbrentt.com/

PREFACE

Bioinorganic chemistry is an interdisciplinary field that bridges the disciplines of chemistry and biology, focusing on the role of metals and inorganic elements in biological systems. This branch of science examines how metal ions interact with biomolecules and their essential functions in various biological processes. Metals such as iron, copper, zinc, magnesium, and manganese are integral to the structure and function of numerous biological molecules, including enzymes, proteins, and nucleic acids.

These metal ions often serve as cofactors that are critical for enzymatic activity, participating in electron transfer, catalysis, and stabilization of protein structures. For instance, the presence of iron in hemoglobin is vital for oxygen transport in the blood, while magnesium is a key component of chlorophyll, enabling photosynthesis in plants. The study of bioinorganic chemistry not only reveals the biochemical roles of these metals but also sheds light on their implications in health and disease.

Deficiencies or excesses of certain metal ions can lead to various health disorders. For example, copper deficiency can result in anemia, while excess iron accumulation can cause oxidative stress and organ damage. Understanding these relationships is crucial for developing therapeutic strategies, such as metal-based drugs, which exploit the unique properties of metal ions to target specific biological pathways.

In recent years, bioinorganic chemistry has expanded to explore the use of nanoparticles and other novel materials in biomedical applications. Metal nanoparticles, including those composed of gold, silver, and platinum, have shown promise in drug delivery, imaging, and photothermal therapy. These advancements are driven by the ability of nanoparticles to interact with biological systems at the molecular level, enabling precise targeting of diseased tissues. Furthermore, bioinorganic chemistry plays a pivotal role in environmental science, where the interactions between metal ions and biological systems impact ecosystem health. The study of metal biogeochemistry is essential for understanding the bioavailability and toxicity of metals in various environments, including soil and aquatic systems. Additionally, bioinorganic chemistry encompasses the investigation of metalloenzymes, which are enzymes that contain metal ions in their active sites. These metalloenzymes are involved in a wide array of biological reactions, including DNA replication, repair, and energy metabolism. The structural and mechanistic studies of metalloenzymes provide valuable insights into their catalytic mechanisms and the roles of metal ions in facilitating these processes. Techniques such as X-ray crystallography, NMR spectroscopy, and mass spectrometry are commonly employed in this field to elucidate the structures and functions of metal-containing biomolecules.

As our understanding of bioinorganic chemistry continues to evolve, researchers are increasingly focused on the design of artificial metalloenzymes and biomimetic systems that mimic natural processes. This area of research holds potential for advancing biotechnology and developing novel therapeutic agents.

Overall, bioinorganic chemistry is a vibrant and rapidly evolving field that integrates principles from various scientific disciplines to unravel the complexities of metal ions in biological systems. It not only enhances our understanding of fundamental biological processes but also paves the way for innovative applications in medicine, environmental science, and materials science. Through the exploration of the intricate relationships between metals and biological systems, bioinorganic chemistry contributes to the advancement of knowledge that is crucial for addressing pressing global challenges, including health and environmental sustainability.

Maria Jose de Leone

Santiago Salinas

CONTENTS

INTRODUCTION

Bioinorganic chemistry is an interdisciplinary field that bridges the areas of biology and inorganic chemistry. It focuses on the role of inorganic elements, particularly metal ions, in biological processes. This branch of chemistry is concerned with understanding how metals function in living organisms, either as essential components of enzymes and proteins or as key players in various biochemical reactions. The study of bioinorganic chemistry sheds light on the interaction between metal ions and biological molecules, offering insights into how these elements contribute to life-sustaining processes such as respiration, photosynthesis, and DNA synthesis.

ROLE OF METALS IN BIOLOGICAL SYSTEMS

Metals play a crucial role in biological systems, often as cofactors in enzymes that catalyze biochemical reactions. Many essential biological processes, such as electron transfer, oxygen transport, and catalysis, involve metal ions. For instance, iron is a key component of hemoglobin, the protein responsible for transporting oxygen in the blood. Similarly, magnesium is central to the structure of chlorophyll, enabling plants to capture sunlight for photosynthesis. Zinc, copper, and manganese are also important, serving as cofactors in various enzymes that mediate critical biological reactions, including DNA synthesis, antioxidant defense, and energy metabolism.

METALLOPROTEINS AND ENZYMES

Metalloproteins are proteins that contain metal ions bound to their structure, and they play a fundamental role in bioinorganic chemistry. These metal ions act as active sites within the proteins, enabling them to carry out specific biological functions. An example of a metalloprotein is cytochrome c, a key component of the electron transport chain in cellular respiration. It contains a heme group with an iron ion, which alternates between oxidation states to transfer electrons. Another example is carbonic anhydrase, a zinc-containing enzyme that facilitates the rapid conversion of carbon dioxide to bicarbonate and protons, essential for maintaining acid-base balance in the body.

Enzymes that contain metal ions, often referred to as metalloenzymes, are central to many biological reactions. These enzymes exploit the unique chemical properties of metals, such as their ability to change oxidation states or bind to small molecules. Metal ions in enzymes can stabilize reaction intermediates, participate in electron transfer, or facilitate the binding of substrates. For example, nitrogenase, a metalloenzyme found in certain bacteria, contains molybdenum and iron and catalyzes the conversion of nitrogen gas (N2) into ammonia (NH3), a crucial step in the nitrogen cycle.

METAL ION TRANSPORT AND STORAGE

In living organisms, metal ions are not free-floating; their transport and storage are tightly regulated. This regulation is essential because free metal ions can be toxic if they accumulate in excess or participate in harmful redox reactions. Specific proteins are responsible for the safe transport and storage of metal ions, ensuring that they are delivered to the appropriate locations within cells.

Ferritin is a well-known example of a metal storage protein. It stores iron in a non-toxic, soluble form and releases it when needed for processes such as hemoglobin synthesis. Another key protein involved in metal transport is transferrin, which binds to iron in the bloodstream and transports it to cells where it is required. Similarly, metallothioneins are small proteins that bind to metal ions like zinc and copper, protecting cells from metal toxicity and oxidative stress.

ESSENTIAL AND TRACE METALS IN BIOLOGICAL SYSTEMS

Bioinorganic chemistry classifies metals into two categories: essential metals and trace metals. Essential metals, such as sodium, potassium, calcium, and magnesium, are required in relatively large amounts for basic cellular functions. Sodium and potassium, for example, play a vital role in maintaining the electrical potential of cell membranes, which is critical for nerve transmission and muscle contraction. Calcium is essential for muscle function, blood clotting, and bone structure.

Trace metals, on the other hand, are required in much smaller amounts but are no less important. These include iron, zinc, copper, manganese, and cobalt, among others. Iron is a critical component of hemoglobin and myoglobin, facilitating oxygen transport and storage. Zinc is involved in DNA replication and gene expression, and copper plays a key role in oxidative stress regulation and iron metabolism. The deficiency or excess of trace metals can lead to a variety of health issues, highlighting their importance in maintaining biological homeostasis.

BIOINORGANIC CHEMISTRY IN MEDICINE

Bioinorganic chemistry has significant implications in the field of medicine, particularly in the development of metal-based drugs. Cisplatin, a platinum-based compound, is one of the most well-known examples of a metal-containing drug used in cancer chemotherapy. Cisplatin binds to DNA in cancer cells, interfering with their ability to replicate and ultimately leading to cell death. This drug has been highly effective in treating various types of cancer, including testicular, ovarian, and lung cancers.

Metal complexes are also used in diagnostic imaging. For example, gadolinium is used as a contrast agent in magnetic resonance imaging (MRI) due to its paramagnetic properties, enhancing the visibility of internal organs and tissues. Additionally, technetium-99m, a radioactive isotope, is widely used in nuclear medicine for imaging and diagnosis of diseases, particularly in cardiac and bone scans.

ENVIRONMENTAL AND INDUSTRIAL APPLICATIONS

Beyond biological systems, bioinorganic chemistry has applications in environmental and industrial fields. Metal ions play a role in bioremediation processes, where microorganisms use metals to detoxify and clean up environmental pollutants. For instance, bacteria that contain metalloenzymes can reduce toxic metal ions, such as chromium(VI), to less harmful forms, contributing to the detoxification of contaminated soils and water.

In industry, bioinorganic chemistry is applied in the development of catalysts that mimic natural metalloenzymes. These catalysts are used to promote chemical reactions in an efficient and environmentally friendly manner. For example, biomimetic catalysts that mimic nitrogenase are being developed to reduce nitrogen into ammonia, offering a potential alternative to the energy-intensive Haber-Bosch process used in fertilizer production.

CONCLUSION

Bioinorganic chemistry offers a unique perspective on the role of inorganic elements, particularly metals, in biological processes. From the fundamental roles of metals in enzymes and proteins to their applications in medicine, environmental science, and industry, bioinorganic chemistry highlights the intricate relationship between inorganic elements and living organisms. Understanding this relationship not only provides insight into how life functions at a molecular level but also offers practical solutions for addressing challenges in health, industry, and the environment. As research in bioinorganic chemistry continues to evolve, it promises to yield further discoveries that could enhance our understanding of biology and improve human health and technology.

CHAPTER 1

BIO-ELEMENTS IN THE PERIODIC TABLE

Bio-elements, or essential elements, are chemical elements that are critical to the structure and function of living organisms. While living organisms are composed primarily of a few elements like carbon, hydrogen, oxygen, and nitrogen, there are numerous other elements that play crucial roles in biological processes.

These elements, often found in the periodic table, participate in biochemical reactions, maintain structural integrity, and enable vital functions like respiration, growth, and reproduction.

1.1 CLASSIFICATION OF BIO-ELEMENTS

Bio-elements can be classified based on their abundance and function in living organisms. These classifications often fall into three categories:

- Major bio-elements (bulk elements) include carbon (C), hydrogen (H), oxygen (O), and nitrogen (N), which make up the majority of an organism's mass. These elements are the building blocks of biomolecules like proteins, carbohydrates, lipids, and nucleic acids.
- Macronutrients such as calcium (Ca), phosphorus (P), potassium (K), sulfur (S), sodium (Na), chlorine (Cl), and magnesium (Mg) are required in significant amounts for essential processes, including muscle contraction, bone formation, and nerve transmission.
- Micronutrients or trace elements, like iron (Fe), zinc (Zn), copper (Cu), manganese (Mn), iodine (I), selenium (Se), and cobalt (Co), are needed in much smaller amounts but are nonetheless critical for enzyme functions, redox reactions, and other biological processes.

1.2 THE ROLE OF MAJOR BIO-ELEMENTS

The major bio-elements—carbon, hydrogen, oxygen, and nitrogen—form the foundation of all living matter. Together, they make up about 96% of the human body's mass and are involved in virtually every biochemical process.

- Carbon (C) is the central element in organic chemistry, serving as the backbone for all biological molecules. Carbon's ability to form stable bonds with other elements, including itself, enables the vast diversity of life forms and biochemical functions.

- Hydrogen (H) is a key component of water and organic molecules. It also participates in energy transfer processes, particularly in cellular respiration and photosynthesis, where it plays a role in proton gradients and ATP synthesis.
- Oxygen (O) is essential for respiration, as it acts as the final electron acceptor in the electron transport chain, a process that generates ATP. Oxygen is also a major component of water and organic compounds.
- Nitrogen (N) is found in amino acids, the building blocks of proteins, and nucleotides, which make up DNA and RNA. It is essential for cell growth, reproduction, and repair.

1.3 MACRONUTRIENTS AND THEIR FUNCTIONS

Macronutrients are elements that organisms require in relatively large amounts. These include calcium, phosphorus, potassium, sulfur, sodium, chlorine, and magnesium, each playing specific roles in maintaining the structural and functional integrity of biological systems.

- Calcium (Ca) is a crucial element for skeletal structure, as it is the primary component of bones and teeth. Additionally, calcium ions are involved in muscle contraction, blood clotting, and signal transduction within cells.
- Phosphorus (P) is a component of nucleic acids (DNA and RNA) and ATP, the energy currency of the cell. It is also involved in the formation of bones and teeth, much like calcium.
- Potassium (K) and sodium (Na) are vital for maintaining the electrical potential across cell membranes, essential for nerve impulse transmission and muscle contraction. Potassium, in particular, is important for maintaining osmotic balance and regulating heart function.
- Magnesium (Mg) plays a role as a cofactor in numerous enzymatic reactions, including those involved in DNA replication and protein synthesis. It is also essential for stabilizing the structure of ATP.
- Sulfur (S) is found in certain amino acids (cysteine and methionine) and vitamins, contributing to the structure of proteins and the formation of important coenzymes like CoA (Coenzyme A).

1.4 TRACE ELEMENTS AND THEIR BIOLOGICAL ROLES

Trace elements, though required in minute quantities, are indispensable for numerous biological functions. These elements often serve as cofactors for enzymes, facilitating various biochemical reactions.

- Iron (Fe) is essential for oxygen transport and storage, as it is a component of hemoglobin and myoglobin. It is also involved in electron transfer processes within the mitochondria during cellular respiration.
- Zinc (Zn) plays a vital role in immune function, DNA synthesis, wound healing, and protein metabolism. It is a structural component of zinc-finger proteins that regulate gene expression.
- Copper (Cu) is involved in the electron transport chain and the synthesis of hemoglobin. It also plays a role in antioxidant defense by neutralizing free radicals.
- Iodine (I) is a key component of thyroid hormones, which regulate metabolism, growth, and development.
- Selenium (Se) functions as an antioxidant, protecting cells from oxidative damage by being a component of enzymes like glutathione peroxidase.
- Cobalt (Co) is an integral part of vitamin B12, essential for red blood cell production and nervous system function.

1.5 THE PERIODIC TABLE AND BIO-ELEMENTS

The periodic table serves as an organizational tool for understanding the distribution and properties of bio-elements. Many of the essential bio-elements come from groups 1, 2, 13, 14, 15, and transition metals. Their position in the periodic table often correlates with their chemical behavior and role in biological systems.

For instance, elements in group 1, like sodium and potassium, are highly reactive metals that play essential roles in cellular ionic balance. Group 2 elements, such as calcium and magnesium, are alkaline earth metals involved in structural and signaling functions. Transition metals like iron, copper, and zinc, found in the d-block, are versatile in redox chemistry and enzyme catalysis.

1.6 DEFICIENCIES AND TOXICITY OF BIO-ELEMENTS

While bio-elements are essential for life, their levels must be tightly regulated. Deficiencies in bio-elements can lead to severe health problems. For instance, iron deficiency leads to anemia, zinc deficiency impairs immune function, and iodine deficiency results in goiter and developmental issues.

Conversely, excess amounts of certain bio-elements can be toxic. For example, excess iron can cause oxidative stress and tissue damage, while elevated levels of copper and selenium can lead to cellular toxicity. This delicate balance underscores the importance of maintaining appropriate levels of bio-elements for optimal health.

1.7 APPLICATIONS OF BIO-ELEMENTS IN MEDICINE

Bio-elements have wide applications in medicine. For example, iron supplements are used to treat anemia, magnesium is given to patients with heart problems, and iodine is supplemented in areas where its deficiency is prevalent. Metals like platinum are used in cancer treatment, particularly in drugs like cisplatin, which interferes with DNA replication in cancer cells. Additionally, diagnostic tools such as MRI use gadolinium, a lanthanide, as a contrast agent to enhance imaging. Radioactive isotopes like technetium-99m are used in nuclear medicine for diagnostic purposes, demonstrating the diverse applications of bio-elements in modern healthcare.

1.8 CONCLUSION

Bio-elements are essential components of life, forming the structural framework of biomolecules, facilitating enzymatic reactions, and maintaining the delicate balance of biological processes. Understanding the role of these elements, their classification, and their position on the periodic table provides valuable insights into their importance in sustaining life. The periodic table not only serves as a map of chemical elements but also as a guide for understanding the intricate relationship between chemistry and biology. By studying bio-elements, we can better appreciate the complexity of life and develop solutions to health and environmental challenges.

CHAPTER 2

PRE-LIFE AND EARLY LIFE FORMS; EXTREMOPHILES

The origin of life is one of the most profound mysteries in science, encompassing the transition from non-living chemical systems to complex, self-replicating life forms.

Life on Earth is believed to have arisen approximately 3.5 to 4 billion years ago, during a period when the planet's environment was vastly different from today. Before life emerged, a prebiotic or "pre-life" stage existed, where complex organic molecules formed under natural conditions, setting the stage for life. Early life forms, once they emerged, were simple microorganisms that thrived in the harsh, primordial conditions of early Earth.

These organisms, which we now refer to as extremophiles, thrived in environments that were too extreme for most modern life, such as boiling hot springs, acidic lakes, and deep-sea hydrothermal vents.

2.1 THE PRE-LIFE STAGE: ABIOGENESIS

The pre-life stage, or abiogenesis, refers to the process by which life originated from non-living matter. Abiogenesis involves the formation of simple organic compounds, which eventually led to the more complex molecules necessary for life, such as nucleotides, amino acids, and lipids. One widely accepted hypothesis for the origin of these molecules is the "primordial soup" theory. In this theory, early Earth had a reducing atmosphere composed of gases like methane (CH_4), ammonia (NH_3), hydrogen (H_2), and water vapor (H_2O).

Lightning, ultraviolet radiation, and volcanic activity provided the energy needed for chemical reactions, producing simple organic molecules in the Earth's oceans.

Experiments by Stanley Miller and Harold Urey in the 1950s demonstrated that organic molecules, such as amino acids, could be synthesized under conditions mimicking the early Earth's atmosphere.

These findings provided the first experimental support for the idea that life's building blocks could have arisen naturally on the prebiotic Earth. Over time, these molecules likely accumulated in the Earth's oceans, forming a "primordial soup" from which more complex molecules could form.

2.2 THE RNA WORLD HYPOTHESIS

One of the key steps in the transition from pre-life to life is the development of molecules capable of storing information and catalyzing chemical reactions. The "RNA world" hypothesis suggests that RNA, a molecule that can both store genetic information and catalyze chemical reactions, was one of the first molecules to take on these dual roles. In this hypothesis, RNA played a central role in early life forms before the evolution of DNA and proteins.

RNA is believed to have been a critical component of early life because it can fold into complex three-dimensional shapes that enable it to catalyze reactions, much like enzymes do in modern cells. Furthermore, RNA can serve as a template for its own replication, suggesting that self-replicating RNA molecules could have been the precursors to the first simple organisms.

2.3 FORMATION OF PROTOCELLS

For life to emerge, these self-replicating molecules needed to be enclosed within a membrane, forming what are known as protocells. Protocells are simple vesicles, composed of lipid bilayers, that could have formed spontaneously in the prebiotic environment. These vesicles would have provided a controlled environment for the concentration and replication of genetic material, protecting the RNA or other early biomolecules from the harsh external environment.

The formation of protocells is a key step in the transition from non-living chemistry to living cells. These early protocells would have resembled simple, primitive versions of modern cells, lacking complex organelles but capable of growing, dividing, and replicating genetic material. Over time, as the genetic material inside these protocells became more complex, the first true cells—prokaryotes—likely emerged, marking the beginning of biological evolution.

2.4 EARLY LIFE FORMS: PROKARYOTES

The earliest known life forms were simple prokaryotic organisms, specifically bacteria and archaea. These single-celled organisms lacked a nucleus and complex internal structures, but they possessed the essential components for life, including a membrane, genetic material, and the machinery needed for metabolism and replication. Fossil evidence suggests that these organisms appeared at least 3.5 billion years ago, thriving in an environment that was much more hostile than it is today.

Prokaryotes dominated the Earth for billions of years, and their metabolic diversity allowed them to exploit a wide range of environments. Some of the earliest prokaryotes were likely chemolithotrophs, organisms that obtain energy by oxidizing inorganic compounds, such as hydrogen sulfide (H_2S) or methane (CH_4).

Others, such as cyanobacteria, developed the ability to perform photosynthesis, producing oxygen as a byproduct and eventually transforming Earth's atmosphere in what is known as the Great Oxygenation Event.

2.5 EXTREMOPHILES: LIFE IN EXTREME CONDITIONS

One of the most fascinating aspects of early life is the ability of organisms to survive and thrive in extreme conditions. Extremophiles are organisms that live in environments that would be deadly to most other life forms. These environments include extreme temperatures, pH levels, salinity, pressure, and radiation. Extremophiles provide valuable insights into the resilience and adaptability of life and may offer clues about the potential for life on other planets.

2.5.1 Thermophiles and Hyperthermophiles

Thermophiles are organisms that thrive at high temperatures, often above 45°C, while hyperthermophiles prefer even hotter environments, with some species living in conditions exceeding 100°C. These organisms are commonly found in hot springs, geysers, and deep-sea hydrothermal vents. The enzymes and proteins of thermophiles and hyperthermophiles are adapted to withstand extreme heat, making them stable and functional in conditions that would denature the proteins of most other organisms.

2.5.2 Halophiles

Halophiles are extremophiles that thrive in environments with high salt concentrations, such as salt lakes, salt mines, and hypersaline seas like the Dead Sea. These organisms have evolved mechanisms to prevent dehydration in salty conditions, such as accumulating compatible solutes inside their cells to balance the osmotic pressure.

2.5.3 Acidophiles and Alkaliphiles

Acidophiles are organisms that live in highly acidic environments, with pH levels as low as 1 or even lower. These organisms can be found in places like acid mine drainage and volcanic craters. Conversely, alkaliphiles thrive in highly alkaline environments, where pH levels can reach 9 or above. These organisms are often found in soda lakes or alkaline soils.

2.5.4 Barophiles (Piezophiles)

Barophiles, also known as piezophiles, are organisms that thrive under high-pressure conditions.

These organisms are typically found in the deep ocean, where the pressure can reach thousands of atmospheres. The cell membranes and proteins of barophiles are specially adapted to function under intense pressure.

2.6 EXTREMOPHILES AND THE SEARCH FOR EXTRATERRESTRIAL LIFE

The discovery of extremophiles on Earth has expanded our understanding of the conditions under which life can exist. The ability of these organisms to thrive in environments once thought uninhabitable suggests that life may be more adaptable and resilient than previously believed. This has important implications for the search for life beyond Earth. Extremophiles serve as models for the types of life that might exist on other planets or moons, such as Mars, Europa, or Enceladus, where conditions are extreme but may still support life.

2.7 CONCLUSION: PRE-LIFE, EARLY LIFE, AND EXTREMOPHILES

The study of pre-life, early life forms, and extremophiles provides a glimpse into the remarkable adaptability of life. From the primordial soup to the development of protocells and the rise of prokaryotic organisms, the transition from chemistry to biology was a gradual and complex process. Early life forms adapted to extreme environments, laying the foundation for the diversity of life we see today. Extremophiles, in particular, demonstrate the potential for life in the most inhospitable environments, both on Earth and potentially elsewhere in the universe. The exploration of these organisms not only deepens our understanding of life's origins but also broadens the scope of where life might exist.

CHAPTER 3

THE ALKALINE AND ALKALINE EARTH METALS

3.1 OVERVIEW

The alkaline and alkaline earth metals are two important groups of elements located in the s-block of the periodic table. These metals are characterized by their high reactivity and form a significant portion of the Earth's crust. The alkaline metals, located in Group 1 of the periodic table, include lithium (Li), sodium (Na), potassium (K), rubidium (Rb), cesium (Cs), and francium (Fr). Alkaline earth metals, on the other hand, belong to Group 2 and include beryllium (Be), magnesium (Mg), calcium (Ca), strontium (Sr), barium (Ba), and radium (Ra). These metals share similar chemical properties due to their similar electron configurations, and they play critical roles in various industrial, biological, and environmental processes.

The alkaline and alkaline earth metals are essential elements that have far-reaching applications in industry, biology, and environmental processes. While they share certain characteristics due to their positions in the periodic table, their distinct properties influence their behavior in chemical reactions, their roles in biological systems, and their industrial uses. Understanding the chemistry and uses of these metals not only provides insight into the building blocks of matter but also highlights their significance in modern life.

3.1.1 Properties of Alkaline Metals

General Characteristics

Alkaline metals are known for their softness, silvery appearance, and high reactivity. These elements have one electron in their outermost s-orbital, which they readily lose to form cations with a +1 charge. This makes them highly reactive, particularly with water, where they form hydroxides and release hydrogen gas. The reactivity of alkaline metals increases down the group, with francium being the most reactive, though it is rarely studied due to its extreme radioactivity.

Physical Properties

Alkaline metals have low densities compared to other metals. For example, lithium, sodium, and potassium are light enough to float on water. Their melting and boiling points decrease as you move down the group. This trend is attributed to the weakening metallic bonds, as the size of the metal atoms increases down the group, reducing the attraction between the metal ions and the delocalized electrons.

Chemical Reactivity

The reactivity of alkaline metals is a key characteristic. They react vigorously with water to form alkaline hydroxides and hydrogen gas. The general reaction can be written as:

$$2M + 2H_2O \rightarrow 2MOH + H_2$$

where M represents an alkaline metal. The reactivity increases down the group, with lithium being the least reactive and cesium being highly reactive, often resulting in explosive reactions with water.

Alkaline metals also react with halogens to form ionic halides, following the reaction:

$$2M + X_2 \rightarrow 2MX$$

where X represents a halogen. The resulting compounds, such as sodium chloride (NaCl), are highly stable and form crystalline structures due to the strong electrostatic attraction between the metal cations and halide anions.

3.1.2 Uses of Alkaline Metals

Alkaline metals have numerous applications in everyday life and industry. Sodium and potassium are essential in biological processes, particularly in maintaining the sodium-potassium balance in cells, which is crucial for nerve impulse transmission and muscle contraction.

Sodium chloride (table salt) is a staple in food seasoning and preservation. Potassium compounds are used in fertilizers to support plant growth. Lithium is widely used in rechargeable batteries, particularly in the production of lithium-ion batteries, which power everything from smartphones to electric vehicles.

3.1.3 Properties of Alkaline Earth Metals

General Characteristics

Alkaline earth metals, located in Group 2 of the periodic table, have two electrons in their outermost s-orbital, which they lose to form cations with a +2 charge.

These metals are harder and denser than their alkaline counterparts. While they are also highly reactive, their reactivity is less than that of alkaline metals due to the higher ionization energy required to remove two electrons. Like alkaline metals, the reactivity of alkaline earth metals increases down the group, with radium being the most reactive.

Physical Properties

Alkaline earth metals have higher melting and boiling points compared to alkaline metals. They are also denser and harder. For example, beryllium and magnesium are lightweight metals used in various industrial applications. Their atomic radii increase as you move down the group, and like alkaline metals, they exhibit a decrease in ionization energy down the group.

Chemical Reactivity

Alkaline earth metals are less reactive than alkaline metals but still react with water, albeit less vigorously. For example, magnesium reacts slowly with cold water but reacts more rapidly when heated. Calcium, strontium, and barium react with water more readily than magnesium, forming hydroxides and releasing hydrogen gas:

$$M + 2H_2O \rightarrow M(OH)_2 + H_2$$

The hydroxides of alkaline earth metals are less soluble in water than those of alkaline metals, but their solubility increases down the group. For instance, barium hydroxide is highly soluble in water, forming a strongly alkaline solution.

Alkaline earth metals also react with oxygen to form oxides, such as magnesium oxide (MgO) and calcium oxide (CaO), which are commonly used in construction and industrial processes. In addition, they react with halogens to form halides, like calcium chloride ($CaCl_2$) and magnesium chloride ($MgCl_2$).

3.1.4 Uses of Alkaline Earth Metals

Alkaline earth metals are widely used in industry, medicine, and agriculture. Calcium is essential for human health, particularly in maintaining strong bones and teeth. Calcium compounds, such as calcium carbonate ($CaCO_3$), are used in construction as a component of cement and concrete. Magnesium is used in lightweight alloys for the automotive and aerospace industries due to its strength and low density. It is also an essential element in photosynthesis, being a central component of the chlorophyll molecule.

Beryllium is used in aerospace materials due to its high melting point and stiffness, while strontium and barium are used in fireworks and signal flares to produce bright colors. Radium, although highly radioactive and dangerous, was once used in luminescent paints, though this practice has largely been discontinued due to health risks.

3.1.5 Comparison Between Alkaline and Alkaline Earth Metals

While both alkaline and alkaline earth metals share some similarities, such as their metallic nature and reactivity, they also have notable differences. Alkaline metals are more reactive than alkaline earth metals, as they require less energy to lose their single valence electron. Alkaline earth metals, having two valence electrons, are less reactive and generally harder and denser. Alkaline metals form highly soluble hydroxides, while the hydroxides of alkaline earth metals are less soluble in water. The chemical reactivity of both groups increases as you move down the periodic table.

3.1.6 Environmental and Biological Significance

Both alkaline and alkaline earth metals play critical roles in the environment and biological systems. Sodium and potassium are vital for maintaining cellular homeostasis in organisms. Calcium is crucial for bone structure, muscle contraction, and cellular signaling in humans and animals. Magnesium, a key component of chlorophyll, is essential for plant photosynthesis. The environmental availability and balance of these metals affect ecosystems, agriculture, and human health.

Excessive exposure to certain alkaline and alkaline earth metals, such as barium or radium, can pose health risks, including toxicity and radioactivity concerns. As a result, monitoring and managing the use and disposal of these metals in industrial processes is crucial to environmental safety.

3.2 ION CHANNELS

Ion channels are pore-forming membrane proteins that allow ions to pass through the channel pore. Their functions include establishing a resting membrane potential, shaping action potentials and other electrical signals by gating the flow of ions across the cell membrane, controlling the flow of ions across secretory and epithelial cells, and regulating cell volume.

Ion channels are present in the membranes of all cells. Ion channels are one of the two classes of ionophoric proteins, the other being ion transporters.

The study of ion channels often involves biophysics, electrophysiology, and pharmacology, while using techniques including voltage clamp, patch clamp, immunohistochemistry, X-ray crystallography, fluoroscopy, and RT-PCR. Their classification as molecules is referred to as channelomics.

3.2.1 Basic Features

There are two distinctive features of ion channels that differentiate them from other types of ion transporter proteins:

- The rate of ion transport through the channel is very high (often 10^6 ions per second or greater).
- Ions pass through channels down their electrochemical gradient, which is a function of ion concentration and membrane potential, "downhill", without the input (or help) of metabolic energy (e.g. ATP, co-transport mechanisms, or active transport mechanisms).

Ion channels are located within the membrane of all excitable cells, and of many intracellular organelles. They are often described as narrow, water-filled tunnels that allow only ions of a certain size and/or charge to pass through. This characteristic is called selective permeability. The archetypal channel pore is just one or two atoms wide at its narrowest point and is selective for specific species of ion, such as sodium or potassium. However, some channels may be permeable to the passage of more than one type of ion, typically sharing a common charge: positive (cations) or negative (anions). Ions often move through the segments of the channel pore in a single file nearly as quickly as the ions move through the free solution. In many ion channels, passage through the pore is governed by a "gate", which may be opened or closed in response to chemical or electrical signals, temperature, or mechanical force.

Ion channels are integral membrane proteins, typically formed as assemblies of several individual proteins. Such "multi-subunit" assemblies usually involve a circular arrangement of identical or homologous proteins closely packed around a water-filled pore through the plane of the membrane or lipid bilayer. For most voltage-gated ion channels, the pore-forming subunit(s) are called the α subunit, while the auxiliary subunits are denoted β, γ, and so on.

3.2.2 Biological Role

Because channels underlie the nerve impulse and because "transmitter-activated" channels mediate conduction across the synapses, channels are especially prominent components of the nervous system. Indeed, numerous toxins that organisms have evolved for shutting down the nervous systems of predators and prey (e.g., the venoms produced by spiders, scorpions, snakes, fish, bees, sea snails, and others) work by modulating ion channel conductance and/or kinetics.

In addition, ion channels are key components in a wide variety of biological processes that involve rapid changes in cells, such as cardiac, skeletal, and smooth muscle contraction, epithelial transport of nutrients and ions, T-cell activation, and pancreatic beta-cell insulin release. In the search for new drugs, ion channels are a frequent target.

3.2.3 Diversity

There are over 300 types of ion channels just in the cells of the inner ear. Ion channels may be classified by the nature of their gating, the species of ions passing through those gates, the number of gates (pores), and localization of proteins.

Further heterogeneity of ion channels arises when channels with different constitutive subunits give rise to a specific kind of current. Absence or mutation of one or more of the contributing types of channel subunits can result in loss of function and, potentially, underlie neurologic diseases.

Classification by Gating

Ion channels may be classified by gating, i.e. what opens and closes the channels. For example, voltage-gated ion channels open or close depending on the voltage gradient across the plasma membrane, while ligand-gated ion channels open or close depending on binding of ligands to the channel.

Voltage-Gated

Voltage-gated ion channels open and close in response to membrane potential.

- **Voltage-Gated Sodium Channels:** This family contains at least 9 members and is largely responsible for action potential creation and propagation. The pore-forming α subunits are very large (up to 4,000 amino acids) and consist of four homologous repeat domains (I-IV) each comprising six transmembrane segments (S1-S6) for a total of 24 transmembrane segments. The members of this family also coassemble with auxiliary β subunits, each spanning the membrane once. Both α and β subunits are extensively glycosylated.
- **Voltage-Gated Calcium Channels:** This family contains 10 members, though these are known to coassemble with $\alpha_2\delta$, β, and γ subunits. These channels play an important role in both linking muscle excitation with contraction as well as neuronal excitation with transmitter release. The α subunits have an overall structural resemblance to those of the sodium channels and are equally large.
  - *Cation Channels of Sperm:* This small family of channels, normally referred to as Catsper channels, is related to the two-pore channels and distantly related to TRP channels.
- **Voltage-Gated Potassium Channels (K_V):** This family contains almost 40 members, which are further divided into 12 subfamilies. These channels are known mainly for their role in repolarizing the cell membrane following action potentials. The α subunits have six transmembrane segments, homologous to a single domain of the sodium channels. Correspondingly, they assemble as tetramers to produce a functioning channel.
- **Some Transient Receptor Potential Channels:** This group of channels, normally referred to simply as TRP channels, is named after their role in Drosophila phototransduction. This family, containing at least 28 members, is incredibly diverse in its method of activation. Some TRP channels seem to be constitutively open, while others are gated by voltage, intracellular Ca^{2+}, pH, redox state, osmolarity, and mechanical stretch. These channels also vary according to the ion(s) they pass,

some being selective for Ca^{2+} while others are less selective, acting as cation channels. This family is subdivided into 6 subfamilies based on homology: classical (TRPC), vanilloid receptors (TRPV), melastatin (TRPM), polycystins (TRPP), mucolipins (TRPML), and ankyrin transmembrane protein 1 (TRPA).

- **Hyperpolarization-Activated Cyclic Nucleotide-Gated Channels:** The opening of these channels is due to hyperpolarization rather than the depolarization required for other cyclic nucleotide-gated channels. These channels are also sensitive to the cyclic nucleotides cAMP and cGMP, which alter the voltage sensitivity of the channel's opening. These channels are permeable to the monovalent cations K^+ and Na^+. There are 4 members of this family, all of which form tetramers of six-transmembrane α subunits. As these channels open under hyperpolarizing conditions, they function as pacemaking channels in the heart, particularly the SA node.
- **Voltage-Gated Proton Channels:** Voltage-gated proton channels open with depolarization, but in a strongly pH-sensitive manner. The result is that these channels open only when the electrochemical gradient is outward, such that their opening will only allow protons to leave cells. Their function thus appears to be acid extrusion from cells. Another important function occurs in phagocytes (e.g. eosinophils, neutrophils, macrophages) during the "respiratory burst." When bacteria or other microbes are engulfed by phagocytes, the enzyme NADPH oxidase assembles in the membrane and begins to produce reactive oxygen species (ROS) that help kill bacteria. NADPH oxidase is electrogenic, moving electrons across the membrane, and proton channels open to allow proton flux to balance the electron movement electrically.

Ligand-Gated (Neurotransmitter)

Also known as ionotropic receptors, this group of channels open in response to specific ligand molecules binding to the extracellular domain of the receptor protein. Ligand binding causes a conformational change in the structure of the channel protein that ultimately leads to the opening of the channel gate and subsequent ion flux across the plasma membrane. Examples of such channels include the cation-permeable nicotinic acetylcholine receptors, ionotropic glutamate-gated receptors, acid-sensing ion channels (ASICs), ATP-gated P2X receptors, and the anion-permeable γ-aminobutyric acid-gated $GABA_A$ receptor.

Ion channels activated by second messengers may also be categorized in this group, although ligands and second messengers are otherwise distinguished from each other.

Lipid-Gated

This group of channels opens in response to specific lipid molecules binding to the channel's transmembrane domain typically near the inner leaflet of the plasma membrane. Phosphatidylinositol 4,5-bisphosphate (PIP_2) and phosphatidic acid (PA) are the best-characterized lipids to gate these channels. Many of the leak potassium channels are gated by lipids including the inward-rectifier potassium channels and two pore domain potassium channels TREK-1 and TRAAK. KCNQ potassium channel family are gated by PIP_2. The voltage activated potassium channel (Kv) is regulated by PA. Its midpoint of activation shifts +50 mV upon PA hydrolysis, near resting membrane potentials. This suggests Kv could be opened by lipid hydrolysis independent of voltage and may qualify this channel as dual lipid and voltage gated channel.

Other Gating

Gating also includes activation and inactivation by second messengers from the inside of the cell membrane - rather than from outside the cell, as in the case for ligands.

- **Some Potassium Channels:**
  - **Inward-Rectifier Potassium Channels:** These channels allow potassium ions to flow into the cell in an "inwardly rectifying" manner: potassium flows more efficiently into than out of the cell. This family is composed of 15 official and 1 unofficial member and is further subdivided into 7 subfamilies based on homology. These channels are affected by intracellular ATP, PIP_2, and G-protein βγ subunits. They are involved in important physiological processes such as pacemaker activity in the heart, insulin release, and potassium uptake in glial cells. They contain only two transmembrane segments, corresponding to the core pore-forming segments of the K_V and K_{Ca} channels. Their α subunits form tetramers.
  - **Calcium-Activated Potassium Channels:** This family of channels is activated by intracellular Ca^{2+} and contains 8 members.
  - **Tandem Pore Domain Potassium Channel:** This family of 15 members form what are known as leak channels, and they display Goldman-Hodgkin-Katz (open) rectification. Contrary to their common name of 'Two-pore-domain potassium channels', these channels have only one pore but two pore domains per subunit.
  - Two-pore channels include ligand-gated and voltage-gated cation channels, so-named because they contain two pore-forming subunits. As their name suggests, they have two pores.
  - Light-gated channels like channelrhodopsin are directly opened by photons.
  - Mechanosensitive ion channels open under the influence of stretch, pressure, shear, and displacement.

- Cyclic nucleotide-gated channels: This superfamily of channels contains two families: the cyclic nucleotide-gated (CNG) channels and the hyperpolarization-activated, cyclic nucleotide-gated (HCN) channels. This grouping is functional rather than evolutionary.
- **Cyclic Nucleotide-Gated Channels:** This family of channels is characterized by activation by either intracellular cAMP or cGMP. These channels are primarily permeable to monovalent cations such as K^+ and Na^+. They are also permeable to Ca^{2+}, though it acts to close them. There are 6 members of this family, which is divided into 2 subfamilies.
- Hyperpolarization-activated cyclic nucleotide-gated channels
- Temperature-Gated Channels: Members of the transient receptor potential ion channel superfamily, such as TRPV1 or TRPM8, are opened either by hot or cold temperatures.

Classification by Type of Ions

- **Chloride Channels:** This superfamily of channels consists of approximately 13 members. They include ClCs, CLICs, Bestrophins and CFTRs. These channels are non-selective for small anions; however chloride is the most abundant anion, and hence they are known as chloride channels.
- **Potassium Channels:**
  - Voltage-gated potassium channels e.g., Kvs, Kirs etc.
  - Calcium-activated potassium channels e.g., BKCa or MaxiK, SK, etc.
  - Inward-rectifier potassium channels
  - Two-pore-domain potassium channels: This family of 15 members form what is known as leak channels, and they display Goldman-Hodgkin-Katz (open) rectification.
- **Sodium Channels:**
  - Voltage-gated sodium channels (NaVs)
  - Epithelial sodium channels (ENaCs)
- **Calcium channels (CaVs)**
- **Phosphate Chanels:** To date, only one phosphate channel, Xenotropic and polytropic retrovirus receptor 1 (XPR1), has been identified in animals. It is a pyrophosphate-gated channel.
- **Proton Channels:**
  - Voltage-gated proton channels
- **Non-Selective Cation Channels:** These non-selectively allow many types of cations, mainly Na^+, K^+ and Ca^{2+}, through the channel.
  - Most transient receptor potential channels

Classification by Cellular Localization

Ion channels are also classified according to their subcellular localization. The plasma membrane accounts for around 2% of the total membrane in the cell, whereas intracellular organelles contain 98% of the cell's membrane. The major intracellular compartments are endoplasmic reticulum, Golgi apparatus, and mitochondria. On the basis of localization, ion channels are classified as:

- **Plasma Membrane Channels:**
  - *Examples:* Voltage-gated potassium channels (Kv), Sodium channels (Nav), Calcium channels (Cav) and Chloride channels (ClC)
- Intracellular channels, which are further classified into different organelles
  - Endoplasmic reticulum channels: RyR, SERCA, ORAi
  - Mitochondrial channels: mPTP, KATP, BK, IK, CLIC5, Kv7.4 at the inner membrane and VDAC and CLIC4 as outer membrane channels.

Other Classifications

Some ion channels are classified by the duration of their response to stimuli:

- **Transient Receptor Potential Channels:** This group of channels, normally referred to simply as TRP channels, is named after their role in *Drosophila* visual phototransduction. This family, containing at least 28 members, is diverse in its mechanisms of activation. Some TRP channels remain constitutively open, while others are gated by voltage, intracellular Ca^{2+}, pH, redox state, osmolarity, and mechanical stretch. These channels also vary according to the ion(s) they pass, some being selective for Ca^{2+} while others are less selective cation channels. This family is subdivided into 6 subfamilies based on homology: canonical TRP (TRPC), vanilloid receptors (TRPV), melastatin (TRPM), polycystins (TRPP), mucolipins (TRPML), and ankyrin transmembrane protein 1 (TRPA).

3.2.4 Detailed Structure

Channels differ with respect to the ion they let pass (for example, Na^+, K^+, Cl^-), the ways in which they may be regulated, the number of subunits of which they are composed and other aspects of structure.

Channels belonging to the largest class, which includes the voltage-gated channels that underlie the nerve impulse, consist of four or sometimes five subunits with six transmembrane helices each. On activation, these helices move about and open the pore. Two of these six helices are separated by a loop that lines the pore and is the primary determinant of ion selectivity and conductance in this channel class and some others.

The existence and mechanism for ion selectivity was first postulated in the late 1960s by Bertil Hille and Clay Armstrong. The idea of the ionic selectivity for potassium channels was that the carbonyl oxygens of the protein backbones of the "selectivity filter" (named by Bertil Hille) could efficiently replace the water molecules that normally shield potassium ions, but that sodium ions were smaller and cannot be completely dehydrated to allow such shielding, and therefore could not pass through. This mechanism was finally confirmed when the first structure of an ion channel was elucidated. A bacterial potassium channel KcsA, consisting of just the selectivity filter, "P" loop, and two transmembrane helices was used as a model to study the permeability and the selectivity of ion channels in the Mackinnon lab. The determination of the molecular structure of KcsA by Roderick MacKinnon using X-ray crystallography won a share of the 2003 Nobel Prize in Chemistry.

Because of their small size and the difficulty of crystallizing integral membrane proteins for X-ray analysis, it is only very recently that scientists have been able to directly examine what channels "look like." Particularly in cases where the crystallography required removing channels from their membranes with detergent, many researchers regard images that have been obtained as tentative. An example is the long-awaited crystal structure of a voltage-gated potassium channel, which was reported in May 2003. One inevitable ambiguity about these structures relates to the strong evidence that channels change conformation as they operate (they open and close, for example), such that the structure in the crystal could represent any one of these operational states. Most of what researchers have deduced about channel operation so far they have established through electrophysiology, biochemistry, gene sequence comparison and mutagenesis.

Channels can have single (CLICs) to multiple transmembrane (K channels, P2X receptors, Na channels) domains which span plasma membrane to form pores. Pore can determine the selectivity of the channel. Gate can be formed either inside or outside the pore region.

3.2.5 Pharmacology

Chemical substances can modulate the activity of ion channels, for example by blocking or activating them.

Ion Channel Blockers

A variety of ion channel blockers (inorganic and organic molecules) can modulate ion channel activity and conductance. Some commonly used blockers include:

- Tetrodotoxin (TTX), used by puffer fish and some types of newts for defense. It blocks sodium channels.

- Saxitoxin is produced by a dinoflagellate also known as "red tide". It blocks voltage-dependent sodium channels.
- Conotoxin is used by cone snails to hunt prey.
- Lidocaine and novocaine belong to a class of local anesthetics which block sodium ion channels.
- Dendrotoxin is produced by mamba snakes, and blocks potassium channels.
- Iberiotoxin is produced by the *Hottentotta tamulus* (Eastern Indian scorpion) and blocks potassium channels.
- Heteropodatoxin is produced by *Heteropoda venatoria* (brown huntsman spider or laya) and blocks potassium channels.

Ion Channel Activators

Several compounds are known to promote the opening or activation of specific ion channels. These are classified by the channel on which they act:

- Calcium channel openers, such as Bay K8644
- Chloride channel openers, such as phenanthroline
- Potassium channel openers, such as minoxidil
- Sodium channel openers, such as DDT

3.2.6 Diseases

There are a number of disorders which disrupt normal functioning of ion channels and have disastrous consequences for the organism. Genetic and autoimmune disorders of ion channels and their modifiers are known as channelopathies.

- Shaker gene mutations cause a defect in the voltage gated ion channels, slowing down the repolarization of the cell.
- Equine hyperkalaemic periodic paralysis as well as human hyperkalaemic periodic paralysis (HyperPP) are caused by a defect in voltage-dependent sodium channels.
- Paramyotonia congenita (PC) and potassium-aggravated myotonias (PAM)
- Generalized epilepsy with febrile seizures plus (GEFS+)
- Episodic ataxia (EA), characterized by sporadic bouts of severe discoordination with or without myokymia, and can be provoked by stress, startle, or heavy exertion such as exercise.
- Familial hemiplegic migraine (FHM)

- Spinocerebellar ataxia type 13
- Long QT syndrome is a ventricular arrhythmia syndrome caused by mutations in one or more of presently ten different genes, most of which are potassium channels and all of which affect cardiac repolarization.
- Brugada syndrome is another ventricular arrhythmia caused by voltage-gated sodium channel gene mutations.
- Polymicrogyria is a developmental brain malformation caused by voltage-gated sodium channel and NMDA receptor gene mutations.
- Cystic fibrosis is caused by mutations in the CFTR gene, which is a chloride channel.
- Mucolipidosis type IV is caused by mutations in the gene encoding the TRPML1 channel
- Mutations in and overexpression of ion channels are important events in cancer cells. In Glioblastoma multiforme, upregulation of gBK potassium channels and ClC-3 chloride channels enables glioblastoma cells to migrate within the brain, which may lead to the diffuse growth patterns of these tumors.

3.3 SODIUM AND POTASSIUM

3.3.1 Sodium

Sodium is a chemical element; it has symbol Na (from Neo-Latin *natrium*) and atomic number 11. It is a soft, silvery-white, highly reactive metal. Sodium is an alkali metal, being in group 1 of the periodic table. Its only stable isotope is ^{23}Na. The free metal does not occur in nature and must be prepared from compounds. Sodium is the sixth most abundant element in the Earth's crust and exists in numerous minerals such as feldspars, sodalite, and halite (NaCl). Many salts of sodium are highly water-soluble: sodium ions have been leached by the action of water from the Earth's minerals over eons, and thus sodium and chlorine are the most common dissolved elements by weight in the oceans.

Sodium was first isolated by Humphry Davy in 1807 by the electrolysis of sodium hydroxide. Among many other useful sodium compounds, sodium hydroxide (lye) is used in soap manufacture, and sodium chloride (edible salt) is a de-icing agent and a nutrient for animals including humans.

Sodium is an essential element for all animals and some plants. Sodium ions are the major cation in the extracellular fluid (ECF) and as such are the major contributor to the ECF osmotic pressure. Animal cells actively pump sodium ions out of the cells by means of the sodium–potassium pump, an enzyme complex embedded in the cell membrane, in order to maintain a roughly ten-times higher concentration of sodium ions outside the cell than inside. In nerve cells, the sudden flow of sodium ions into the cell through voltage-gated sodium channels enables transmission of a nerve impulse in a process called the action potential.

Characteristics

Physical

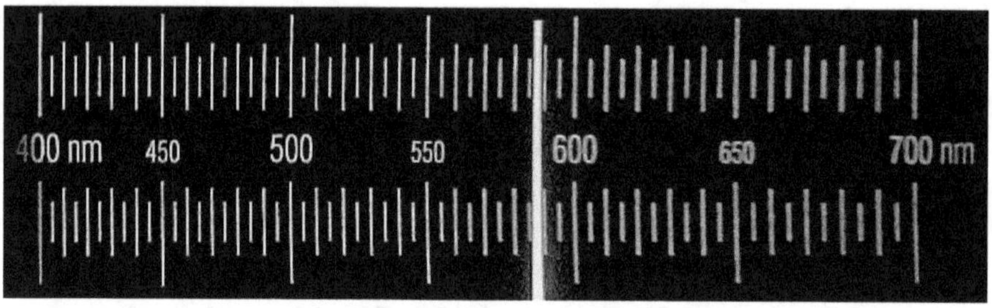

Fig: Emission Spectrum for Sodium, Showing the D Line

Sodium at standard temperature and pressure is a soft silvery metal that combines with oxygen in the air, forming sodium oxides. Bulk sodium is usually stored in oil or an inert gas.

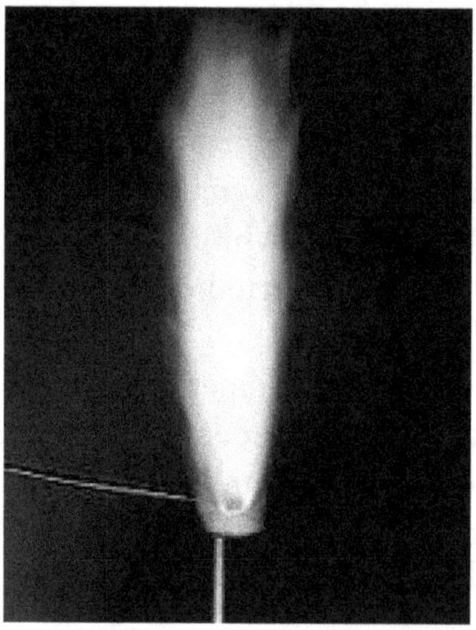

Fig: A Positive Flame Test for Sodium has a Bright Yellow Color

Sodium metal can be easily cut with a knife. It is a good conductor of electricity and heat. Due to having low atomic mass and large atomic radius, sodium is third-least dense of all elemental metals and is one of only three metals that can float on water, the other two being lithium and potassium.

The melting (98 °C) and boiling (883 °C) points of sodium are lower than those of lithium but higher than those of the heavier alkali metals potassium, rubidium, and caesium, following periodic trends down the group. These properties change dramatically at elevated pressures: at 1.5 Mbar, the color changes from silvery metallic to black; at 1.9 Mbar the material becomes transparent with a red color; and at 3 Mbar, sodium is a clear and transparent solid. All of these high-pressure allotropes are insulators and electrides.

In a flame test, sodium and its compounds glow yellow because the excited 3s electrons of sodium emit a photon when they fall from 3p to 3s; the wavelength of this photon corresponds to the D line at about 589.3 nm. Spin-orbit interactions involving the electron in the 3p orbital split the D line into two, at 589.0 and 589.6 nm; hyperfine structures involving both orbitals cause many more lines.

Isotopes

Twenty isotopes of sodium are known, but only ^{23}Na is stable. ^{23}Na is created in the carbon-burning process in stars by fusing two carbon atoms together; this requires temperatures above 600 megakelvins and a star of at least three solar masses. Two radioactive, cosmogenic isotopes are the byproduct of cosmic ray spallation: ^{22}Na has a half-life of 2.6 years and ^{24}Na, a half-life of 15 hours; all other isotopes have a half-life of less than one minute. Two nuclear isomers have been discovered, the longer-lived one being ^{24m}Na with a half-life of around 20.2 milliseconds. Acute neutron radiation, as from a nuclear criticality accident, converts some of the stable ^{23}Na in human blood to ^{24}Na; the neutron radiation dosage of a victim can be calculated by measuring the concentration of ^{24}Na relative to ^{23}Na.

Chemistry

Sodium atoms have 11 electrons, one more than the stable configuration of the noble gas neon. The first and second ionization energies are 495.8 kJ/mol and 4562 kJ/mol, respectively. As a result, sodium usually forms ionic compounds involving the Na^+ cation.

Metallic Sodium

Metallic sodium is generally less reactive than potassium and more reactive than lithium. Sodium metal is highly reducing, with the standard reduction potential for the Na^+/Na couple being −2.71 volts, though potassium and lithium have even more negative potentials.

Salts and Oxides

Sodium compounds are of immense commercial importance, being particularly central to industries producing glass, paper, soap, and textiles. The most important sodium compounds are table salt (NaCl), soda ash (Na_2CO_3), baking soda ($NaHCO_3$), caustic soda (NaOH), sodium nitrate ($NaNO_3$), di- and tri-sodium phosphates, sodium thiosulfate ($Na_2S_2O_3 \cdot 5H_2O$), and borax ($Na_2B_4O_7 \cdot 10H_2O$). In compounds, sodium is usually ionically bonded to water and anions and is viewed as a hard Lewis acid.

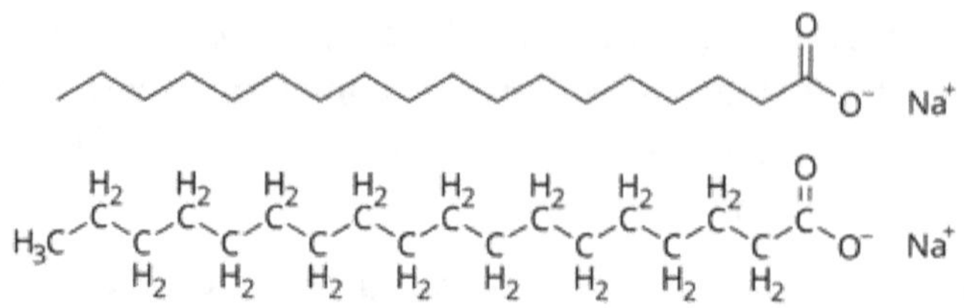

Fig: Two Equivalent Images of the Chemical Structure of Sodium Stearate, A Typical Soap

Most soaps are sodium salts of fatty acids. Sodium soaps have a higher melting temperature (and seem "harder") than potassium soaps.

Like all the alkali metals, sodium reacts exothermically with water. The reaction produces caustic soda (sodium hydroxide) and flammable hydrogen gas. When burned in air, it forms primarily sodium peroxide with some sodium oxide.

Aqueous Solutions

Sodium tends to form water-soluble compounds, such as halides, sulfates, nitrates, carboxylates and carbonates. The main aqueous species are the aquo complexes $[Na(H_2O)_n]^+$, where n = 4–8; with n = 6 indicated from X-ray diffraction data and computer simulations. Direct precipitation of sodium salts from aqueous solutions is rare because sodium salts typically have a high affinity for water. An exception is sodium bismuthate ($NaBiO_3$), which is insoluble in cold water and decomposes in hot water. Because of the high solubility of its compounds, sodium salts are usually isolated as solids by evaporation or by precipitation with an organic antisolvent, such as ethanol; for example, only 0.35 g/L of sodium chloride will dissolve in ethanol. Crown ethers, like 15-crown-5, may be used as a phase-transfer catalyst.

Sodium content of samples is determined by atomic absorption spectrophotometry or by potentiometry using ion-selective electrodes.

Electrides and Sodides

Like the other alkali metals, sodium dissolves in ammonia and some amines to give deeply colored solutions; evaporation of these solutions leaves a shiny film of metallic sodium. The solutions contain the coordination complex $[Na(NH_3)_6]^+$, with the positive charge counterbalanced by electrons as anions; cryptands permit the isolation of these complexes as crystalline solids. Sodium forms complexes with crown ethers, cryptands and other ligands. For example, 15-crown-5 has a high affinity for sodium because the cavity size of 15-crown-5 is 1.7–2.2 Å, which is enough to fit the sodium ion (1.9 Å). Cryptands, like crown ethers and other ionophores, also have a high affinity for the sodium ion; derivatives of the alkalide Na^- are obtainable by the addition of cryptands to solutions of sodium in ammonia via disproportionation.

Organosodium Compounds

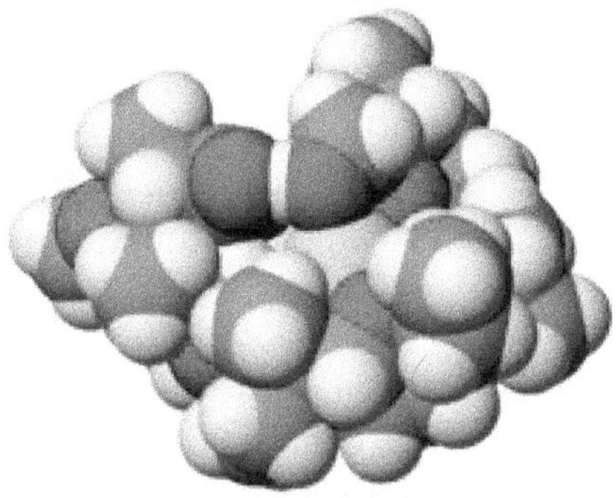

Fig: The Structure of the Complex of Sodium (Na^+, Shown in Yellow) and the Antibiotic Monensin-A

Many organosodium compounds have been prepared. Because of the high polarity of the C-Na bonds, they behave like sources of carbanions (salts with organic anions). Some well-known derivatives include sodium cyclopentadienide (NaC_5H_5) and trityl sodium ($(C_6H_5)_3CNa$). Sodium naphthalene, $Na^+[C_{10}H_8\bullet]^-$, a strong reducing agent, forms upon mixing Na and naphthalene in ethereal solutions.

Intermetallic Compounds

Sodium forms alloys with many metals, such as potassium, calcium, lead, and the group 11 and 12 elements. Sodium and potassium form KNa_2 and NaK. NaK is 40–90% potassium and it is liquid at ambient temperature. It is an excellent thermal and electrical conductor. Sodium-calcium alloys are by-products of the electrolytic production of sodium from a binary salt mixture of NaCl-$CaCl_2$ and ternary mixture NaCl-$CaCl_2$-$BaCl_2$. Calcium is only partially miscible with sodium, and the 1–2% of it dissolved in the sodium obtained from said mixtures can be precipitated by cooling to 120 °C and filtering.

In a liquid state, sodium is completely miscible with lead. There are several methods to make sodium-lead alloys. One is to melt them together and another is to deposit sodium electrolytically on molten lead cathodes. $NaPb_3$, NaPb, Na_9Pb_4, Na_5Pb_2, and $Na_{15}Pb_4$ are some of the known sodium-lead alloys. Sodium also forms alloys with gold ($NaAu_2$) and silver ($NaAg_2$). Group 12 metals (zinc, cadmium and mercury) are known to make alloys with sodium. $NaZn_{13}$ and $NaCd_2$ are alloys of zinc and cadmium. Sodium and mercury form NaHg, $NaHg_4$, $NaHg_2$, Na_3Hg_2, and Na_3Hg.

History

Because of its importance in human health, salt has long been an important commodity. In medieval Europe, a compound of sodium with the Latin name of *sodanum* was used as a headache remedy. The name sodium is thought to originate from the Arabic *suda*, meaning headache, as the headache-alleviating properties of sodium carbonate or soda were well known in early times.

Although sodium, sometimes called *soda*, had long been recognized in compounds, the metal itself was not isolated until 1807 by Sir Humphry Davy through the electrolysis of sodium hydroxide. In 1809, the German physicist and chemist Ludwig Wilhelm Gilbert proposed the names *Natronium* for Humphry Davy's "sodium" and *Kalium* for Davy's "potassium".

The chemical abbreviation for sodium was first published in 1814 by Jöns Jakob Berzelius in his system of atomic symbols, and is an abbreviation of the element's Neo-Latin name *natrium*, which refers to the Egyptian *natron*, a natural mineral salt mainly consisting of hydrated sodium carbonate. Natron historically had several important industrial and household uses, later eclipsed by other sodium compounds.

Sodium imparts an intense yellow color to flames. As early as 1860, Kirchhoff and Bunsen noted the high sensitivity of a sodium flame test, and stated in Annalen der Physik und Chemie:

- In a corner of our 60 m^3 room farthest away from the apparatus, we exploded 3 mg of sodium chlorate with milk sugar while observing the nonluminous flame before the slit. After a while, it glowed a bright yellow and showed a strong sodium line that disappeared only after 10 minutes. From the weight of the sodium salt and the volume of air in the room, we easily calculate that one part by weight of air could not contain more than 1/20 millionth weight of sodium.

Occurrence

The Earth's crust contains 2.27% sodium, making it the sixth most abundant element on Earth and the fourth most abundant metal, behind aluminium, iron, calcium, and magnesium and ahead of potassium.Sodium's estimated oceanic abundance is 10.8 grams per liter. Because of its high reactivity, it is never found as a pure element. It is found in many minerals, some very soluble, such as halite and natron, others much less soluble, such as amphibole and zeolite. The insolubility of certain sodium minerals such as cryolite and feldspar arises from their polymeric anions, which in the case of feldspar is a polysilicate. In the universe, sodium is the 15th most abundant element with a 20,000 parts-per-billion abundance, making sodium 0.002% of the total atoms in the universe.

Astronomical Observations

Atomic sodium has a very strong spectral line in the yellow-orange part of the spectrum (the same line as is used in sodium-vapour street lights). This appears as an absorption line in many types of stars, including the Sun. The line was first studied in 1814 by Joseph von Fraunhofer during his investigation of the lines in the solar spectrum, now known as the Fraunhofer lines. Fraunhofer named it the "D" line, although it is now known to actually be a group of closely spaced lines split by a fine and hyperfine structure.

The strength of the D line allows its detection in many other astronomical environments. In stars, it is seen in any whose surfaces are cool enough for sodium to exist in atomic form (rather than ionised). This corresponds to stars of roughly F-type and cooler. Many other stars appear to have a sodium absorption line, but this is actually caused by gas in the foreground interstellar medium. The two can be distinguished via high-resolution spectroscopy, because interstellar lines are much narrower than those broadened by stellar rotation.

Sodium has also been detected in numerous Solar System environments, including the exospheres of Mercury and the Moon, and numerous other bodies. Some comets have a sodium tail, which was first detected in observations of Comet Hale–Bopp in 1997. Sodium has even been detected in the atmospheres of some extrasolar planets via transit spectroscopy.

Commercial Production

Employed in rather specialized applications, about 100,000 tonnes of metallic sodium are produced annually. Metallic sodium was first produced commercially in the late nineteenth century by carbothermal reduction of sodium carbonate at 1100 °C, as the first step of the Deville process for the production of aluminium:

$$Na_2CO_3 + 2\,C \rightarrow 2\,Na + 3\,CO$$

The high demand for aluminium created the need for the production of sodium. The introduction of the Hall–Héroult process for the production of aluminium by electrolysing a molten salt bath ended the need for large quantities of sodium. A related process based on the reduction of sodium hydroxide was developed in 1886.

Sodium is now produced commercially through the electrolysis of molten sodium chloride (common salt), based on a process patented in 1924. This is done in a Downs cell in which the NaCl is mixed with calcium chloride to lower the melting point below 700 °C. As calcium is less electropositive than sodium, no calcium will be deposited at the cathode. This method is less expensive than the previous Castner process (the electrolysis of sodium hydroxide). If sodium of high purity is required, it can be distilled once or several times.

The market for sodium is volatile due to the difficulty in its storage and shipping; it must be stored under a dry inert gas atmosphere or anhydrous mineral oil to prevent the formation of a surface layer of sodium oxide or sodium superoxide.

Uses

Though metallic sodium has some important uses, the major applications for sodium use compounds; millions of tons of sodium chloride, hydroxide, and carbonate are produced annually. Sodium chloride is extensively used for anti-icing and de-icing and as a preservative; examples of the uses of sodium bicarbonate include baking, as a raising agent, and sodablasting. Along with potassium, many important medicines have sodium added to improve their bioavailability; though potassium is the better ion in most cases, sodium is chosen for its lower price and atomic weight. Sodium hydride is used as a base for various reactions (such as the aldol reaction) in organic chemistry.

Metallic sodium is used mainly for the production of sodium borohydride, sodium azide, indigo, and triphenylphosphine. A once-common use was the making of tetraethyllead and titanium metal; because of the move away from TEL and new titanium production methods, the production of sodium declined after 1970. Sodium is also used as an alloying metal, an anti-scaling agent, and as a reducing agent for metals when other materials are ineffective.

Note the free element is not used as a scaling agent, ions in the water are exchanged for sodium ions. Sodium plasma ("vapor") lamps are often used for street lighting in cities, shedding light that ranges from yellow-orange to peach as the pressure increases. By itself or with potassium, sodium is a desiccant; it gives an intense blue coloration with benzophenone when the desiccate is dry.

In organic synthesis, sodium is used in various reactions such as the Birch reduction, and the sodium fusion test is conducted to qualitatively analyse compounds. Sodium reacts with alcohols and gives alkoxides, and when sodium is dissolved in ammonia solution, it can be used to reduce alkynes to trans-alkenes. Lasers emitting light at the sodium D line are used to create artificial laser guide stars that assist in the adaptive optics for land-based visible-light telescopes.

Heat Transfer

Liquid sodium is used as a heat transfer fluid in sodium-cooled fast reactors because it has the high thermal conductivity and low neutron absorption cross section required to achieve a high neutron flux in the reactor. The high boiling point of sodium allows the reactor to operate at ambient (normal) pressure, but drawbacks include its opacity, which hinders visual maintenance, and its strongly reducing properties. Sodium will explode in contact with water, although it will only burn gently in air.

Radioactive sodium-24 may be produced by neutron bombardment during operation, posing a slight radiation hazard; the radioactivity stops within a few days after removal from the reactor. If a reactor needs to be shut down frequently,

NaK is used. Because NaK is a liquid at room temperature, the coolant does not solidify in the pipes.

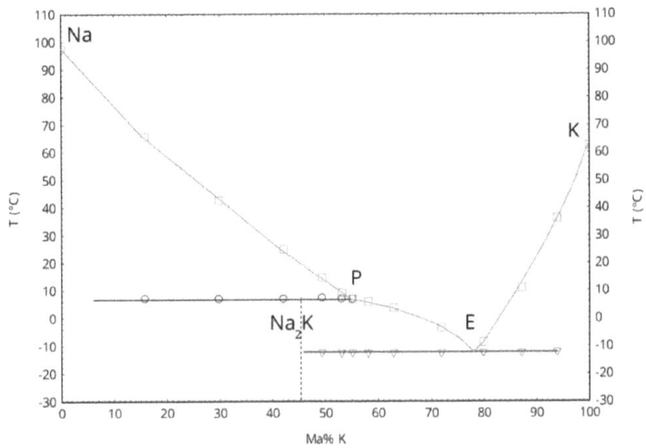

Fig: NaK Phase Diagram, Showing the Melting Point of Sodium as a Function of Potassium Concentration. NaK with 77% Potassium is Eutectic and has the Lowest Melting Point of the NaK Alloys at −12.6 °C.

In this case, the pyrophoricity of potassium requires extra precautions to prevent and detect leaks. Another heat transfer application is poppet valves in high-performance internal combustion engines; the valve stems are partially filled with sodium and work as a heat pipe to cool the valves.

Biological Role

Biological Role in Humans

In humans, sodium is an essential mineral that regulates blood volume, blood pressure, osmotic equilibrium and pH. The minimum physiological requirement for sodium is estimated to range from about 120 milligrams per day in newborns to 500 milligrams per day over the age of 10.

Diet

Sodium chloride, also known as *edible salt* or *table salt* (chemical formula NaCl), is the principal source of sodium (Na) in the diet, and is used as seasoning and preservative in such commodities as pickled preserves and jerky; for Americans, most sodium chloride comes from processed foods. Other sources of sodium are its natural occurrence in food and such food additives as monosodium glutamate (MSG), sodium nitrite, sodium saccharin, baking soda (sodium bicarbonate), and sodium benzoate. The U.S. Institute of Medicine set its tolerable upper intake level for sodium at 2.3 grams per day, but the average person in the United States consumes 3.4 grams per day. The American Heart Association recommends no more than 1.5 g of sodium per day.

The Committee to Review the Dietary Reference Intakes for Sodium and Potassium, which is part of the National Academies of Sciences, Engineering, and Medicine, has determined that there isn't enough evidence from research studies to establish Estimated Average Requirement (EAR) and Recommended Dietary Allowance (RDA) values for sodium. As a result, the committee has established Adequate Intake (AI) levels instead, as follows. The sodium AI for infants of 0–6 months is established at 110 mg/day, 7–12 months: 370 mg/day; for children 1–3 years: 800 mg/day, 4–8 years: 1,000 mg/day; for adolescents: 9–13 years – 1,200 mg/day, 14–18 years 1,500 mg/day; for adults regardless of their age or sex: 1,500 mg/day.

Sodium chloride (NaCl) contains approximately 39.34% of elemental sodium (Na) the total mass. This means that 1 gram of sodium chloride contains approximately 393.4 mg of elemental sodium.

For example, to find out how much sodium chloride contains 1500 mg of elemental sodium (the value of 1500 mg sodium is the adequate intake (AI) for an adult), we can use the proportion:

$$393.4\ mg\ Na : 1000\ mg\ NaCl = 1500\ mg\ Na : x\ mg\ NaCl$$

Solving for x gives us the amount of sodium chloride that contains 1500 mg of elemental sodium:

$$x = (1500\ mg\ Na \times 1000\ mg\ NaCl) / 393.4\ mg\ Na = 3812.91\ mg$$

This mean that 3812.91 mg of sodium chloride contain 1500 mg of elemental sodium.

High Sodium Consumption

High sodium consumption is unhealthy, and can lead to alteration in the mechanical performance of the heart. High sodium consumption is also associated with chronic kidney disease, high blood pressure, cardiovascular diseases, and stroke.

High Blood Pressure

There is a strong correlation between higher sodium intake and higher blood pressure. Studies have found that lowering sodium intake by 2 g per day tends to lower systolic blood pressure by about two to four mm Hg. It has been estimated that such a decrease in sodium intake would lead to 9–17% fewer cases of hypertension.

Hypertension causes 7.6 million premature deaths worldwide each year. Since edible salt contains about 39.3% sodium—the rest being chlorine and trace chemicals; thus, 2.3 g sodium is about 5.9 g, or 5.3 ml, of salt—about one US teaspoon.

One scientific review found that people with or without hypertension who excreted less than 3 grams of sodium per day in their urine (and therefore were taking in less than 3 g/d) had a *higher* risk of death, stroke, or heart attack than those excreting 4 to 5 grams per day. Levels of 7 g per day or more in people with hypertension were associated with higher mortality and cardiovascular events, but this was not found to be true for people without hypertension.

The US FDA states that adults with hypertension and prehypertension should reduce daily sodium intake to 1.5 g.

Physiology

The renin–angiotensin system regulates the amount of fluid and sodium concentration in the body. Reduction of blood pressure and sodium concentration in the kidney result in the production of renin, which in turn produces aldosterone and angiotensin, which stimulates the reabsorption of sodium back into the bloodstream. When the concentration of sodium increases, the production of renin decreases, and the sodium concentration returns to normal. The sodium ion (Na^+) is an important electrolyte in neuron function, and in osmoregulation between cells and the extracellular fluid.

This is accomplished in all animals by Na^+/K^+-ATPase, an active transporter pumping ions against the gradient, and sodium/potassium channels. The difference in extracellular and intracellular ion concentration, maintained by the sodium-potassium pump, produce electrical signals in the form of action potentials that supports cardiac muscle contraction and promote long distance communication between neurons. Sodium is the most prevalent metallic ion in extracellular fluid.

In humans, unusually low or high sodium levels in the blood is recognized in medicine as hyponatremia and hypernatremia. These conditions may be caused by genetic factors, ageing, or prolonged vomiting or diarrhea.

Biological Role in Plants

In C4 plants, sodium is a micronutrient that aids metabolism, specifically in regeneration of phosphoenolpyruvate and synthesis of chlorophyll. In others, it substitutes for potassium in several roles, such as maintaining turgor pressure and aiding in the opening and closing of stomata. Excess sodium in the soil can limit the uptake of water by decreasing the water potential, which may result in plant wilting; excess concentrations in the cytoplasm can lead to enzyme inhibition, which in turn causes necrosis and chlorosis.

In response, some plants have developed mechanisms to limit sodium uptake in the roots, to store it in cell vacuoles, and restrict salt transport from roots to leaves. Excess sodium may also be stored in old plant tissue, limiting the damage to new growth. Halophytes have adapted to be able to flourish in sodium rich environments.

Safety and Precautions

Sodium forms flammable hydrogen and caustic sodium hydroxide on contact with water; ingestion and contact with moisture on skin, eyes or mucous membranes can cause severe burns. Sodium spontaneously explodes in the presence of water due to the formation of hydrogen (highly explosive) and sodium hydroxide (which dissolves in the water, liberating more surface). However, sodium exposed to air and ignited or reaching autoignition (reported to occur when a molten pool of sodium reaches about 290 °C, 554 °F) displays a relatively mild fire.

In the case of massive (non-molten) pieces of sodium, the reaction with oxygen eventually becomes slow due to formation of a protective layer. Fire extinguishers based on water accelerate sodium fires. Those based on carbon dioxide and bromochlorodifluoromethane should not be used on sodium fire. Metal fires are Class D, but not all Class D extinguishers are effective when used to extinguish sodium fires. An effective extinguishing agent for sodium fires is Met-L-X. Other effective agents include Lith-X, which has graphite powder and an organophosphate flame retardant, and dry sand.

Sodium fires are prevented in nuclear reactors by isolating sodium from oxygen with surrounding pipes containing inert gas. Pool-type sodium fires are prevented using diverse design measures called catch pan systems. They collect leaking sodium into a leak-recovery tank where it is isolated from oxygen.

Liquid sodium fires are more dangerous to handle than solid sodium fires, particularly if there is insufficient experience with the safe handling of molten sodium. In a technical report for the United States Fire Administration, R. J. Gordon writes (emphasis in original)

- Molten sodium is extremely dangerous because it is much more reactive than a solid mass. In the liquid form, every sodium atom is free and mobile to instantaneously combine with any available oxygen atom or other oxidizer, and any gaseous by-product will be created as a rapidly expanding gas bubble within the molten mass. Even a minute amount of water can create this type of reaction. Any amount of water introduced into a pool of molten sodium is likely to cause a violent explosion inside the liquid mass, releasing the hydrogen as a rapidly expanding gas and causing the molten sodium to erupt from the container.
- When molten sodium is involved in a fire, the combustion occurs at the surface of the liquid. An inert gas, such as nitrogen or argon, can be used to form an inert layer over the pool of burning liquid sodium, but the gas must be applied very gently and contained over the surface. Except for soda ash, most of the powdered agents that are used to

extinguish small fires in solid pieces or shallow pools will sink to the bottom of a molten mass of burning sodium – the sodium will float to the top and continue to burn. If the burning sodium is in a container, it may be feasible to extinguish the fire by placing a lid on the container to exclude oxygen.

3.3.2 Potassium

Potassium is a chemical element; it has symbol K and atomic number 19. It is a silvery white metal that is soft enough to easily cut with a knife. Potassium metal reacts rapidly with atmospheric oxygen to form flaky white potassium peroxide in only seconds of exposure. It was first isolated from potash, the ashes of plants, from which its name derives. In the periodic table, potassium is one of the alkali metals, all of which have a single valence electron in the outer electron shell, which is easily removed to create an ion with a positive charge (which combines with anions to form salts). In nature, potassium occurs only in ionic salts. Elemental potassium reacts vigorously with water, generating sufficient heat to ignite hydrogen emitted in the reaction, and burning with a lilac-colored flame. It is found dissolved in seawater (which is 0.04% potassium by weight), and occurs in many minerals such as orthoclase, a common constituent of granites and other igneous rocks.

Potassium is chemically very similar to sodium, the previous element in group 1 of the periodic table. They have a similar first ionization energy, which allows for each atom to give up its sole outer electron. It was first suggested in 1702 that they were distinct elements that combine with the same anions to make similar salts, which was demonstrated in 1807 when elemental potassium was first isolated via electrolysis. Naturally occurring potassium is composed of three isotopes, of which ^{40}K is radioactive. Traces of ^{40}K are found in all potassium, and it is the most common radioisotope in the human body.

Potassium ions are vital for the functioning of all living cells. The transfer of potassium ions across nerve cell membranes is necessary for normal nerve transmission; potassium deficiency and excess can each result in numerous signs and symptoms, including an abnormal heart rhythm and various electrocardiographic abnormalities. Fresh fruits and vegetables are good dietary sources of potassium. The body responds to the influx of dietary potassium, which raises serum potassium levels, by shifting potassium from outside to inside cells and increasing potassium excretion by the kidneys.

Most industrial applications of potassium exploit the high solubility of its compounds in water, such as saltwater soap. Heavy crop production rapidly depletes the soil of potassium, and this can be remedied with agricultural fertilizers containing potassium, accounting for 95% of global potassium chemical production.

Properties

Physical

Potassium is the second least dense metal after lithium. It is a soft solid with a low melting point, and can be easily cut with a knife. Potassium is silvery in appearance, but it begins to tarnish toward gray immediately on exposure to air. In a flame test, potassium and its compounds emit a lilac color with a peak emission wavelength of 766.5 nanometers.

Neutral potassium atoms have 19 electrons, one more than the configuration of the noble gas argon. Because of its low first ionization energy of 418.8 kJ/mol, the potassium atom is much more likely to lose the last electron and acquire a positive charge, although negatively charged alkalide K^- ions are not impossible. In contrast, the second ionization energy is very high (3052 kJ/mol).

Chemical

Potassium reacts with oxygen, water, and carbon dioxide components in air. With oxygen it forms potassium peroxide. With water potassium forms potassium hydroxide (KOH). The reaction of potassium with water can be violently exothermic, especially since the coproduced hydrogen gas can ignite. Because of this, potassium and the liquid sodium-potassium (NaK) alloy are potent desiccants, although they are no longer used as such.

Compounds

Four oxides of potassium are well studied: potassium oxide (K_2O), potassium peroxide (K_2O_2), potassium superoxide (KO_2) and potassium ozonide (KO_3). The binary potassium-oxygen compounds react with water forming KOH.

KOH is a strong base. Illustrating its hydrophilic character, as much as 1.21 kg of KOH can dissolve in a single liter of water. Anhydrous KOH is rarely encountered. KOH reacts readily with carbon dioxide (CO_2) to produce potassium carbonate (K_2CO_3), and in principle could be used to remove traces of the gas from air. Like the closely related sodium hydroxide, KOH reacts with fats to produce soaps.

In general, potassium compounds are ionic and, owing to the high hydration energy of the K^+ ion, have excellent water solubility. The main species in water solution are the aquo complexes $[K(H_2O)_n]^+$ where $n = 6$ and 7.

Potassium heptafluorotantalate ($K_2[TaF_7]$) is an intermediate in the purification of tantalum from the otherwise persistent contaminant of niobium.

Organopotassium compounds illustrate nonionic compounds of potassium. They feature highly polar covalent K–C bonds. Examples include benzyl potassium $KCH_2C_6H_5$. Potassium intercalates into graphite to give a variety of graphite intercalation compounds, including KC_8.

Isotopes

There are 25 known isotopes of potassium, three of which occur naturally: ^{39}K (93.3%), ^{40}K (0.0117%), and ^{41}K (6.7%) (by mole fraction). Naturally occurring ^{40}K has a half-life of 1.250×10^9 years. It decays to stable ^{40}Ar by electron capture or positron emission (11.2%) or to stable ^{40}Ca by beta decay (88.8%). The decay of ^{40}K to ^{40}Ar is the basis of a common method for dating rocks. The conventional K-Ar dating method depends on the assumption that the rocks contained no argon at the time of formation and that all the subsequent radiogenic argon (^{40}Ar) was quantitatively retained. Minerals are dated by measurement of the concentration of potassium and the amount of radiogenic ^{40}Ar that has accumulated. The minerals best suited for dating include biotite, muscovite, metamorphic hornblende, and volcanic feldspar; whole rock samples from volcanic flows and shallow instrusives can also be dated if they are unaltered. Apart from dating, potassium isotopes have been used as tracers in studies of weathering and for nutrient cycling studies because potassium is a macronutrient required for life on Earth.

^{40}K occurs in natural potassium (and thus in some commercial salt substitutes) in sufficient quantity that large bags of those substitutes can be used as a radioactive source for classroom demonstrations. ^{40}K is the radioisotope with the largest abundance in the human body. In healthy animals and people, ^{40}K represents the largest source of radioactivity, greater even than ^{14}C. In a human body of 70 kg, about 4,400 nuclei of ^{40}K decay per second. The activity of natural potassium is 31 Bq/g.

History

Potash

Potash is primarily a mixture of potassium salts because plants have little or no sodium content, and the rest of a plant's major mineral content consists of calcium salts of relatively low solubility in water. While potash has been used since ancient times, its composition was not understood. Georg Ernst Stahl obtained experimental evidence that led him to suggest the fundamental difference of sodium and potassium salts in 1702, and Henri Louis Duhamel du Monceau was able to prove this difference in 1736. The exact chemical composition of potassium and sodium compounds, and the status as chemical element of potassium and sodium, was not known then, and thus Antoine Lavoisier did not include the alkali in his list of chemical elements in 1789. For a long time the only significant applications for potash were the production of glass, bleach, soap and gunpowder as potassium nitrate. Potassium soaps from animal fats and vegetable oils were especially prized because they tend to be more water-soluble and of softer texture, and are therefore known as soft soaps. The discovery by Justus Liebig in 1840 that potassium is a necessary element for plants and that most types of soil lack potassium caused a steep rise in demand for potassium salts. Wood-ash from fir trees was initially used as a potassium salt source for fertilizer, but, with the discovery in 1868 of mineral deposits containing potassium chloride near Staßfurt, Germany, the production of potassium-containing fertilizers began at an industrial scale. Other potash deposits were discovered, and by the 1960s Canada became the dominant producer.

Metal

Potassium *metal* was first isolated in 1807 by Humphry Davy, who derived it by electrolysis of molten caustic potash (KOH) with the newly discovered voltaic pile. Potassium was the first metal that was isolated by electrolysis. Later in the same year, Davy reported extraction of the metal sodium from a mineral derivative (caustic soda, NaOH, or lye) rather than a plant salt, by a similar technique, demonstrating that the elements, and thus the salts, are different. Although the production of potassium and sodium metal should have shown that both are elements, it took some time before this view was universally accepted.

Because of the sensitivity of potassium to water and air, air-free techniques are normally employed for handling the element. It is unreactive toward nitrogen and saturated hydrocarbons such as mineral oil or kerosene. It readily dissolves in liquid ammonia, up to 480 g per 1000 g of ammonia at 0 °C.

Depending on the concentration, the ammonia solutions are blue to yellow, and their electrical conductivity is similar to that of liquid metals. Potassium slowly reacts with ammonia to form KNH_2, but this reaction is accelerated by minute amounts of transition metal salts. Because it can reduce the salts to the metal, potassium is often used as the reductant in the preparation of finely divided metals from their salts by the Rieke method. Illustrative is the preparation of magnesium:

$$MgCl_2 + 2\,K \rightarrow Mg + 2\,KCl$$

Occurrence

Potassium is formed in supernovae by nucleosynthesis from lighter atoms. Potassium is principally created in Type II supernovae via an explosive oxygen-burning process. These are nuclear fusion reactions, not to be confused with chemical burning of potassium in oxygen. ^{40}K is also formed in s-process nucleosynthesis and the neon burning process.

Potassium is the 20th most abundant element in the solar system and the 17th most abundant element by weight in the Earth. It makes up about 2.6% of the weight of the Earth's crust and is the seventh most abundant element in the crust. The potassium concentration in seawater is 0.39 g/L (0.039 wt/v%), about one twenty-seventh the concentration of sodium.

Geology

Elemental potassium does not occur in nature because of its high reactivity. It reacts violently with water and also reacts with oxygen. Orthoclase (potassium feldspar) is a common rock-forming mineral. Granite for example contains 5% potassium, which is well above the average in the Earth's crust. Sylvite (KCl), carnallite ($KCl{\cdot}MgCl_2{\cdot}6H_2O$), kainite ($MgSO_4{\cdot}KCl{\cdot}3H_2O$) and langbeinite ($MgSO_4{\cdot}K_2SO_4$) are the minerals found in large evaporite deposits worldwide. The deposits often show layers starting with the least soluble at the bottom and the most soluble on top. Deposits of niter (potassium nitrate) are formed by decomposition of organic material in contact with atmosphere, mostly in caves; because of the good water solubility of niter the formation of larger deposits requires special environmental conditions.

Commercial Production

Mining

Potassium salts such as carnallite, langbeinite, polyhalite, and sylvite form extensive evaporite deposits in ancient lake bottoms and seabeds, making extraction of potassium salts in these environments commercially viable. The principal source of potassium – potash – is mined in Canada, Russia, Belarus, Kazakhstan, Germany, Israel, the U.S., Jordan, and other places around the world. The first mined deposits were located near Staßfurt, Germany, but the deposits span from Great Britain over Germany into Poland. They are located in the Zechstein and were deposited in the Middle to Late Permian. The largest deposits ever found lie 1,000 meters (3,300 feet) below the surface of the Canadian province of Saskatchewan. The deposits are located in the Elk Point Group produced in the Middle Devonian. Saskatchewan, where several large mines have operated since the 1960s pioneered the technique of freezing of wet sands (the Blairmore formation) to drive mine shafts through them. The main potash mining company in Saskatchewan until its merge was the Potash Corporation of Saskatchewan, now Nutrien. The water of the Dead Sea is used by Israel and Jordan as a source of potash, while the concentration in normal oceans is too low for commercial production at current prices.

Chemical Extraction

Several methods are used to separate potassium salts from sodium and magnesium compounds. The most-used method is fractional precipitation using the solubility differences of the salts. Electrostatic separation of the ground salt mixture is also used in some mines. The resulting sodium and magnesium waste is either stored underground or piled up in slag heaps. Most of the mined potassium mineral ends up as potassium chloride after processing. The mineral industry refers to potassium chloride either as potash, muriate of potash, or simply MOP.

Pure potassium metal can be isolated by electrolysis of its hydroxide in a process that has changed little since it was first used by Humphry Davy in 1807. Although the electrolysis process was developed and used in industrial scale in the 1920s, the thermal method by reacting sodium with potassium chloride in a chemical equilibrium reaction became the dominant method in the 1950s.

$$Na + KCl \rightarrow NaCl + K$$

The production of sodium potassium alloys is accomplished by changing the reaction time and the amount of sodium used in the reaction. The Griesheimer process employing the reaction of potassium fluoride with calcium carbide was also used to produce potassium.

$$2\,KF + CaC_2 \rightarrow 2\,K + CaF_2 + 2\,C$$

Reagent-grade potassium metal costs about $10.00/pound ($22/kg) in 2010 when purchased by the tonne. Lower purity metal is considerably cheaper. The market is volatile because long-term storage of the metal is difficult. It must be stored in a dry inert gas atmosphere or anhydrous mineral oil to prevent the formation of a surface layer of potassium superoxide, a pressure-sensitive explosive that detonates when scratched. The resulting explosion often starts a fire difficult to extinguish.

Cation Identification

Potassium is now quantified by ionization techniques, but at one time it was quantitated by gravimetric analysis. Reagents used to precipitate potassium salts include sodium tetraphenylborate, hexachloroplatinic acid, and sodium cobaltinitrite into respectively potassium tetraphenylborate, potassium hexachloroplatinate, and potassium cobaltinitrite. The reaction with sodium cobaltinitrite is illustrative:

$$3\,K^+ + Na_3\left[Co(NO_2)_6\right] \rightarrow K_3\left[Co(NO_2)_6\right] + 3\,Na^+$$

The potassium cobaltinitrite is obtained as a yellow solid.

Commercial Uses

Fertilizer

Potassium ions are an essential component of plant nutrition and are found in most soil types. They are used as a fertilizer in agriculture, horticulture, and hydroponic culture in the form of chloride (KCl), sulfate (K_2SO_4), or nitrate (KNO_3), representing the 'K' in 'NPK'. Agricultural fertilizers consume 95% of global potassium chemical production, and about 90% of this potassium is supplied as KCl. The potassium content of most plants ranges from 0.5% to 2% of the harvested weight of crops, conventionally expressed as amount of K_2O. Modern high-yield agriculture depends upon fertilizers to replace the potassium lost at harvest. Most agricultural fertilizers contain potassium chloride, while potassium sulfate is used for chloride-sensitive crops or crops needing higher sulfur content. The sulfate is produced mostly by decomposition of the complex minerals kainite ($MgSO_4 \cdot KCl \cdot 3H_2O$) and langbeinite ($MgSO_4 \cdot K_2SO_4$). Only a very few fertilizers contain potassium nitrate. In 2005, about 93% of world potassium production was consumed by the fertilizer industry. Furthermore, potassium can play a key role in nutrient cycling by controlling litter composition.

Medical Use

- **Potassium Citrate:** Potassium citrate is used to treat a kidney stone condition called renal tubular acidosis.
- **Potassium Chloride:** Potassium, in the form of potassium chloride is used as a medication to treat and prevent low blood potassium. Low blood potassium may occur due to vomiting, diarrhea, or certain medications. It is given by slow injection into a vein or by mouth.

Food Additives

Potassium sodium tartrate ($KNaC_4H_4O_6$, Rochelle salt) is a main constituent of some varieties of baking powder; it is also used in the silvering of mirrors. Potassium bromate ($KBrO_3$) is a strong oxidizer (E924), used to improve dough strength and rise height. Potassium bisulfite ($KHSO_3$) is used as a food preservative, for example in wine and beer-making (but not in meats). It is also used to bleach textiles and straw, and in the tanning of leathers.

Industrial

Major potassium chemicals are potassium hydroxide, potassium carbonate, potassium sulfate, and potassium chloride. Megatons of these compounds are produced annually.

KOH is a strong base, which is used in industry to neutralize strong and weak acids, to control pH and to manufacture potassium salts. It is also used to saponify fats and oils, in industrial cleaners, and in hydrolysis reactions, for example of esters.

Potassium nitrate (KNO_3) or saltpeter is obtained from natural sources such as guano and evaporites or manufactured via the Haber process; it is the oxidant in gunpowder (black powder) and an important agricultural fertilizer. Potassium cyanide (KCN) is used industrially to dissolve copper and precious metals, in particular silver and gold, by forming complexes. Its applications include gold mining, electroplating, and electroforming of these metals; it is also used in organic synthesis to make nitriles. Potassium carbonate (K_2CO_3 or potash) is used in the manufacture of glass, soap, color TV tubes, fluorescent lamps, textile dyes and pigments. Potassium permanganate ($KMnO_4$) is an oxidizing, bleaching and purification substance and is used for production of saccharin. Potassium chlorate ($KClO_3$) is added to matches and explosives. Potassium bromide (KBr) was formerly used as a sedative and in photography.

While potassium chromate (K_2CrO_4) is used in the manufacture of a host of different commercial products such as inks, dyes, wood stains (by reacting with the tannic acid in wood), explosives, fireworks, fly paper, and safety matches, as well as in the tanning of leather, all of these uses are due to the chemistry of the chromate ion rather than to that of the potassium ion.

Niche Uses

There are thousands of uses of various potassium compounds. One example is potassium superoxide, KO_2, an orange solid that acts as a portable source of oxygen and a carbon dioxide absorber. It is widely used in respiration systems in mines, submarines and spacecraft as it takes less volume than the gaseous oxygen.

$$4\,KO_2 + 2\,CO_2 \rightarrow 2\,K_2CO_3 + 3\,O_2$$

Another example is potassium cobaltinitrite, $K_3[Co(NO_2)_6]$, which is used as artist's pigment under the name of Aureolin or Cobalt Yellow.

The stable isotopes of potassium can be laser cooled and used to probe fundamental and technological problems in quantum physics. The two bosonic isotopes possess convenient Feshbach resonances to enable studies requiring tunable interactions, while ^{40K} is one of only two stable fermions amongst the alkali metals.

Laboratory Uses

An alloy of sodium and potassium, NaK is a liquid used as a heat-transfer medium and a desiccant for producing dry and air-free solvents. It can also be used in reactive distillation. The ternary alloy of 12% Na, 47% K and 41% Cs has the lowest melting point of −78 °C of any metallic compound. Metallic potassium is used in several types of magnetometers.

Biological Role

Potassium is the eighth or ninth most common element by mass (0.2%) in the human body, so that a 60 kg adult contains a total of about 120 g of potassium. The body has about as much potassium as sulfur and chlorine, and only calcium and phosphorus are more abundant (with the exception of the ubiquitous CHON elements). Potassium ions are present in a wide variety of proteins and enzymes.

Biochemical Function

Potassium levels influence multiple physiological processes, including

- Resting cellular-membrane potential and the propagation of action potentials in neuronal, muscular, and cardiac tissue. Due to the electrostatic and chemical properties, K^+ ions are larger than Na^+ ions, and ion channels and pumps in cell membranes can differentiate between the two ions, actively pumping or passively passing one of the two ions while blocking the other.
- Hormone secretion and action.
- Vascular tone.
- Systemic blood pressure control.
- Gastrointestinal motility.
- Acid–base homeostasis.
- Glucose and insulin metabolism.
- Mineralocorticoid action.
- Renal concentrating ability.
- Fluid and electrolyte balance.
- Local cortical monoaminergic norepinephrine, serotonin, and dopamine levels, and through them, sleep/wake balance, and spontaneous activity.

Homeostasis

Potassium homeostasis denotes the maintenance of the total body potassium content, plasma potassium level, and the ratio of the intracellular to extracellular potassium concentrations within narrow limits, in the face of pulsatile intake (meals), obligatory renal excretion, and shifts between intracellular and extracellular compartments.

Plasma Levels

Plasma potassium is normally kept at 3.5 to 5.5 millimoles (mmol) [or milliequivalents (mEq)] per liter by multiple mechanisms. Levels outside this range are associated with an increasing rate of death from multiple causes, and some cardiac, kidney, and lung diseases progress more rapidly if serum potassium levels are not maintained within the normal range.

An average meal of 40–50 mmol presents the body with more potassium than is present in all plasma (20–25 mmol). This surge causes the plasma potassium to rise up to 10% before clearance by renal and extrarenal mechanisms.

Hypokalemia, a deficiency of potassium in the plasma, can be fatal if severe. Common causes are increased gastrointestinal loss (vomiting, diarrhea), and increased renal loss (diuresis). Deficiency symptoms include muscle weakness, paralytic ileus, ECG abnormalities, decreased reflex response; and in severe cases, respiratory paralysis, alkalosis, and cardiac arrhythmia.

Control Mechanisms

Potassium content in the plasma is tightly controlled by four basic mechanisms, which have various names and classifications. These are:

- A reactive negative-feedback system,
- A reactive feed-forward system,
- A predictive or circadian system, and
- An internal or cell membrane transport system.

Collectively, the first three are sometimes termed the "external potassium homeostasis system"; and the first two, the "reactive potassium homeostasis system".

- The reactive negative-feedback system refers to the system that induces renal secretion of potassium in response to a rise in the plasma potassium (potassium ingestion, shift out of cells, or intravenous infusion.)
- The reactive feed-forward system refers to an incompletely understood system that induces renal potassium secretion in response to potassium ingestion prior to any rise in the plasma potassium. This is probably initiated by gut cell potassium receptors that detect ingested potassium and trigger vagal afferent signals to the pituitary gland.
- The predictive or circadian system increases renal secretion of potassium during mealtime hours (e.g. daytime for humans, nighttime for rodents) independent of the presence, amount, or absence of potassium ingestion. It is mediated by a circadian oscillator in the suprachiasmatic nucleus of the brain (central clock), which causes the kidney (peripheral clock) to secrete potassium in this rhythmic circadian fashion.
- The ion transport system moves potassium across the cell membrane using two mechanisms. One is active and pumps sodium out of, and potassium into, the cell. The other is passive and allows potassium to leak out of the cell. Potassium and sodium cations influence fluid distribution between intracellular and extracellular compartments by osmotic forces. The movement of potassium and sodium through the cell membrane is mediated by the Na^+/K^+-ATPase pump. This ion pump uses ATP to pump three sodium ions out of the cell and two potassium ions into the cell, creating an electrochemical gradient and electromotive

force across the cell membrane. The highly selective potassium ion channels (which are tetramers) are crucial for hyperpolarization inside neurons after an action potential is triggered, to cite one example. The most recently discovered potassium ion channel is KirBac3.1, which makes a total of five potassium ion channels (KcsA, KirBac1.1, KirBac3.1, KvAP, and MthK) with a determined structure. All five are from prokaryotic species.

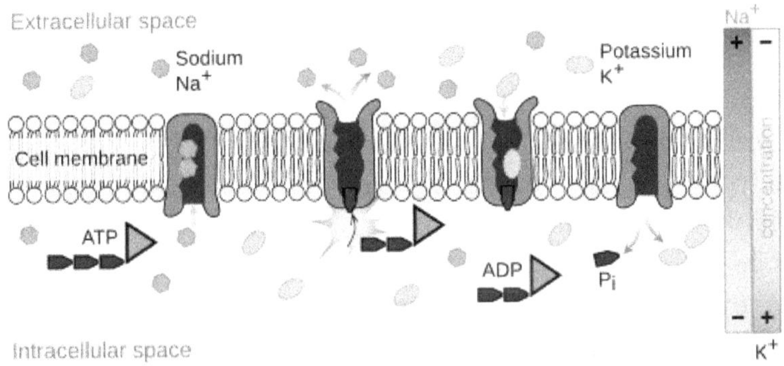

Fig: The Action of the Sodium-Potassium Pump is an Example of Primary Active Transport. The Two Carrier Proteins Embedded in the Cell Membrane on the Left are using ATP to Move Sodium out of the Cell Against the Concentration Gradient; The Two Proteins on the Right are Using Secondary Active Transport to Move Potassium into the Cell. This Process Results in Reconstitution of ATP

Renal Filtration, Reabsorption, and Excretion

Renal handling of potassium is closely connected to sodium handling. Potassium is the major cation (positive ion) inside animal cells (150 mmol/L, 4.8 g/L), while sodium is the major cation of extracellular fluid (150 mmol/L, 3.345 g/L). In the kidneys, about 180 liters of plasma is filtered through the glomeruli and into the renal tubules per day. This filtering involves about 600 mg of sodium and 33 mg of potassium. Since only 1–10 mg of sodium and 1–4 mg of potassium are likely to be replaced by diet, renal filtering must efficiently reabsorb the remainder from the plasma.

Sodium is reabsorbed to maintain extracellular volume, osmotic pressure, and serum sodium concentration within narrow limits. Potassium is reabsorbed to maintain serum potassium concentration within narrow limits. Sodium pumps in the renal tubules operate to reabsorb sodium. Potassium must be conserved, but because the amount of potassium in the blood plasma is very small and the pool of potassium in the cells is about 30 times as large, the situation is not so critical for potassium. Since potassium is moved passively in counter flow to sodium in response to an apparent (but not actual) Donnan equilibrium, the urine can never sink below the concentration of potassium in serum except sometimes by actively excreting water at the end of the processing. Potassium is excreted twice and reabsorbed three times before the urine reaches the collecting tubules. At that point, urine usually has about the same potassium concentration as plasma. At the end of the processing, potassium is secreted one more time if the serum levels are too high. With no potassium intake, it is excreted at about 200 mg per day until, in about a week, potassium in the serum declines to a mildly deficient level of 3.0–3.5 mmol/L. If potassium is still withheld, the concentration continues to fall until a severe deficiency causes eventual death.

The potassium moves passively through pores in the cell membrane. When ions move through ion transporters (pumps) there is a gate in the pumps on both sides of the cell membrane and only one gate can be open at once. As a result, approximately 100 ions are forced through per second. Ion channels have only one gate, and there only one kind of ion can stream through, at 10 million to 100 million ions per second. Calcium is required to open the pores, although calcium may work in reverse by blocking at least one of the pores. Carbonyl groups inside the pore on the amino acids mimic the water hydration that takes place in water solution by the nature of the electrostatic charges on four carbonyl groups inside the pore.

Dietary Recommendations

North America:

- The U.S. National Academy of Medicine (NAM), on behalf of both the U.S. and Canada, sets Dietary Reference Intakes, including Estimated Average Requirements (EARs) and Recommended Dietary Allowances (RDAs), or Adequate Intakes (AIs) for when there is not sufficient information to set EARs and RDAs.
- For both males and females under 9 years of age, the AIs for potassium are: 400 mg of potassium for 0 to 6-month-old infants, 860 mg of potassium for 7 to 12-month-old infants, 2,000 mg of potassium for 1 to 3-year-old children, and 2,300 mg of potassium for 4 to 8-year-old children.
- For males 9 years of age and older, the AIs for potassium are: 2,500 mg of potassium for 9 to 13-year-old males, 3,000 mg of potassium for 14 to 18-year-old males, and 3,400 mg for males that are 19 years of age and older.
- For females 9 years of age and older, the AIs for potassium are: 2,300 mg of potassium for 9 to 18-year-old females, and 2,600 mg of potassium for females that are 19 years of age and older.
- For pregnant and lactating females, the AIs for potassium are: 2,600 mg

of potassium for 14 to 18-year-old pregnant females, 2,900 mg for pregnant females that are 19 years of age and older; furthermore, 2,500 mg of potassium for 14 to 18-year-old lactating females, and 2,800 mg for lactating females that are 19 years of age and older. As for safety, the NAM also sets tolerable upper intake levels (ULs) for vitamins and minerals, but for potassium the evidence was insufficient, so no UL was established.

- As of 2004, most Americans adults consume less than 3,000 mg.

Europe:

- Likewise, in the European Union, in particular in Germany, and Italy, insufficient potassium intake is somewhat common.
- The National Health Service in the United Kingdom recommends a similar intake, saying that "adults (19 to 64 years) need need 3500 mg per day" and that excess amounts may cause health problems such as stomach pain and diarrhea.

Food Sources

Potassium is present in all fruits, vegetables, meat and fish. Foods with high potassium concentrations include yam, parsley, dried apricots, milk, chocolate, all nuts (especially almonds and pistachios), potatoes, bamboo shoots, bananas, avocados, coconut water, soybeans, and bran.

The United States Department of Agriculture also lists tomato paste, orange juice, beet greens, white beans, plantains, and many other dietary sources of potassium, ranked in descending order according to potassium content. A day's worth of potassium is in 5 plantains or 11 bananas.

Deficient Intake

Diets low in potassium can lead to hypertension and hypokalemia.

Supplementation

Supplements of potassium are most widely used in conjunction with diuretics that block reabsorption of sodium and water upstream from the distal tubule (thiazides and loop diuretics), because this promotes increased distal tubular potassium secretion, with resultant increased potassium excretion. A variety of prescription and over-the counter supplements are available. Potassium chloride may be dissolved in water, but the salty/bitter taste makes liquid supplements unpalatable. Potassium is also available in tablets or capsules, which are formulated to allow potassium to leach slowly out of a matrix, since very high concentrations of potassium ion that occur adjacent to a solid tablet can injure the gastric or intestinal mucosa. For this reason, non-prescription potassium pills are limited by law in the US to a maximum of 99 mg of potassium. Potassium supplementation can also be combined with other metabolites, such as citrate or chloride, to achieve specific clinical effects.

Potassium supplements may be employed to mitigate the impact of hypertension, thereby reducing cardiovascular risk. Potassium chloride and potassium bicarbonate may be useful to control mild hypertension. In 2020, potassium was the 33rd most commonly prescribed medication in the U.S., with more than 17 million prescriptions. Potassium supplementation has been shown to reduce both systolic and diastolic blood pressure in individuals with essential hypertension.

Additionally, potassium supplements may be employed with the aim of preventing the formation of kidney stones, a condition that can lead to renal complications if left untreated. Low potassium levels can lead to decreased calcium reabsorption in the kidneys, increasing the risk of elevated urine calcium and the formation of kidney stones. By maintaining adequate potassium levels, this risk can be reduced.

The mechanism of action of potassium involves various types of transporters and channels that facilitate its movement across cell membranes. This process can lead to an increase in the pumping of hydrogen ions. This, in turn, can escalate the production of gastric acid, potentially contributing to the development of gastric ulcers.

Potassium has a role in bone health. It contributes to the acid-base equilibrium in the body and helps protect bone tissue. Potassium salts produce an alkaline component that can aid in maintaining bone health.

For individuals with diabetes, potassium supplementation may be necessary, particularly for those with type 2 diabetes. Potassium is essential for the secretion of insulin by pancreatic beta cells, which helps regulate glucose levels. Without sufficient potassium, insulin secretion is compromised, leading to hyperglycemia and worsening diabetes.

Excessive potassium intake can have adverse effects, such as gastrointestinal discomfort and disturbances in heart rhythm.

Potassium supplementation can have side effects on ulceration, particularly in relation to peptic ulcer disease. Potassium channels have the potential to increase gastric acid secretion, which can lead to an increased risk of ulcerations. Medications used for peptic ulcer disease, known as "proton pump inhibitors", work by inhibiting potassium pumps that activate the H/K ATPase. This inhibition helps to reduce the secretion of hydrochloric acid into the parietal cell, thereby decreasing acidic synthesis and lowering the risk of ulcers. Nicorandil, a drug used for the treatment of ischemic heart disease, can stimulate nitrate and potassium ATP channels, and as a result, it has been associated with side effects such as

GI, oral, and anal ulcers. Prolonged and chronic use of potassium supplements has been linked to more severe side effects, including ulcers outside of the gastrointestinal (GI) tract. Close monitoring is necessary for patients who are also taking angiotensinogen-converting enzyme inhibitors, angiotensin receptor blockers, or potassium-sparing diuretics.

Detection by Taste Buds

Potassium can be detected by taste because it triggers three of the five types of taste sensations, according to concentration. Dilute solutions of potassium ions taste sweet, allowing moderate concentrations in milk and juices, while higher concentrations become increasingly bitter/alkaline, and finally also salty to the taste. The combined bitterness and saltiness of high-potassium solutions makes high-dose potassium supplementation by liquid drinks a palatability challenge.

Precautions

Potassium metal can react violently with water producing KOH and hydrogen gas.

$$2\,K(s) + 2\,H_2O(l) \rightarrow 2\,KOH(aq) + H_2(g)\uparrow$$

This reaction is exothermic and releases sufficient heat to ignite the resulting hydrogen in the presence of oxygen. Finely powdered potassium ignites in air at room temperature.

The bulk metal ignites in air if heated. Because its density is 0.89 g/cm^3, burning potassium floats in water that exposes it to atmospheric oxygen. Many common fire extinguishing agents, including water, either are ineffective or make a potassium fire worse. Nitrogen, argon, sodium chloride (table salt), sodium carbonate (soda ash), and silicon dioxide (sand) are effective if they are dry. Some Class D dry powder extinguishers designed for metal fires are also effective. These agents deprive the fire of oxygen and cool the potassium metal.

During storage, potassium forms peroxides and superoxides. These peroxides may react violently with organic compounds such as oils. Both peroxides and superoxides may react explosively with metallic potassium.

Because potassium reacts with water vapor in the air, it is usually stored under anhydrous mineral oil or kerosene. Unlike lithium and sodium, potassium should not be stored under oil for longer than six months, unless in an inert (oxygen-free) atmosphere, or under vacuum. After prolonged storage in air dangerous shock-sensitive peroxides can form on the metal and under the lid of the container, and can detonate upon opening.

Ingestion of large amounts of potassium compounds can lead to hyperkalemia, strongly influencing the cardiovascular system. Potassium chloride is used in the U.S. for lethal injection executions.

3.4 MAGNESIUM

Magnesium is a chemical element; it has symbol Mg and atomic number 12. It is a shiny gray metal having a low density, low melting point and high chemical reactivity. Like the other alkaline earth metals (group 2 of the periodic table) it occurs naturally only in combination with other elements and it almost always has an oxidation state of +2. It reacts readily with air to form a thin passivation coating of magnesium oxide that inhibits further corrosion of the metal. The free metal burns with a brilliant-white light. The metal is obtained mainly by electrolysis of magnesium salts obtained from brine. It is less dense than aluminium and is used primarily as a component in strong and lightweight alloys that contain aluminium.

In the cosmos, magnesium is produced in large, aging stars by the sequential addition of three helium nuclei to a carbon nucleus. When such stars explode as supernovas, much of the magnesium is expelled into the interstellar medium where it may recycle into new star systems. Magnesium is the eighth most abundant element in the Earth's crust and the fourth most common element in the Earth (after iron, oxygen and silicon), making up 13% of the planet's mass and a large fraction of the planet's mantle. It is the third most abundant element dissolved in seawater, after sodium and chlorine.

This element is the eleventh most abundant element by mass in the human body and is essential to all cells and some 300 enzymes. Magnesium ions interact with polyphosphate compounds such as ATP, DNA, and RNA. Hundreds of enzymes require magnesium ions to function. Magnesium compounds are used medicinally as common laxatives and antacids (such as milk of magnesia), and to stabilize abnormal nerve excitation or blood vessel spasm in such conditions as eclampsia.

3.4.1 Characteristics

Physical Properties

Elemental magnesium is a gray-white lightweight metal, two-thirds the density of aluminium. Magnesium has the lowest melting (923 K (650 °C)) and the lowest boiling point (1,363 K (1,090 °C)) of all the alkaline earth metals. Pure polycrystalline magnesium is brittle and easily fractures along shear bands. It becomes much more malleable when alloyed with small amounts of other metals, such as 1% aluminium. The malleability of polycrystalline magnesium can also be significantly improved by reducing its grain size to ca. 1 micron or less. When finely powdered, magnesium reacts with water to produce hydrogen gas:

$$Mg(s) + 2\,H_2O(g) \rightarrow Mg(OH)_2(aq) + H_2(g) + 1203.6\,kJ/mol$$

However, this reaction is much less dramatic than the reactions of the alkali metals with water, because the magnesium hydroxide builds up on the surface of the magnesium metal and inhibits further reaction.

Chemical Properties

Oxidation

The principal property of magnesium metal is its reducing power. One hint is that it tarnishes slightly when exposed to air, although, unlike the heavier alkaline earth metals, an oxygen-free environment is unnecessary for storage because magnesium is protected by a thin layer of oxide that is fairly impermeable and difficult to remove.

Direct reaction of magnesium with air or oxygen at ambient pressure forms only the "normal" oxide MgO. However, this oxide may be combined with hydrogen peroxide to form magnesium peroxide, MgO_2, and at low temperature the peroxide may be further reacted with ozone to form magnesium superoxide $Mg(O_2)_2$.

Magnesium reacts with nitrogen in the solid state if it is powdered and heated to just below the melting point, forming Magnesium nitride Mg_3N_2.

Magnesium reacts with water at room temperature, though it reacts much more slowly than calcium, a similar group 2 metal. When submerged in water, hydrogen bubbles form slowly on the surface of the metal; this reaction happens much more rapidly with powdered magnesium. The reaction also occurs faster with higher temperatures. Magnesium's reversible reaction with water can be harnessed to store energy and run a magnesium-based engine. Magnesium also reacts exothermically with most acids such as hydrochloric acid (HCl), producing magnesium chloride and hydrogen gas, similar to the HCl reaction with aluminium, zinc, and many other metals. Although it is difficult to ignite in mass or bulk, magnesium metal will ignite.

Magnesium may also be used as an igniter for thermite, a mixture of aluminium and iron oxide powder that ignites only at a very high temperature.

Organic Chemistry

Organomagnesium compounds are widespread in organic chemistry. They are commonly found as Grignard reagents, formed by reaction of magnesium with haloalkanes. Examples of Grignard reagents are phenylmagnesium bromide and ethylmagnesium bromide. The Grignard reagents function as a common nucleophile, attacking the electrophilic group such as the carbon atom that is present within the polar bond of a carbonyl group. A prominent organomagnesium reagent beyond Grignard reagents is magnesium anthracene, which is used as a source of highly active magnesium. The related butadiene-magnesium adduct serves as a source for the butadiene dianion. Complexes of dimagnesium(I) have been observed.

Detection in Solution

The presence of magnesium ions can be detected by the addition of ammonium chloride, ammonium hydroxide and monosodium phosphate to an aqueous or dilute HCl solution of the salt. The formation of a white precipitate indicates the presence of magnesium ions.

Azo violet dye can also be used, turning deep blue in the presence of an alkaline solution of magnesium salt. The color is due to the adsorption of azo violet by $Mg(OH)_2$.

3.4.2 Forms

Alloys

As of 2013, magnesium alloys consumption was less than one million tonnes per year, compared with 50 million tonnes of aluminium alloys. Their use has been historically limited by the tendency of Mg alloys to corrode, creep at high temperatures, and combust.

Corrosion

In magnesium alloys, the presence of iron, nickel, copper, or cobalt strongly activates corrosion. In more than trace amounts, these metals precipitate as intermetallic compounds, and the precipitate locales function as active cathodic sites that reduce water, causing the loss of magnesium. Controlling the quantity of these metals improves corrosion resistance. Sufficient manganese overcomes the corrosive effects of iron. This requires precise control over composition, increasing costs. Adding a cathodic poison captures atomic hydrogen within the structure of a metal. This prevents the formation of free hydrogen gas, an essential factor of corrosive chemical processes. The addition of about one in three hundred parts arsenic reduces the corrosion rate of magnesium in a salt solution by a factor of nearly ten.

High-Temperature Creep and Flammability

Magnesium's tendency to creep (gradually deform) at high temperatures is greatly reduced by alloying with zinc and rare-earth elements. Flammability is significantly reduced by a small amount of calcium in the alloy. By using rare-earth elements, it may be possible to manufacture magnesium alloys that are able to not catch fire at higher temperatures compared to magnesium's liquidus and in some cases potentially pushing it close to magnesium's boiling point.

Compounds

Magnesium forms a variety of compounds important to industry and biology, including magnesium carbonate, magnesium chloride, magnesium citrate, magnesium hydroxide (milk of magnesia), magnesium oxide, magnesium sulfate, and magnesium sulfate heptahydrate (Epsom salts). As recently as 2020, magnesium hydride was under investigation as a way to store hydrogen.

Isotopes

Magnesium has three stable isotopes: ^{24}Mg, ^{25}Mg and ^{26}Mg. All are present in significant amounts in nature. About 79% of Mg is ^{24}Mg. The isotope ^{28}Mg is radioactive and in the 1950s to 1970s was produced by several nuclear power plants for use in scientific experiments. This isotope has a relatively short half-life (21 hours) and its use was limited by shipping times.

The nuclide ^{26}Mg has found application in isotopic geology, similar to that of aluminium. ^{26}Mg is a radiogenic daughter product of ^{26}Al, which has a half-life of 717,000 years. Excessive quantities of stable ^{26}Mg have been observed in the Ca-Al-rich inclusions of some carbonaceous chondrite meteorites. This anomalous abundance is attributed to the decay of its parent ^{26}Al in the inclusions, and researchers conclude that such meteorites were formed in the solar nebula before the ^{26}Al had decayed.

These are among the oldest objects in the Solar System and contain preserved information about its early history.

It is conventional to plot $^{26}Mg/^{24}Mg$ against an Al/Mg ratio. In an isochron dating plot, the Al/Mg ratio plotted is $^{27}Al/^{24}Mg$. The slope of the isochron has no age significance, but indicates the initial $^{26}Al/^{27}Al$ ratio in the sample at the time when the systems were separated from a common reservoir.

3.4.3 Production

Occurrence

Magnesium is the eighth-most-abundant element in the Earth's crust by mass and tied in seventh place with iron in molarity. It is found in large deposits of magnesite, dolomite, and other minerals, and in mineral waters, where magnesium ion is soluble.

Although magnesium is found in more than 60 minerals, only dolomite, magnesite, brucite, carnallite, talc, and olivine are of commercial importance.

The Mg^{2+} cation is the second-most-abundant cation in seawater (about 1/8 the mass of sodium ions in a given sample), which makes seawater and sea salt attractive commercial sources for Mg. To extract the magnesium, calcium hydroxide is added to the seawater to precipitate magnesium hydroxide.

$$MgCl_2 + Ca(OH)_2 \rightarrow Mg(OH)_2 + CaCl_2$$

Magnesium hydroxide (brucite) is poorly soluble in water and can be collected by filtration. It reacts with hydrochloric acid to magnesium chloride.

$$Mg(OH)_2 + 2\ HCl \rightarrow MgCl_2 + 2H_2O$$

From magnesium chloride, electrolysis produces magnesium.

Production Quantities

World production was approximately 1,100 kt in 2017, with the bulk being produced in China (930 kt) and Russia (60 kt). The United States was in the 20th century the major world supplier of this metal, supplying 45% of world production even as recently as 1995. Since the Chinese mastery of the Pidgeon process the US market share is at 7%, with a single US producer left as of 2013: US Magnesium, a Renco Group company located on the shores of the Great Salt Lake.

In September 2021, China took steps to reduce production of magnesium as a result of a government initiative to reduce energy availability for manufacturing industries, leading to a significant price increase.

Pidgeon and Bolzano Processes

The Pidgeon process and the Bolzano process are similar. In both, magnesium oxide is the precursor to magnesium metal. The magnesium oxide is produced as a solid solution with calcium oxide by calcining the mineral dolomite, which is a solid solution of calcium and magnesium carbonates:

$$CaCO_3 \cdot MgCO_3 \rightarrow MgO \cdot CaO + 2\,CO_2$$

Reduction occurs at high temperatures with silicon. A ferrosilicon alloy is used rather than pure silicon as it is more economical. The iron component has no bearing on the reaction, having the simplified equation:

$$MgO \cdot CaO + Si \rightarrow 2\,Mg + Ca_2SiO_4$$

The calcium oxide combines with silicon as the oxygen scavenger, yielding the very stable calcium silicate. The Mg/Ca ratio of the precursors can be adjusted by the addition of MgO or CaO.

The Pidgeon and the Bolzano process differ in the details of the heating and the configuration of the reactor. Both generate gaseous Mg that is condensed and collected. The Pidgeon process dominates the worldwide production. The Pidgeon method is less technologically complex and because of distillation/vapour deposition conditions, a high purity product is easily achievable. China is almost completely reliant on the silicothermic Pidgeon process.

Dow Process

Besides the Pigeon process, the second most used process for magnesium production is electrolysis. This is a two step process. The first step is to prepare feedstock containing magnesium chloride and the second step is to dissociate the compound in electrolytic cells as magnesium metal and chlorine gas. The basic reaction is as follows:

$$MgCl_2 \rightarrow Mg(g) + Cl_2(g)$$

The temperatures at which this reaction is operated is between 680 and 750 °C. The magnesium chloride can be obtained using the Dow process, a process that mixes sea water and dolomite in a flocculator or by dehydration of magnesium chloride brines. The electrolytic cells are partially submerged in a molten salt electrolyte to which the produced magnesium chloride is added in concentrations between 6-18%. This process does have its share of disadvantages including production of harmful chlorine gas and the overall reaction being very energy intensive, creating environmental risks. The Pidgeon process is more advantageous regarding its simplicity, shorter construction period, low power consumption and overall good magnesium quality compared to the electrolysis method.

In the United States, magnesium was once obtained principally with the Dow process in Corpus Christi TX, by electrolysis of fused magnesium chloride from brine and sea water. A saline solution containing Mg^{2+}ions is first treated with lime (calcium oxide) and the precipitated magnesium hydroxide is collected:

$$Mg^{2+}\ (aq)\ + CaO(s)\ + H_2O(l)\ \rightarrow Ca^{2+}\ (aq)\ + Mg(OH)_2(s)$$

The hydroxide is then converted to magnesium chloride by treatment with hydrochloric acid and heating of the product to eliminate water:

$$Mg(OH)_2 + 2\ HCl\ \rightarrow\ MgCl_2 + 2\ H_2O$$

The salt is then electrolyzed in the molten state. At the cathode, the Mg^{2+}ion is reduced by two electrons to magnesium metal:

$$Mg^{2+} + 2e^- \rightarrow Mg$$

At the anode, each pair of Cl^- ions is oxidized to chlorine gas, releasing two electrons to complete the circuit:

$$2Cl^- \rightarrow Cl_2(g) + 2e^-$$

Carbothermic Process

The carbothermic route to magnesium has been recognized as a low energy, yet high productivity path to magnesium extraction. The chemistry is as follows:

$$C + MgO \rightarrow CO + Mg$$

A disadvantage of this method is that slow cooling the vapour can cause the reaction to quickly revert. To prevent this from happening, the magnesium can be dissolved directly in a suitable metal solvent before reversion starts happening. Rapid quenching of the vapour can also be performed to prevent reversion.

YSZ Process

A newer process, solid oxide membrane technology, involves the electrolytic reduction of MgO. At the cathode, Mg^{2+}ion is reduced by two electrons to magnesium metal. The electrolyte is yttria-stabilized zirconia (YSZ). The anode is a liquid metal. At the YSZ/liquid metal anode O^{2-} is oxidized. A layer of graphite borders the liquid metal anode, and at this interface carbon and oxygen react to form carbon monoxide. When silver is used as the liquid metal anode, there is no reductant carbon or hydrogen needed, and only oxygen gas is evolved at the anode. It was reported in 2011 that this method provides a 40% reduction in cost per pound over the electrolytic reduction method.

Rieke Process

Rieke et al. developed a "general approach for preparing highly reactive metal powders by reducing metal salts in ethereal or hydrocarbon solvents using alkali metals as reducing agents" now known as the Rieke process. Rieke finalized the identification of Rieke metals in 1989, one of which was Rieke-magnesium, first produced in 1974.

3.4.4 History

The name magnesium originates from the Greek word for locations related to the tribe of the Magnetes, either a district in Thessaly called Magnesia or Magnesia ad Sipylum, now in Turkey. It is related to magnetite and manganese, which also originated from this area, and required differentiation as separate substances.

In 1618, a farmer at Epsom in England attempted to give his cows water from a local well. The cows refused to drink because of the water's bitter taste, but the farmer noticed that the water seemed to heal scratches and rashes. The substance obtained by evaporating the water became known as Epsom salts and its fame spread. It was eventually recognized as hydrated magnesium sulfate, $MgSO_4·7H_2O$.

The metal itself was first isolated by Sir Humphry Davy in England in 1808. He used electrolysis on a mixture of magnesia and mercuric oxide. Antoine Bussy prepared it in coherent form in 1831. Davy's first suggestion for a name was 'magnium', but the name magnesium is now used in most European languages.

3.4.5 Uses

Magnesium Metal

Magnesium is the third-most-commonly-used structural metal, following iron and aluminium. The main applications of magnesium are, in order: aluminium alloys, die-casting (alloyed with zinc), removing sulfur in the production of iron and steel, and the production of titanium in the Kroll process.

Magnesium is used in lightweight materials and alloys. For example, when infused with silicon carbide nanoparticles, it has extremely high specific strength.

Historically, magnesium was one of the main aerospace construction metals and was used for German military aircraft as early as World War I and extensively for

German aircraft in World War II. The Germans coined the name "Elektron" for magnesium alloy, a term which is still used today. In the commercial aerospace industry, magnesium was generally restricted to engine-related components, due to fire and corrosion hazards. Magnesium alloy use in aerospace is increasing in the 21st century, driven by the importance of fuel economy. Magnesium alloys can act as replacements for aluminium and steel alloys in structural applications.

Aircraft

- Wright Aeronautical used a magnesium crankcase in the WWII-era Wright R-3350 Duplex Cyclone aviation engine. This presented a serious problem for the earliest models of the Boeing B-29 Superfortress heavy bomber when an in-flight engine fire ignited the engine crankcase. The resulting combustion was as hot as 5,600 °F (3,100 °C) and could sever the wing spar from the fuselage.

Automotive

- Mercedes-Benz used the alloy Elektron in the bodywork of an early model Mercedes-Benz 300 SLR; these cars competed in the 1955 World Sportscar Championship including a win at the Mille Miglia, and at Le Mans where one was involved in the 1955 Le Mans disaster when spectators were showered with burning fragments of elektron.
- Porsche used magnesium alloy frames in the 917/053 that won Le Mans in 1971, and continues to use magnesium alloys for its engine blocks due to the weight advantage.
- Volkswagen Group has used magnesium in its engine components for many years.
- Mitsubishi Motors uses magnesium for its paddle shifters.
- BMW used magnesium alloy blocks in their N52 engine, including an aluminium alloy insert for the cylinder walls and cooling jackets surrounded by a high-temperature magnesium alloy AJ62A. The engine was used worldwide between 2005 and 2011 in various 1, 3, 5, 6, and 7 series models; as well as the Z4, X1, X3, and X5.
- Chevrolet used the magnesium alloy AE44 in the 2006 Corvette Z06.

Both AJ62A and AE44 are recent developments in high-temperature low-creep magnesium alloys. The general strategy for such alloys is to form intermetallic precipitates at the grain boundaries, for example by adding mischmetal or calcium.

Electronics

Because of low density and good mechanical and electrical properties, magnesium is used for manufacturing of mobile phones, laptop and tablet computers, cameras, and other electronic components. It was used as a premium feature because of its light weight in some 2020 laptops.

Source of Light

When burning in air, magnesium produces a brilliant white light that includes strong ultraviolet wavelengths. Magnesium powder (flash powder) was used for subject illumination in the early days of photography. Later, magnesium filament was used in electrically ignited single-use photography flashbulbs. Magnesium powder is used in fireworks and marine flares where a brilliant white light is required. It was also used for various theatrical effects, such as lightning, pistol flashes, and supernatural appearances.

Magnesium is flammable, burning at a temperature of approximately 3,100 °C (3,370 K; 5,610 °F), and the autoignition temperature of magnesium ribbon is approximately 473 °C (746 K; 883 °F). Magnesium's high combustion temperature makes it a useful tool for starting emergency fires. Other uses include flash photography, flares, pyrotechnics, fireworks sparklers, and trick birthday candles. Magnesium is also often used to ignite thermite or other materials that require a high ignition temperature. Magnesium continues to be used as an incendiary element in warfare.

Flame temperatures of magnesium and magnesium alloys can reach 3,100 °C (5,610 °F), although flame height above the burning metal is usually less than 300 mm (12 in). Once ignited, such fires are difficult to extinguish because they resist several substances commonly used to put out fires; combustion continues in nitrogen (forming magnesium nitride), in carbon dioxide (forming magnesium oxide and carbon), and in water (forming magnesium oxide and hydrogen, which also combusts due to heat in the presence of additional oxygen). This property was used in incendiary weapons during the firebombing of cities in World War II, where the only practical civil defense was to smother a burning flare under dry sand to exclude atmosphere from the combustion.

Chemical Reagent

In the form of turnings or ribbons, to prepare Grignard reagents, which are useful in organic synthesis.

Other

- As an additive agent in conventional propellants and the production of nodular graphite in cast iron.
- As a reducing agent to separate uranium and other metals from their salts.

- As a sacrificial (galvanic) anode to protect boats, underground tanks, pipelines, buried structures, and water heaters.
- Alloyed with zinc to produce the zinc sheet used in photoengraving plates in the printing industry, dry-cell battery walls, and roofing.
- Alloyed with aluminium with aluminium-magnesium alloys being used mainly for beverage cans, sports equipment such as golf clubs, fishing reels, and archery bows and arrows.
- Many car and aircraft manufacturers have made engine and body parts from magnesium.
- Magnesium batteries have been commercialized as primary batteries, and are an active topic of research for rechargeable batteries.

Compounds

- Magnesium compounds, primarily magnesium oxide (MgO), are used as a refractory material in furnace linings for producing iron, steel, nonferrous metals, glass, and cement. Magnesium oxide and other magnesium compounds are also used in the agricultural, chemical, and construction industries. Magnesium oxide from calcination is used as an electrical insulator in fire-resistant cables.
- Magnesium reacts with haloalkanes to give Grignard reagents, which are used for a wide variety of organic reactions forming carbon–carbon bonds.
- Magnesium salts are included in various foods, fertilizers (magnesium is a component of chlorophyll), and microbe culture media.
- Magnesium sulfite is used in the manufacture of paper (sulfite process).
- Magnesium phosphate is used to fireproof wood used in construction.
- Magnesium hexafluorosilicate is used for moth-proofing textiles.

3.4.6 Biological Roles

Mechanism of Action

The important interaction between phosphate and magnesium ions makes magnesium essential to the basic nucleic acid chemistry of all cells of all known living organisms.

More than 300 enzymes require magnesium ions for their catalytic action, including all enzymes using or synthesizing ATP and those that use other nucleotides to synthesize DNA and RNA. The ATP molecule is normally found in a chelate with a magnesium ion.

Nutrition

Diet

Spices, nuts, cereals, cocoa and vegetables are good sources of magnesium. Green leafy vegetables such as spinach are also rich in magnesium.

Dietary Recommendations

In the UK, the recommended daily values for magnesium are 300 mg for men and 270 mg for women. In the U.S. the Recommended Dietary Allowances (RDAs) are 400 mg for men ages 19–30 and 420 mg for older; for women 310 mg for ages 19–30 and 320 mg for older.

Supplementation

Numerous pharmaceutical preparations of magnesium and dietary supplements are available. In two human trials magnesium oxide, one of the most common forms in magnesium dietary supplements because of its high magnesium content per weight, was less bioavailable than magnesium citrate, chloride, lactate or aspartate.

Metabolism

An adult body has 22–26 grams of magnesium, with 60% in the skeleton, 39% intracellular (20% in skeletal muscle), and 1% extracellular. Serum levels are typically 0.7–1.0 mmol/L or 1.8–2.4 mEq/L. Serum magnesium levels may be normal even when intracellular magnesium is deficient. The mechanisms for maintaining the magnesium level in the serum are varying gastrointestinal absorption and renal excretion. Intracellular magnesium is correlated with intracellular potassium. Increased magnesium lowers calcium and can either prevent hypercalcemia or cause hypocalcemia depending on the initial level. Both low and high protein intake conditions inhibit magnesium absorption, as does the amount of phosphate, phytate, and fat in the gut. Unabsorbed dietary magnesium is excreted in feces; absorbed magnesium is excreted in urine and sweat.

Detection in Serum and Plasma

Magnesium status may be assessed by measuring serum and erythrocyte magnesium concentrations coupled with urinary and fecal magnesium content, but intravenous magnesium loading tests are more accurate and practical. A retention of 20% or more of the injected amount indicates deficiency. As of 2004, no biomarker has been established for magnesium.

Magnesium concentrations in plasma or serum may be monitored for efficacy and safety in those receiving the drug therapeutically, to confirm the diagnosis in potential poisoning victims, or to assist in the forensic investigation in a case of fatal overdose. The newborn children of mothers who received parenteral magnesium sulfate during labor may exhibit toxicity with normal serum magnesium levels.

Deficiency

Low plasma magnesium (hypomagnesemia) is common: it is found in 2.5–15% of the general population. From 2005 to 2006, 48 percent of the United States population consumed less magnesium than recommended in the Dietary Reference Intake. Other causes are increased renal or gastrointestinal loss, an increased intracellular shift, and proton-pump inhibitor antacid therapy. Most are asymptomatic, but symptoms referable to neuromuscular, cardiovascular, and metabolic dysfunction may occur. Alcoholism is often associated with magnesium deficiency. Chronically low serum magnesium levels are associated with metabolic syndrome, diabetes mellitus type 2, fasciculation, and hypertension.

Therapy

Intravenous magnesium is recommended by the ACC/AHA/ESC 2006 Guidelines for Management of Patients With Ventricular Arrhythmias and the Prevention of Sudden Cardiac Death for patients with ventricular arrhythmia associated with torsades de pointes who present with long QT syndrome; and for the treatment of patients with digoxin induced arrhythmias.

- Intravenous magnesium sulfate is used for the management of pre-eclampsia and eclampsia.
- Hypomagnesemia, including that caused by alcoholism, is reversible by oral or parenteral magnesium administration depending on the degree of deficiency.
- There is limited evidence that magnesium supplementation may play a role in the prevention and treatment of migraine.

Sorted by type of magnesium salt, other therapeutic applications include:

- Magnesium sulfate, as the heptahydrate called Epsom salts, is used as bath salts, a laxative, and a highly soluble fertilizer.
- Magnesium hydroxide, suspended in water, is used in milk of magnesia antacids and laxatives.
- Magnesium chloride, oxide, gluconate, malate, orotate, glycinate, ascorbate and citrate are all used as oral magnesium supplements.
- Magnesium borate, magnesium salicylate, and magnesium sulfate are used as antiseptics.
- Magnesium bromide is used as a mild sedative (this action is due to the bromide, not the magnesium).

- Magnesium stearate is a slightly flammable white powder with lubricating properties. In pharmaceutical technology, it is used in pharmacological manufacture to prevent tablets from sticking to the equipment while compressing the ingredients into tablet form.
- Magnesium carbonate powder is used by athletes such as gymnasts, weightlifters, and climbers to eliminate palm sweat, prevent sticking, and improve the grip on gymnastic apparatus, lifting bars, and climbing rocks.

Overdose

Overdose from dietary sources alone is unlikely because excess magnesium in the blood is promptly filtered by the kidneys, and overdose is more likely in the presence of impaired renal function. Overdose is not unlikely in case of excessive intake of supplements. Indeed, megadose therapy has caused death in a young child, and severe hypermagnesemia in a woman and a young girl who had healthy kidneys. The most common symptoms of overdose are nausea, vomiting, and diarrhea; other symptoms include hypotension, confusion, slowed heart and respiratory rates, deficiencies of other minerals, coma, cardiac arrhythmia, and death from cardiac arrest.

Function in Plants

Plants require magnesium to synthesize chlorophyll, essential for photosynthesis. Magnesium in the center of the porphyrin ring in chlorophyll functions in a manner similar to the iron in the center of the porphyrin ring in heme. Magnesium deficiency in plants causes late-season yellowing between leaf veins, especially in older leaves, and can be corrected by either applying epsom salts (which is rapidly leached), or crushed dolomitic limestone, to the soil.

3.4.7 Safety Precautions

Magnesium metal and its alloys can be explosive hazards; they are highly flammable in their pure form when molten or in powder or ribbon form. Burning or molten magnesium reacts violently with water. When working with powdered magnesium, safety glasses with eye protection and UV filters (such as welders use) are employed because burning magnesium produces ultraviolet light that can permanently damage the retina of a human eye.

Magnesium is capable of reducing water and releasing highly flammable hydrogen gas:

$$Mg(s)+2H_2O(l) \rightarrow Mg(OH)_2(s)+H_2(g)$$

Therefore, water cannot extinguish magnesium fires. The hydrogen gas produced intensifies the fire. Dry sand is an effective smothering agent, but only on relatively level and flat surfaces. Magnesium reacts with carbon dioxide exothermically to form magnesium oxide and carbon:

$$2Mg(s)+CO_2(g)\rightarrow 2\ MgO(s)+C(s)$$

Hence, carbon dioxide fuels rather than extinguishes magnesium fires. Burning magnesium can be quenched by using a Class D dry chemical fire extinguisher, or by covering the fire with sand or magnesium foundry flux to remove its air source.

3.5 CALCIUM

Calcium is a chemical element; it has symbol Ca and atomic number 20. As an alkaline earth metal, calcium is a reactive metal that forms a dark oxide-nitride layer when exposed to air. Its physical and chemical properties are most similar to its heavier homologues strontium and barium. It is the fifth most abundant element in Earth's crust, and the third most abundant metal, after iron and aluminium. The most common calcium compound on Earth is calcium carbonate, found in limestone and the fossilised remnants of early sea life; gypsum, anhydrite, fluorite, and apatite are also sources of calcium. The name derives from Latin *calx* "lime", which was obtained from heating limestone.

Some calcium compounds were known to the ancients, though their chemistry was unknown until the seventeenth century. Pure calcium was isolated in 1808 via electrolysis of its oxide by Humphry Davy, who named the element. Calcium compounds are widely used in many industries: in foods and pharmaceuticals for calcium supplementation, in the paper industry as bleaches, as components in cement and electrical insulators, and in the manufacture of soaps. On the other hand, the metal in pure form has few applications due to its high reactivity; still, in small quantities it is often used as an alloying component in steelmaking, and sometimes, as a calcium–lead alloy, in making automotive batteries.

Calcium is the most abundant metal and the fifth-most abundant element in the human body. As electrolytes, calcium ions (Ca^{2+}) play a vital role in the physiological and biochemical processes of organisms and cells: in signal transduction pathways where they act as a second messenger; in neurotransmitter release from neurons; in contraction of all muscle cell types; as cofactors in many enzymes; and in fertilization. Calcium ions outside cells are important for maintaining the potential difference across excitable cell membranes, protein synthesis, and bone formation.

3.5.1 Characteristics

Classification

Calcium is a very ductile silvery metal (sometimes described as pale yellow) whose properties are very similar to the heavier elements in its group, strontium, barium, and radium. A calcium atom has twenty electrons, with electron configuration

[Ar]$4s^2$. Like the other elements placed in group 2 of the periodic table, calcium has two valence electrons in the outermost s-orbital, which are very easily lost in chemical reactions to form a dipositive ion with the stable electron configuration of a noble gas, in this case argon.

Hence, calcium is almost always divalent in its compounds, which are usually ionic. Hypothetical univalent salts of calcium would be stable with respect to their elements, but not to disproportionation to the divalent salts and calcium metal, because the enthalpy of formation of MX_2 is much higher than those of the hypothetical MX. This occurs because of the much greater lattice energy afforded by the more highly charged Ca^{2+} cation compared to the hypothetical Ca^{+} cation.

Calcium, strontium, barium, and radium are always considered to be alkaline earth metals; the lighter beryllium and magnesium, also in group 2 of the periodic table, are often included as well. Nevertheless, beryllium and magnesium differ significantly from the other members of the group in their physical and chemical behavior: they behave more like aluminium and zinc respectively and have some of the weaker metallic character of the post-transition metals, which is why the traditional definition of the term "alkaline earth metal" excludes them.

Physical Properties

Calcium metal melts at 842 °C and boils at 1494 °C; these values are higher than those for magnesium and strontium, the neighbouring group 2 metals. It crystallises in the face-centered cubic arrangement like strontium and barium; above 443 °C (716 K), it changes to a body-centered cubic. Its density of 1.526 g/cm^3 (at 20 °C) is the lowest in its group.

Calcium is harder than lead but can be cut with a knife with effort. While calcium is a poorer conductor of electricity than copper or aluminium by volume, it is a better conductor by mass than both due to its very low density. While calcium is infeasible as a conductor for most terrestrial applications as it reacts quickly with atmospheric oxygen, its use as such in space has been considered.

Chemical Properties

The chemistry of calcium is that of a typical heavy alkaline earth metal. For example, calcium spontaneously reacts with water more quickly than magnesium and less quickly than strontium to produce calcium hydroxide and hydrogen gas. It also reacts with the oxygen and nitrogen in air to form a mixture of calcium oxide and calcium nitride.

When finely divided, it spontaneously burns in air to produce the nitride. Bulk calcium is less reactive: it quickly forms a hydration coating in moist air, but below 30% relative humidity it may be stored indefinitely at room temperature.

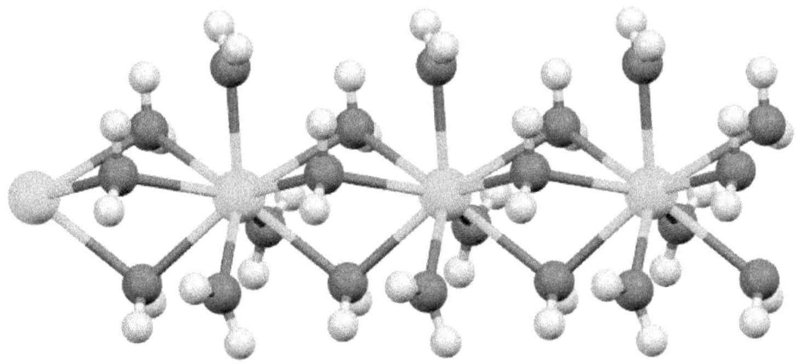

Fig: Structure of the Polymeric $[Ca(H_2O)_6]^{2+}$ Center in Hydrated Calcium Chloride, Illustrating the High Coordination Number Typical for Calcium Complexes

Besides the simple oxide CaO, calcium peroxide, CaO_2, can be made by direct oxidation of calcium metal under a high pressure of oxygen, and there is some evidence for a yellow superoxide $Ca(O_2)_2$.Calcium hydroxide, $Ca(OH)_2$, is a strong base, though not as strong as the hydroxides of strontium, barium or the alkali metals. All four dihalides of calcium are known. Calcium carbonate ($CaCO_3$) and calcium sulfate ($CaSO_4$) are particularly abundant minerals. Like strontium and barium, as well as the alkali metals and the divalent lanthanides europium and ytterbium, calcium metal dissolves directly in liquid ammonia to give a dark blue solution.

Due to the large size of the calcium ion (Ca^{2+}), high coordination numbers are common, up to 24 in some intermetallic compounds such as $CaZn_{13}$. Calcium is readily complexed by oxygen chelates such as EDTA and polyphosphates, which are useful in analytic chemistry and removing calcium ions from hard water. In the absence of steric hindrance, smaller group 2 cations tend to form stronger complexes, but when large polydentate macrocycles are involved the trend is reversed.

Though calcium is in the same group as magnesium and organomagnesium compounds are very widely used throughout chemistry, organocalcium compounds are not similarly widespread because they are more difficult to make and more reactive, though they have recently been investigated as possible catalysts. Organocalcium compounds tend to be more similar to organoytterbium compounds due to the similar ionic radii of Yb^{2+} (102 pm) and Ca^{2+} (100 pm).

Most of these compounds can only be prepared at low temperatures; bulky ligands tend to favor stability. For example, calcium dicyclopentadienyl, $Ca(C_5H_5)_2$, must be made by directly reacting calcium metal with mercurocene or cyclopentadiene itself; replacing the C_5H_5 ligand with the bulkier $C_5(CH_3)_5$ ligand on the other hand increases the compound's solubility, volatility, and kinetic stability.

Isotopes

Natural calcium is a mixture of five stable isotopes (^{40}Ca, ^{42}Ca, ^{43}Ca, ^{44}Ca, and ^{46}Ca) and one isotope with a half-life so long that it is for all practical purposes stable (^{48}Ca, with a half-life of about 4.3×10^{19} years). Calcium is the first (lightest) element to have six naturally occurring isotopes.

By far the most common isotope of calcium in nature is ^{40}Ca, which makes up 96.941% of all natural calcium. It is produced in the silicon-burning process from fusion of alpha particles and is the heaviest stable nuclide with equal proton and neutron numbers; its occurrence is also supplemented slowly by the decay of primordial ^{40}K. Adding another alpha particle leads to unstable ^{44}Ti, which decays via two successive electron captures to stable ^{44}Ca; this makes up 2.806% of all natural calcium and is the second-most common isotope.

The other four natural isotopes, ^{42}Ca, ^{43}Ca, ^{46}Ca, and ^{48}Ca, are significantly rarer, each comprising less than 1% of all natural calcium. The four lighter isotopes are mainly products of the oxygen-burning and silicon-burning processes, leaving the two heavier ones to be produced via neutron capture processes. ^{46}Ca is mostly produced in a "hot" s-process, as its formation requires a rather high neutron flux to allow short-lived ^{45}Ca to capture a neutron. ^{48}Ca is produced by electron capture in the r-process in type Ia supernovae, where high neutron excess and low enough entropy ensures its survival.

^{46}Ca and ^{48}Ca are the first "classically stable" nuclides with a 6-neutron or 8-neutron excess respectively. Although extremely neutron-rich for such a light element, ^{48}Ca is very stable because it is a doubly magic nucleus, having 20 protons and 28 neutrons arranged in closed shells. Its beta decay to ^{48}Sc is very hindered because of the gross mismatch of nuclear spin: ^{48}Ca has zero nuclear spin, being even–even, while ^{48}Sc has spin 6+, so the decay is forbidden by the conservation of angular momentum. While two excited states of ^{48}Sc are available for decay as well, they are also forbidden due to their high spins. As a result, when ^{48}Ca does decay, it does so by double beta decay to ^{48}Ti instead, being the lightest nuclide known to undergo double beta decay.

^{46}Ca can also theoretically undergo double beta decay to ^{46}Ti, but this has never been observed. The most common isotope ^{40}Ca is also doubly magic and could undergo double electron capture to ^{40}Ar, but this has likewise never been observed. Calcium is the only element with two primordial doubly magic isotopes. The experimental lower limits for the half-lives of ^{40}Ca and ^{46}Ca are 5.9×10^{21} years and 2.8×10^{15} years respectively.

Apart from the practically stable ^{48}Ca, the longest lived radioisotope of calcium is ^{41}Ca. It decays by electron capture to stable ^{41}K with a half-life of about 10^5 years.

Its existence in the early Solar System as an extinct radionuclide has been inferred from excesses of ^{41}K: traces of ^{41}Ca also still exist today, as it is a cosmogenic nuclide, continuously produced through neutron activation of natural ^{40}Ca.

Many other calcium radioisotopes are known, ranging from ^{35}Ca to ^{60}Ca. They are all much shorter-lived than ^{41}Ca, the most stable being ^{45}Ca (half-life 163 days) and ^{47}Ca (half-life 4.54 days). Isotopes lighter than ^{42}Ca usually undergo beta plus decay to isotopes of potassium, and those heavier than ^{44}Ca usually undergo beta minus decay to isotopes of scandium, though near the nuclear drip lines, proton emission and neutron emission begin to be significant decay modes as well.

Like other elements, a variety of processes alter the relative abundance of calcium isotopes. The best studied of these processes is the mass-dependent fractionation of calcium isotopes that accompanies the precipitation of calcium minerals such as calcite, aragonite and apatite from solution. Lighter isotopes are preferentially incorporated into these minerals, leaving the surrounding solution enriched in heavier isotopes at a magnitude of roughly 0.025% per atomic mass unit (amu) at room temperature. Mass-dependent differences in calcium isotope composition are conventionally expressed by the ratio of two isotopes (usually $^{44}Ca/^{40}Ca$) in a sample compared to the same ratio in a standard reference material. $^{44}Ca/^{40}Ca$ varies by about 1- 2‰ among organisms on Earth.

3.5.2 History

Calcium compounds were known for millennia, though their chemical makeup was not understood until the 17th century. Lime as a building material and as plaster for statues was used as far back as around 7000 BC. The first dated lime kiln dates back to 2500 BC and was found in Khafajah, Mesopotamia.

About the same time, dehydrated gypsum ($CaSO_4·2H_2O$) was being used in the Great Pyramid of Giza. This material would later be used for the plaster in the tomb of Tutankhamun. The ancient Romans instead used lime mortars made by heating limestone ($CaCO_3$). The name "calcium" itself derives from the Latin word *calx* "lime".

Vitruvius noted that the lime that resulted was lighter than the original limestone, attributing this to the boiling of the water. In 1755, Joseph Black proved that this was due to the loss of carbon dioxide, which as a gas had not been recognised by the ancient Romans.

In 1789, Antoine Lavoisier suspected that lime might be an oxide of a fundamental chemical element. In his table of the elements, Lavoisier listed five "salifiable earths" (i.e., ores that could be made to react with acids to produce salts (*salis* = salt, in Latin): *chaux* (calcium oxide), *magnésie* (magnesia, magnesium oxide), *baryte* (barium sulfate), *alumine* (alumina, aluminium oxide), and *silice* (silica, silicon dioxide)). About these "elements", Lavoisier reasoned:

- We are probably only acquainted as yet with a part of the metallic substances existing in nature, as all those which have a stronger affinity to oxygen than carbon possesses, are incapable, hitherto, of being reduced to a metallic state, and consequently, being only presented to our observation under the form of oxyds, are confounded with earths. It is extremely probable that barytes, which we have just now arranged with earths, is in this situation; for in many experiments it exhibits properties nearly approaching to those of metallic bodies. It is even possible that all the substances we call earths may be only metallic oxyds, irreducible by any hitherto known process.

Calcium, along with its congeners magnesium, strontium, and barium, was first isolated by Humphry Davy in 1808. Following the work of Jöns Jakob Berzelius and Magnus Martin af Pontin on electrolysis, Davy isolated calcium and magnesium by putting a mixture of the respective metal oxides with mercury(II) oxide on a platinum plate which was used as the anode, the cathode being a platinum wire partially submerged into mercury. Electrolysis then gave calcium–mercury and magnesium–mercury amalgams, and distilling off the mercury gave the metal. However, pure calcium cannot be prepared in bulk by this method and a workable commercial process for its production was not found until over a century later.

3.5.3 Occurrence and Production

At 3%, calcium is the fifth most abundant element in the Earth's crust, and the third most abundant metal behind aluminium and iron. It is also the fourth most abundant element in the lunar highlands. Sedimentary calcium carbonate deposits pervade the Earth's surface as fossilized remains of past marine life; they occur in two forms, the rhombohedral calcite (more common) and the orthorhombic aragonite (forming in more temperate seas). Minerals of the first type include limestone, dolomite, marble, chalk, and iceland spar; aragonite beds make up the Bahamas, the Florida Keys, and the Red Sea basins. Corals, sea shells, and pearls are mostly made up of calcium carbonate. Among the other important minerals of calcium are gypsum ($CaSO_4 \cdot 2H_2O$), anhydrite ($CaSO_4$), fluorite (CaF_2), and apatite ($[Ca_5(PO_4)_3X]$, X = OH, Cl, or F).gre

The major producers of calcium are China (about 10000 to 12000 tonnes per year), Russia (about 6000 to 8000 tonnes per year), and the United States (about 2000 to 4000 tonnes per year). Canada and France are also among the minor producers. In 2005, about 24000 tonnes of calcium were produced; about half of the world's extracted calcium is used by the United States, with about 80% of the output used each year. In Russia and China, Davy's method of electrolysis is still used, but is instead applied to molten calcium chloride. Since calcium is less reactive than strontium or barium, the oxide–nitride coating that results in air is stable and lathe machining and other standard metallurgical techniques are suitable for calcium. In the United States and Canada, calcium is instead produced by reducing lime with aluminium at high temperatures.

Geochemical Cycling

Calcium cycling provides a link between tectonics, climate, and the carbon cycle. In the simplest terms, mountain-building exposes calcium-bearing rocks such as basalt and granodiorite to chemical weathering and releases Ca^{2+} into surface water. These ions are transported to the ocean where they react with dissolved CO_2 to form limestone ($CaCO_3$), which in turn settles to the sea floor where it is incorporated into new rocks. Dissolved CO_2, along with carbonate and bicarbonate ions, are termed "dissolved inorganic carbon" (DIC).

The actual reaction is more complicated and involves the bicarbonate ion (HCO^-_3) that forms when CO_2 reacts with water at seawater pH:

$$Ca^2 + 2HCO\bar{3} --> CaCO3 \downarrow + CO_2 + H_2O$$

At seawater pH, most of the dissolved CO_2 is immediately converted back into $HCO^-{}_3$. The reaction results in a net transport of one molecule of CO_2 from the ocean/atmosphere into the lithosphere. The result is that each Ca^{2+} ion released by chemical weathering ultimately removes one CO_2 molecule from the surficial system (atmosphere, ocean, soils and living organisms), storing it in carbonate rocks where it is likely to stay for hundreds of millions of years. The weathering of calcium from rocks thus scrubs CO_2 from the ocean and atmosphere, exerting a strong long-term effect on climate.

3.5.4 Applications

The largest use of metallic calcium is in steelmaking, due to its strong chemical affinity for oxygen and sulfur. Its oxides and sulfides, once formed, give liquid lime aluminate and sulfide inclusions in steel which float out; on treatment, these inclusions disperse throughout the steel and become small and spherical, improving castability, cleanliness and general mechanical properties.

Calcium is also used in maintenance-free automotive batteries, in which the use of 0.1% calcium–lead alloys instead of the usual antimony–lead alloys leads to lower water loss and lower self-discharging.

Due to the risk of expansion and cracking, aluminium is sometimes also incorporated into these alloys. These lead–calcium alloys are also used in casting, replacing lead–antimony alloys.

Calcium is also used to strengthen aluminium alloys used for bearings, for the control of graphitic carbon in cast iron, and to remove bismuth impurities from lead. Calcium metal is found in some drain cleaners, where it functions to generate heat and calcium hydroxide that saponifies the fats and liquefies the proteins (for example, those in hair) that block drains.

Besides metallurgy, the reactivity of calcium is exploited to remove nitrogen from high-purity argon gas and as a getter for oxygen and nitrogen. It is also used as a reducing agent in the production of chromium, zirconium, thorium, vanadium and uranium. It can also be used to store hydrogen gas, as it reacts with hydrogen to form solid calcium hydride, from which the hydrogen can easily be re-extracted.

Calcium isotope fractionation during mineral formation has led to several applications of calcium isotopes. In particular, the 1997 observation by Skulan and DePaolo that calcium minerals are isotopically lighter than the solutions from which the minerals precipitate is the basis of analogous applications in medicine and in paleoceanography. In animals with skeletons mineralized with calcium, the calcium isotopic composition of soft tissues reflects the relative rate of formation and dissolution of skeletal mineral.

In humans, changes in the calcium isotopic composition of urine have been shown to be related to changes in bone mineral balance. When the rate of bone formation exceeds the rate of bone resorption, the $^{44}Ca/^{40}Ca$ ratio in soft tissue rises and vice versa. Because of this relationship, calcium isotopic measurements of urine or blood may be useful in the early detection of metabolic bone diseases like osteoporosis.

A similar system exists in seawater, where $^{44}Ca/^{40}Ca$ tends to rise when the rate of removal of Ca^{2+} by mineral precipitation exceeds the input of new calcium into the ocean. In 1997, Skulan and DePaolo presented the first evidence of change in seawater $^{44}Ca/^{40}Ca$ over geologic time, along with a theoretical explanation of these changes. More recent papers have confirmed this observation, demonstrating that seawater Ca^{2+} concentration is not constant, and that the ocean is never in a "steady state" with respect to calcium input and output. This has important climatological implications, as the marine calcium cycle is closely tied to the carbon cycle.

Many calcium compounds are used in food, as pharmaceuticals, and in medicine, among others. For example, calcium and phosphorus are supplemented in foods through the addition of calcium lactate, calcium diphosphate, and tricalcium phosphate.

The last is also used as a polishing agent in toothpaste and in antacids. Calcium lactobionate is a white powder that is used as a suspending agent for pharmaceuticals. In baking, calcium phosphate is used as a leavening agent. Calcium sulfite is used as a bleach in papermaking and as a disinfectant, calcium silicate is used as a reinforcing agent in rubber, and calcium acetate is a component of liming rosin and is used to make metallic soaps and synthetic resins. Calcium is on the World Health Organization's List of Essential Medicines.

3.5.5 Food Sources

Foods rich in calcium include dairy products such as yogurt and cheese, sardines, salmon, soy products, kale, and fortified breakfast cereals. Because of concerns for long-term adverse side effects, including calcification of arteries and kidney stones, both the U.S. Institute of Medicine (IOM) and the European Food Safety Authority (EFSA) set Tolerable Upper Intake Levels (ULs) for combined dietary and supplemental calcium. From the IOM, people of ages 9–18 years are not to exceed 3 g/day combined intake; for ages 19–50, not to exceed 2.5 g/day; for ages 51 and older, not to exceed 2 g/day. EFSA set the UL for all adults at 2.5 g/day, but decided the information for children and adolescents was not sufficient to determine ULs.

3.5.6 Biological and Pathological Role

Function

Calcium is an essential element needed in large quantities. The Ca^{2+} ion acts as an electrolyte and is vital to the health of the muscular, circulatory, and digestive systems; is indispensable to the building of bone in the form of hydroxyapatite; and supports synthesis and function of blood cells. For example, it regulates the contraction of muscles, nerve conduction, and the clotting of blood. As a result, intra- and extracellular calcium levels are tightly regulated by the body. Calcium can play this role because the Ca^{2+} ion forms stable coordination complexes with many organic compounds, especially proteins; it also forms compounds with a wide range of solubilities, enabling the formation of the skeleton.

Binding

Calcium ions may be complexed by proteins through binding the carboxyl groups of glutamic acid or aspartic acid residues; through interacting with phosphorylated serine, tyrosine, or threonine residues; or by being chelated by γ-carboxylated amino acid residues. Trypsin, a digestive enzyme, uses the first method; osteocalcin, a bone matrix protein, uses the third. Some other bone matrix proteins such as osteopontin and bone sialoprotein use both the first and the second. Direct activation of enzymes by binding calcium is common; some other enzymes are activated by noncovalent association with direct calcium-binding enzymes. Calcium also binds to the phospholipid layer of the cell membrane, anchoring proteins associated with the cell surface.

Solubility

As an example of the wide range of solubility of calcium compounds, monocalcium phosphate is very soluble in water, 85% of extracellular calcium is as dicalcium phosphate with a solubility of 2.00 mM, and the hydroxyapatite of bones in an organic matrix is tricalcium phosphate with a solubility of 1000 μM.

Nutrition

Calcium is a common constituent of multivitamin dietary supplements, but the composition of calcium complexes in supplements may affect its bioavailability which varies by solubility of the salt involved: calcium citrate, malate, and lactate are highly bioavailable, while the oxalate is less. Other calcium preparations include calcium carbonate, calcium citrate malate, and calcium gluconate. The intestine absorbs about one-third of calcium eaten as the free ion, and plasma calcium level is then regulated by the kidneys.

Hormonal Regulation of Bone Formation and Serum Levels

Parathyroid hormone and vitamin D promote the formation of bone by allowing and enhancing the deposition of calcium ions there, allowing rapid bone turnover without affecting bone mass or mineral content. When plasma calcium levels fall, cell surface receptors are activated and the secretion of parathyroid hormone occurs; it then proceeds to stimulate the entry of calcium into the plasma pool by taking it from targeted kidney, gut, and bone cells, with the bone-forming action of parathyroid hormone being antagonised by calcitonin, whose secretion increases with increasing plasma calcium levels.

Abnormal Serum Levels

Excess intake of calcium may cause hypercalcemia. However, because calcium is absorbed rather inefficiently by the intestines, high serum calcium is more likely caused by excessive secretion of parathyroid hormone (PTH) or possibly by excessive intake of vitamin D, both of which facilitate calcium absorption. All these conditions result in excess calcium salts being deposited in the heart, blood vessels, or kidneys. Symptoms include anorexia, nausea, vomiting, memory loss, confusion, muscle weakness, increased urination, dehydration, and metabolic bone disease.

Chronic hypercalcaemia typically leads to calcification of soft tissue and its serious consequences: for example, calcification can cause loss of elasticity of vascular walls and disruption of laminar blood flow—and thence to plaque rupture and thrombosis. Conversely, inadequate calcium or vitamin D intakes may result in hypocalcemia, often caused also by inadequate secretion of parathyroid hormone or defective PTH receptors in cells. Symptoms include neuromuscular excitability, which potentially causes tetany and disruption of conductivity in cardiac tissue.

Bone Disease

As calcium is required for bone development, many bone diseases can be traced to the organic matrix or the hydroxyapatite in molecular structure or organization of bone. Osteoporosis is a reduction in mineral content of bone per unit volume, and can be treated by supplementation of calcium, vitamin D, and bisphosphonates.

Inadequate amounts of calcium, vitamin D, or phosphates can lead to softening of bones, called osteomalacia.

3.5.7 Safety

Metallic Calcium

Because calcium reacts exothermically with water and acids, calcium metal coming into contact with bodily moisture results in severe corrosive irritation. When swallowed, calcium metal has the same effect on the mouth, oesophagus, and stomach, and can be fatal. However, long-term exposure is not known to have distinct adverse effects.

CHAPTER 4

IRON: GENERAL FEATURES OF ITS INORGANIC CHEMISTRY AND BIOCHEMISTRY

4.1 GENERAL, AND AQUEOUS CHEMISTRY

General and aqueous chemistry form the backbone of many scientific and industrial processes. Understanding the principles of chemical reactions, solubility, acids and bases, redox reactions, and the behavior of substances in water is essential for various fields, including environmental science, biology, medicine, and industry. As one of the most important solvents on Earth, water plays a central role in these processes, making aqueous chemistry a critical area of study for understanding both natural phenomena and human-made systems.

4.1.1 Introduction to General Chemistry

General chemistry is the branch of science that studies the composition, structure, properties, and changes of matter. It lays the foundation for understanding how substances interact at the molecular and atomic levels. General chemistry encompasses the study of the periodic table, chemical bonds, reactions, states of matter, and thermodynamics. The core principles of chemistry are essential for advancing in more specialized fields such as organic, inorganic, physical, and analytical chemistry.

Atoms, molecules, and ions are the basic building blocks of matter. Atoms consist of protons, neutrons, and electrons, which interact in ways that define the chemical properties of substances. The periodic table organizes elements based on their atomic structure and properties, serving as a guide for understanding chemical behavior.

4.1.2 States of Matter

Matter exists in different states—solid, liquid, gas, and plasma—depending on the temperature and pressure conditions. In solids, atoms or molecules are closely packed, leading to a defined shape and volume. Liquids have a fixed volume but can change shape to fit their container, while gases expand to fill any space available. The transition between these states occurs through processes like melting, freezing, condensation, and evaporation. Understanding these states and transitions is critical for studying chemical reactions, as the phase of a substance can affect its reactivity and behavior.

4.1.3 Chemical Reactions

A chemical reaction occurs when substances, known as reactants, undergo a transformation to form new substances, called products. Reactions can be categorized into different types, such as synthesis, decomposition, single replacement, and double replacement. Balancing chemical equations is fundamental in chemistry, as it ensures the conservation of mass, where the number of atoms of each element remains constant before and after the reaction.

Reactions often involve the exchange of energy, either releasing it in the form of heat (exothermic reactions) or absorbing it (endothermic reactions). Thermodynamics, a crucial aspect of general chemistry, studies these energy changes. The first law of thermodynamics, also known as the law of energy conservation, states that energy cannot be created or destroyed, only transformed. Understanding energy flow in reactions helps explain phenomena such as combustion, photosynthesis, and cellular respiration.

4.1.4 Aqueous Chemistry

Aqueous chemistry focuses on reactions that occur in water, one of the most common solvents. Water's polar nature, with a partial negative charge near the oxygen atom and a partial positive charge near the hydrogen atoms, makes it an excellent solvent for ionic and polar covalent compounds. When substances dissolve in water, they dissociate into ions or molecules, facilitating reactions.

The concept of solubility is crucial in aqueous chemistry. Solubility refers to the maximum amount of a substance that can dissolve in a solvent at a given temperature. Various factors, such as temperature, pressure, and the nature of the solute and solvent, influence solubility. For example, table salt (sodium chloride) dissolves easily in water, while oil does not because oil is nonpolar, and water is polar.

4.1.5 Acids and Bases in Aqueous Solutions

One of the most important aspects of aqueous chemistry is the study of acids and bases. Acids are substances that increase the concentration of hydrogen ions (H^+) in water, while bases increase the concentration of hydroxide ions (OH^-). The pH scale measures the acidity or basicity of a solution, with a pH of 7 being neutral, values below 7 indicating acidic conditions, and values above 7 indicating basic conditions. Acid-base reactions, also known as neutralization reactions, occur when an acid and a base react to form water and a salt. For example:

$$HCl(aq) + NaOH(aq) \rightarrow NaCl(aq) + H2O(l)$$

The behavior of weak and strong acids and bases differs significantly. Strong acids, like hydrochloric acid (HCl), completely dissociate in water, releasing all their hydrogen ions. Weak acids, like acetic acid (CH_3COOH), only partially dissociate, leading to an equilibrium between the acid and its ions in solution.

4.1.6 Buffer Solutions

Buffer solutions are an essential concept in aqueous chemistry, particularly in biological systems. A buffer is a solution that resists changes in pH when small amounts of acid or base are added. Buffers are typically made of a weak acid and its conjugate base or a weak base and its conjugate acid. For example, a mixture of acetic acid and sodium acetate acts as a buffer solution.

Buffers play a crucial role in maintaining pH stability in biological systems, such as blood, where the pH needs to remain within a narrow range for proper cellular function. In aqueous chemistry, understanding how buffers work helps explain processes like enzyme activity, metabolic reactions, and environmental chemistry.

4.1.7 Precipitation Reactions

Precipitation reactions occur when two soluble ionic compounds react in an aqueous solution to form an insoluble product, known as a precipitate. For example, when solutions of silver nitrate ($AgNO_3$) and sodium chloride (NaCl) are mixed, silver chloride (AgCl) precipitates out of the solution:

$$AgNO3(aq) + NaCl(aq) \rightarrow AgCl(s) + NaNO3(aq)$$

4.1.8 Redox Reactions in Aqueous Solutions

Reduction-oxidation (redox) reactions involve the transfer of electrons between substances. In a redox reaction, one substance is oxidized (loses electrons) while another is reduced (gains electrons). These reactions are crucial in many natural and industrial processes, including corrosion, combustion, and electrochemistry.

In aqueous solutions, redox reactions are often involved in electrochemical cells, where chemical energy is converted into electrical energy or vice versa. For example, in a galvanic cell, a spontaneous redox reaction generates electrical current. Electrolysis, the process of using electrical energy to drive a non-spontaneous reaction, is another important application of redox chemistry.

4.1.9 Water as a Universal Solvent

Water's ability to dissolve a wide variety of substances makes it known as the "universal solvent." This property is due to its polar nature and its ability to form hydrogen bonds with solutes. Water can dissolve many ionic and polar covalent compounds, facilitating biological reactions, nutrient transport, and chemical processes in both natural and industrial settings. In environmental chemistry, water's solvent properties play a key role in the movement of nutrients, pollutants, and minerals. However, this also means that water can easily become contaminated by harmful substances, making water purification a crucial aspect of environmental science.

4.1.10 The Role of Aqueous Chemistry in Environmental Processes

Aqueous chemistry is essential in understanding environmental processes such as acid rain, water treatment, and aquatic ecosystems. Acid rain, caused by the dissolution of sulfur and nitrogen oxides in water, leads to environmental damage by lowering the pH of soils and water bodies. Understanding the chemistry behind this process helps in developing strategies to mitigate its effects.

Water treatment processes, such as coagulation, filtration, and disinfection, rely on aqueous chemistry principles to remove contaminants and make water safe for consumption. Similarly, understanding how substances dissolve and react in water is crucial for studying nutrient cycles and pollution in aquatic ecosystems.

4.2 MOBILIZATION, TRANSPORT, DELIVERY, AND MINERALIZATION OF IRON

Iron is an essential element for virtually all living organisms due to its role in critical biological processes such as oxygen transport, DNA synthesis, and energy production. However, due to its reactivity, iron poses challenges in terms of its mobilization, transport, delivery, and storage. In biological systems, iron exists in two main oxidation states: ferrous iron (Fe^{2+}) and ferric iron (Fe^{3+}), each of which has different solubility properties. The management of iron within organisms must therefore be tightly regulated to prevent toxicity while ensuring its availability for essential processes.

The mobilization, transport, delivery, and mineralization of iron are highly regulated processes essential for maintaining iron homeostasis in biological systems. Iron's role in critical biological functions, such as oxygen transport, energy production, and DNA synthesis, makes its proper regulation crucial. Understanding these processes not only provides insight into normal physiology but also helps in diagnosing and treating disorders of iron metabolism, such as anemia and iron overload conditions. The interplay between iron mobilization, transport, storage, and delivery ensures that iron is available where it is needed while minimizing its potential toxicity.

4.2.1 Iron Mobilization

Iron mobilization refers to the process by which iron is extracted from various storage forms within the body to be made bioavailable for metabolic activities.

Most iron in the body is stored in the form of ferritin, a protein that acts as an iron buffer, safely storing iron as Fe^{3+} in cells until it is required. When iron is needed, specialized enzymes reduce Fe^{3+} to Fe^{2+}, the more soluble and bioavailable form of iron.

In the human body, dietary iron is mobilized from food sources such as meat, fish, and leafy vegetables. Iron in animal products is usually present as heme iron, which is more easily absorbed by the body, while plant-based sources provide non-heme iron, which is less readily absorbed. For efficient mobilization of non-heme iron, gastric acid in the stomach plays a critical role by reducing Fe^{3+} to Fe^{2+}, aided by dietary factors such as vitamin C, which enhances iron absorption. Conversely, certain substances like phytates and polyphenols can inhibit iron absorption.

Once iron is mobilized from storage forms or dietary intake, it needs to be transported safely to cells for utilization in various biochemical processes.

4.2.2 Iron Transport

Iron transport in biological systems is a carefully regulated process to prevent free iron from participating in harmful oxidative reactions. In the bloodstream, iron is bound to transferrin, a glycoprotein that carries Fe^{3+} to various tissues. Transferrin has two binding sites for Fe^{3+} and can circulate in either a fully loaded (diferric), partially loaded (monoferric), or unloaded (apotransferrin) state. This ensures that free iron, which can generate free radicals through the Fenton reaction, is kept in a non-toxic bound form.

The process of iron transport begins in the gut, where dietary iron is absorbed by enterocytes (intestinal cells) in the duodenum. Non-heme iron must be reduced to Fe^{2+} by the enzyme duodenal cytochrome b before being transported into enterocytes by the divalent metal transporter 1 (DMT1). Inside the enterocytes, Fe^{2+} is either stored as ferritin or transported into the bloodstream by the iron exporter ferroportin. Once in the bloodstream, Fe^{2+} is oxidized back to Fe^{3+} by the enzyme hephaestin or ceruloplasmin, allowing it to bind to transferrin.

Transferrin then carries iron through the bloodstream to various tissues, including the liver, spleen, and bone marrow. Iron uptake by cells occurs via transferrin receptors, which are present on the surface of most cell types. Cells internalize transferrin-bound iron through receptor-mediated endocytosis, a process in which the transferrin-iron complex is engulfed by the cell and brought into an endosome. Inside the endosome, Fe^{3+} is reduced to Fe^{2+}, released into the cytoplasm, and used for essential functions such as hemoglobin synthesis or stored in ferritin for later use.

4.2.3 Iron Delivery and Utilization

Once iron has been transported to the cells, it must be delivered to its site of utilization. In erythroid cells (red blood cell precursors), iron is primarily used for the synthesis of hemoglobin, the protein responsible for oxygen transport in the blood. In the mitochondria of these cells, iron is incorporated into heme, a prosthetic group that binds oxygen in hemoglobin. Iron-sulfur clusters, another important iron-containing structure, are also synthesized in mitochondria and are crucial for a variety of cellular processes, including electron transport and DNA repair.

The delivery of iron to the mitochondria is facilitated by specialized transport proteins, including mitoferrin. Once inside the mitochondria, iron is either incorporated into heme or used to form iron-sulfur clusters. These clusters are essential components of the mitochondrial electron transport chain, where they facilitate the transfer of electrons during cellular respiration.

In non-erythroid cells, iron has other functions, such as acting as a cofactor for enzymes involved in DNA synthesis and repair, as well as various metabolic processes. For instance, iron is a critical component of ribonucleotide reductase, the enzyme that catalyzes the formation of deoxyribonucleotides from ribonucleotides, a key step in DNA synthesis. Additionally, iron-containing enzymes like catalase help protect cells from oxidative damage by breaking down hydrogen peroxide into water and oxygen.

4.2.4 Mineralization of Iron

Iron mineralization refers to the process of converting iron into insoluble forms for long-term storage within cells. Ferritin is the primary protein responsible for storing iron in a non-toxic form. Each ferritin molecule can store up to 4500 iron atoms in its hollow core as Fe^{3+}. The mineralization of iron in ferritin occurs through a carefully regulated process in which Fe^{2+} is oxidized to Fe^{3+} and then stored as ferric hydroxide ($Fe(OH)_3$). This process prevents iron from participating in harmful redox reactions that can generate free radicals and cause oxidative damage to cells.

When iron is needed, ferritin releases Fe^{3+}, which is subsequently reduced to Fe^{2+} for transport and utilization. This dynamic process of iron storage and release ensures that the body has a readily available supply of iron while minimizing the risk of iron toxicity.

In addition to ferritin, hemosiderin is another iron storage complex found in the body. Hemosiderin is an insoluble form of iron storage that accumulates in tissues when iron levels are excessively high, such as in conditions of iron overload or repeated blood transfusions.

Unlike ferritin, hemosiderin is less readily accessible for iron mobilization, making it a less efficient storage form.

4.2.5 Regulation of Iron Homeostasis

The regulation of iron levels within the body is critical for maintaining health, as both iron deficiency and iron overload can have serious consequences. The liver plays a central role in regulating systemic iron homeostasis through the production of the hormone hepcidin. Hepcidin controls the amount of iron that enters the bloodstream by regulating the activity of ferroportin, the only known iron exporter in cells. When iron levels are high, hepcidin is secreted by the liver and binds to ferroportin, causing it to be internalized and degraded. This blocks iron from being released into the bloodstream, reducing iron absorption from the diet and iron mobilization from storage sites.

In conditions of iron deficiency, hepcidin production is suppressed, allowing ferroportin to remain active and promoting iron absorption and mobilization. Disorders of iron regulation, such as hereditary hemochromatosis (iron overload) or anemia of chronic disease (iron deficiency), result from disruptions in hepcidin production or ferroportin function.

4.[illegible] Regulation of Iron Homeostasis

[illegible]

CHAPTER 5

OXYGEN TRANSPORT AND THE RESPIRATORY CHAIN

5.1 OXYGEN AND OXYGEN TRANSPORT BY HAEMOGLOBIN AND MYOGLOBIN

5.1.1 Oxygen

Oxygen is a chemical element; it has symbol O and atomic number 8. It is a member of the chalcogen group in the periodic table, a highly reactive nonmetal, and an oxidizing agent that readily forms oxides with most elements as well as with other compounds. Oxygen is the most abundant element in Earth's crust, and after hydrogen and helium, it is the third-most abundant element in the universe. At standard temperature and pressure, two atoms of the element bind to form dioxygen, a colorless and odorless diatomic gas with the formula O_2. Diatomic oxygen gas currently constitutes 20.95% of the Earth's atmosphere, though this has changed considerably over long periods of time. Oxygen makes up almost half of the Earth's crust in the form of oxides.

All plants, animals, and fungi need oxygen for cellular respiration, which extracts energy by the reaction of oxygen with molecules derived from food and produces carbon dioxide as a waste product. In tetrapods breathing brings oxygen into the lungs where gas exchange takes place, carbon dioxide diffuses out of the blood, and oxygen diffuses into the blood. The body's circulatory system transports the oxygen to the cells, where cellular respiration takes place.

Many major classes of organic molecules in living organisms contain oxygen atoms, such as proteins, nucleic acids, carbohydrates, and fats, as do the major constituent inorganic compounds of animal shells, teeth, and bone. Most of the mass of living organisms is oxygen as a component of water, the major constituent of lifeforms. Oxygen is continuously replenished in Earth's atmosphere by photosynthesis, which uses the energy of sunlight to produce oxygen from water and carbon dioxide. Oxygen is too chemically reactive to remain a free element in air without being continuously replenished by the photosynthetic action of living organisms. Another form (allotrope) of oxygen, ozone (O_3), strongly absorbs ultraviolet UVB radiation and the high-altitude ozone layer helps protect the biosphere from ultraviolet radiation. However, ozone present at the surface is a byproduct of smog and thus a pollutant.

Oxygen was isolated by Michael Sendivogius before 1604, but it is commonly believed that the element was discovered independently by Carl Wilhelm Scheele, in Uppsala, in 1773 or earlier, and Joseph Priestley in Wiltshire, in 1774. Priority is often given for Priestley because his work was published first. Priestley, however, called oxygen "dephlogisticated air", and did not recognize it as a chemical element. The name *oxygen* was coined in 1777 by Antoine Lavoisier, who first recognized oxygen as a chemical element and correctly characterized the role it plays in combustion.

Common uses of oxygen include production of steel, plastics and textiles, brazing, welding and cutting of steels and other metals, rocket propellant, oxygen therapy, and life support systems in aircraft, submarines, spaceflight and diving.

Early Experiments

One of the first known experiments on the relationship between combustion and air was conducted by the 2nd century BCE Greek writer on mechanics, Philo of Byzantium. In his work *Pneumatica*, Philo observed that inverting a vessel over a burning candle and surrounding the vessel's neck with water resulted in some water rising into the neck. Philo incorrectly surmised that parts of the air in the vessel were converted into the classical element fire and thus were able to escape through pores in the glass. Many centuries later Leonardo da Vinci built on Philo's work by observing that a portion of air is consumed during combustion and respiration.

In the late 17th century, Robert Boyle proved that air is necessary for combustion. English chemist John Mayow (1641–1679) refined this work by showing that fire requires only a part of air that he called *spiritus nitroaereus.*

In one experiment, he found that placing either a mouse or a lit candle in a closed container over water caused the water to rise and replace one-fourteenth of the air's volume before extinguishing the subjects. From this, he surmised that nitroaereus is consumed in both respiration and combustion.

Mayow observed that antimony increased in weight when heated, and inferred that the nitroaereus must have combined with it. He also thought that the lungs separate nitroaereus from air and pass it into the blood and that animal heat and muscle movement result from the reaction of nitroaereus with certain substances in the body.

Accounts of these and other experiments and ideas were published in 1668 in his work *Tractatus duo* in the tract "De respiratione".

Phlogiston Theory

Robert Hooke, Ole Borch, Mikhail Lomonosov, and Pierre Bayen all produced oxygen in experiments in the 17th and the 18th century but none of them recognized it as a chemical element. This may have been in part due to the prevalence of the philosophy of combustion and corrosion called the *phlogiston theory*, which was then the favored explanation of those processes.

Established in 1667 by the German alchemist J. J. Becher, and modified by the chemist Georg Ernst Stahl by 1731, phlogiston theory stated that all combustible materials were made of two parts. One part, called phlogiston, was given off when the substance containing it was burned, while the dephlogisticated part was thought to be its true form, or calx.

Highly combustible materials that leave little residue, such as wood or coal, were thought to be made mostly of phlogiston; non-combustible substances that corrode, such as iron, contained very little. Air did not play a role in phlogiston theory, nor were any initial quantitative experiments conducted to test the idea; instead, it was based on observations of what happens when something burns, that most common objects appear to become lighter and seem to lose something in the process.

Discovery

Polish alchemist, philosopher, and physician Michael Sendivogius in his work *De Lapide Philosophorum Tractatus duodecim e naturae fonte et manuali experientia depromti* ["Twelve Treatises on the Philosopher's Stone drawn from the source of nature and manual experience"] (1604) described a substance contained in air, referring to it as 'cibus vitae' (food of life,) and according to Polish historian Roman Bugaj, this substance is identical with oxygen. Sendivogius, during his experiments performed between 1598 and 1604, properly recognized that the substance is equivalent to the gaseous byproduct released by the thermal decomposition of potassium nitrate. In Bugaj's view, the isolation of oxygen and the proper association of the substance to that part of air which is required for life, provides sufficient evidence for the discovery of oxygen by Sendivogius.

This discovery of Sendivogius was however frequently denied by the generations of scientists and chemists which succeeded him.

It is also commonly claimed that oxygen was first discovered by Swedish pharmacist Carl Wilhelm Scheele. He had produced oxygen gas by heating mercuric oxide (HgO) and various nitrates in 1771–72.

Scheele called the gas "fire air" because it was then the only known agent to support combustion. He wrote an account of this discovery in a manuscript titled *Treatise on Air and Fire*, which he sent to his publisher in 1775. That document was published in 1777.

In the meantime, on August 1, 1774, an experiment conducted by the British clergyman Joseph Priestley focused sunlight on mercuric oxide contained in a glass tube, which liberated a gas he named "dephlogisticated air". He noted that candles burned brighter in the gas and that a mouse was more active and lived longer while breathing it. After breathing the gas himself, Priestley wrote: "The feeling of it to my lungs was not sensibly different from that of common air, but I fancied that my breast felt peculiarly light and easy for some time afterwards." Priestley published his findings in 1775 in a paper titled "An Account of Further Discoveries in Air", which was included in the second volume of his book titled *Experiments and Observations on Different Kinds of Air*. Because he published his findings first, Priestley is usually given priority in the discovery.

The French chemist Antoine Laurent Lavoisier later claimed to have discovered the new substance independently. Priestley visited Lavoisier in October 1774 and told him about his experiment and how he liberated the new gas. Scheele had also dispatched a letter to Lavoisier on September 30, 1774, which described his discovery of the previously unknown substance, but Lavoisier never acknowledged receiving it (a copy of the letter was found in Scheele's belongings after his death).

Lavoisier's Contribution

Lavoisier conducted the first adequate quantitative experiments on oxidation and gave the first correct explanation of how combustion works. He used these and similar experiments, all started in 1774, to discredit the phlogiston theory and to prove that the substance discovered by Priestley and Scheele was a chemical element.

In one experiment, Lavoisier observed that there was no overall increase in weight when tin and air were heated in a closed container. He noted that air rushed in when he opened the container, which indicated that part of the trapped air had been consumed. He also noted that the tin had increased in weight and that increase was the same as the weight of the air that rushed back in. This and other experiments on combustion were documented in his book *Sur la combustion en général*, which was published in 1777. In that work, he proved that air is a mixture of two gases; 'vital air', which is essential to combustion and respiration, and *azote*, which did not support either. *Azote* later became *nitrogen* in English, although it has kept the earlier name in French and several other European languages.

Later History

John Dalton's original atomic hypothesis presumed that all elements were monatomic and that the atoms in compounds would normally have the simplest atomic ratios with respect to one another. For example, Dalton assumed that water's formula was HO, leading to the conclusion that the atomic mass of oxygen was 8 times that of hydrogen, instead of the modern value of about 16. In 1805, Joseph Louis Gay-Lussac and Alexander von Humboldt showed that water is formed of two volumes of hydrogen and one volume of oxygen; and by 1811 Amedeo Avogadro had arrived at the correct interpretation of water's composition, based on what is now called Avogadro's law and the diatomic elemental molecules in those gases.

The first commercial method of producing oxygen was chemical, the so-called Brin process involving a reversible reaction of barium oxide. It was invented in 1852 and commercialized in 1884, but was displaced by newer methods in early 20th century.

By the late 19th century scientists realized that air could be liquefied and its components isolated by compressing and cooling it. Using a cascade method, Swiss chemist and physicist Raoul Pierre Pictet evaporated liquid sulfur dioxide in order to liquefy carbon dioxide, which in turn was evaporated to cool oxygen gas enough to liquefy it. He sent a telegram on December 22, 1877, to the French Academy of Sciences in Paris announcing his discovery of liquid oxygen. Just two days later, French physicist Louis Paul Cailletet announced his own method of liquefying molecular oxygen.

Only a few drops of the liquid were produced in each case and no meaningful analysis could be conducted. Oxygen was liquefied in a stable state for the first time on March 29, 1883, by Polish scientists from Jagiellonian University, Zygmunt Wróblewski and Karol Olszewski.

In 1891 Scottish chemist James Dewar was able to produce enough liquid oxygen for study. The first commercially viable process for producing liquid oxygen was independently developed in 1895 by German engineer Carl von Linde and British engineer William Hampson.

Both men lowered the temperature of air until it liquefied and then distilled the component gases by boiling them off one at a time and capturing them separately. Later, in 1901, oxyacetylene welding was demonstrated for the first time by burning a mixture of acetylene and compressed O_2. This method of welding and cutting metal later became common.

In 1923, the American scientist Robert H. Goddard became the first person to develop a rocket engine that burned liquid fuel; the engine used gasoline for fuel and liquid oxygen as the oxidizer. Goddard successfully flew a small liquid-fueled rocket 56 m at 97 km/h on March 16, 1926, in Auburn, Massachusetts, US.

In academic laboratories, oxygen can be prepared by heating together potassium chlorate mixed with a small proportion of manganese dioxide.

Oxygen levels in the atmosphere are trending slightly downward globally, possibly because of fossil-fuel burning.

Characteristics

Properties and molecular structure

At standard temperature and pressure, oxygen is a colorless, odorless, and tasteless gas with the molecular formula O_2, referred to as dioxygen.

As *dioxygen*, two oxygen atoms are chemically bound to each other. The bond can be variously described based on level of theory, but is reasonably and simply described as a covalent double bond that results from the filling of molecular orbitals formed from the atomic orbitals of the individual oxygen atoms, the filling of which results in a bond order of two. More specifically, the double bond is the result of sequential, low-to-high energy, or Aufbau, filling of orbitals, and the resulting cancellation of contributions from the 2s electrons, after sequential filling of the low σ and σ^* orbitals; σ overlap of the two atomic 2p orbitals that lie along the O–O molecular axis and π overlap of two pairs of atomic 2p orbitals perpendicular to the O–O molecular axis, and then cancellation of contributions from the remaining two 2p electrons after their partial filling of the π^* orbitals.

This combination of cancellations and σ and π overlaps results in dioxygen's double-bond character and reactivity, and a triplet electronic ground state. An electron configuration with two unpaired electrons, as is found in dioxygen orbitals that are of equal energy—i.e., degenerate—is a configuration termed a spin triplet state. Hence, the ground state of the O_2 molecule is referred to as triplet oxygen. The highest-energy, partially filled orbitals are antibonding, and so their filling weakens the bond order from three to two. Because of its unpaired electrons, triplet oxygen reacts only slowly with most organic molecules, which have paired electron spins; this prevents spontaneous combustion.

In the triplet form, O_2 molecules are paramagnetic. That is, they impart magnetic character to oxygen when it is in the presence of a magnetic field, because of the spin magnetic moments of the unpaired electrons in the molecule, and the negative exchange energy between neighboring O_2 molecules. Liquid oxygen is so magnetic that, in laboratory demonstrations, a bridge of liquid oxygen may be supported against its own weight between the poles of a powerful magnet.

Singlet oxygen is a name given to several higher-energy species of molecular O_2 in which all the electron spins are paired. It is much more reactive with common organic molecules than is normal (triplet) molecular oxygen. In nature, singlet oxygen is commonly formed from water during photosynthesis, using the energy of sunlight. It is also produced in the troposphere by the photolysis of ozone by light of short wavelength and by the immune system as a source of active oxygen. Carotenoids in photosynthetic organisms (and possibly animals) play a major role in absorbing energy from singlet oxygen and converting it to the unexcited ground state before it can cause harm to tissues.

Allotropes

The common allotrope of elemental oxygen on Earth is called dioxygen, O_2, the major part of the Earth's atmospheric oxygen. O_2 has a bond length of 121 pm and a bond energy of 498 kJ/mol. O_2 is used by complex forms of life, such as animals, in cellular respiration. Other aspects of O_2 are covered in the remainder of this article.

Trioxygen (O_3) is usually known as ozone and is a very reactive allotrope of oxygen that is damaging to lung tissue. Ozone is produced in the upper atmosphere when O_2 combines with atomic oxygen made by the splitting of O_2 by ultraviolet (UV) radiation. Since ozone absorbs strongly in the UV region of the spectrum, the ozone layer of the upper atmosphere functions as a protective radiation shield for the planet. Near the Earth's surface, it is a pollutant formed as a by-product of automobile exhaust. At low earth orbit altitudes, sufficient atomic oxygen is present to cause corrosion of spacecraft.

The metastable molecule tetraoxygen (O_4) was discovered in 2001, and was assumed to exist in one of the six phases of solid oxygen. It was proven in 2006 that this phase, created by pressurizing O_2 to 20 GPa, is in fact a rhombohedral O_8 cluster. This cluster has the potential to be a much more powerful oxidizer than either O_2 or O_3 and may therefore be used in rocket fuel. A metallic phase was discovered in 1990 when solid oxygen is subjected to a pressure of above 96 GPa and it was shown in 1998 that at very low temperatures, this phase becomes superconducting.

Physical Properties

Oxygen dissolves more readily in water than nitrogen, and in freshwater more readily than in seawater. Water in equilibrium with air contains approximately 1 molecule of dissolved O_2 for every 2 molecules of N_2 (1:2), compared with an atmospheric ratio of approximately 1:4. The solubility of oxygen in water is temperature-dependent, and about twice as much (14.6 mg/L) dissolves at 0°C than at 20°C (7.6mg/L). At 25°C and 1 standard atmosphere (101.3kPa) of air, freshwater can dissolve about 6.04 milliliters (mL) of oxygen per liter, and seawater contains about 4.95mL per liter. At 5°C the solubility increases to 9.0mL (50% more than at 25°C) per liter for freshwater and 7.2mL (45% more) per liter for sea water.

Oxygen condenses at 90.20K (-182.95°C, -297.31°F) and freezes at 54.36K (-218.79°C, -361.82°F). Both liquid and solid O_2 are clear substances with a light sky-blue color caused by absorption in the red (in contrast with the blue color of the sky, which is due to Rayleigh scattering of blue light). High-purity liquid O_2 is usually obtained by the fractional distillation of liquefied air. Liquid oxygen may also be condensed from air using liquid nitrogen as a coolant.

Liquid oxygen is a highly reactive substance and must be segregated from combustible materials.

The spectroscopy of molecular oxygen is associated with the atmospheric processes of aurora and airglow. The absorption in the Herzberg continuum and Schumann–Runge bands in the ultraviolet produces atomic oxygen that is important in the chemistry of the middle atmosphere. Excited-state singlet molecular oxygen is responsible for red chemiluminescence in solution.

Isotopes and Stellar Origin

Naturally occurring oxygen is composed of three stable isotopes, ^{16}O, ^{17}O, and ^{18}O, with ^{16}O being the most abundant (99.762% natural abundance).

Most ^{16}O is synthesized at the end of the helium fusion process in massive stars but some is made in the neon burning process. ^{17}O is primarily made by the burning of hydrogen into helium during the CNO cycle, making it a common isotope in the hydrogen burning zones of stars. Most ^{18}O is produced when ^{14}N (made abundant from CNO burning) captures a ^{4}He nucleus, making ^{18}O common in the helium-rich zones of evolved, massive stars.

Fifteen radioisotopes have been characterized, ranging from ^{11}O to ^{28}O. The most stable are ^{15}O with a half-life of 122.24 seconds and ^{14}O with a half-life of 70.606 seconds. All of the remaining radioactive isotopes have half-lives that are less than 27 seconds and the majority of these have half-lives that are less than 83 milliseconds. The most common decay mode of the isotopes lighter than ^{16}O is β^+ decay to yield nitrogen, and the most common mode for the isotopes heavier than ^{18}O is beta decay to yield fluorine.

Occurrence

Oxygen is the most abundant chemical element by mass in the Earth's biosphere, air, sea and land. Oxygen is the third most abundant chemical element in the universe, after hydrogen and helium. About 0.9% of the Sun's mass is oxygen. Oxygen constitutes 49.2% of the Earth's crust by mass as part of oxide compounds such as silicon dioxide and is the most abundant element by mass in the Earth's crust. It is also the major component of the world's oceans (88.8% by mass). Oxygen gas is the second most common component of the Earth's atmosphere, taking up 20.8% of its volume and 23.1% of its mass (some 10^{15} tonnes). Earth is unusual among the planets of the Solar System in having such a high concentration of oxygen gas in its atmosphere: Mars (with 0.1% O_2 by volume) and Venus have much less. The O_2 surrounding those planets is produced solely by the action of ultraviolet radiation on oxygen-containing molecules such as carbon dioxide.

The unusually high concentration of oxygen gas on Earth is the result of the oxygen cycle. This biogeochemical cycle describes the movement of oxygen within and between its three main reservoirs on Earth: the atmosphere, the biosphere, and the lithosphere. The main driving factor of the oxygen cycle is photosynthesis, which is responsible for modern Earth's atmosphere. Photosynthesis releases oxygen into the atmosphere, while respiration, decay, and combustion remove it from the atmosphere. In the present equilibrium, production and consumption occur at the same rate.

Free oxygen also occurs in solution in the world's water bodies. The increased solubility of O_2 at lower temperatures has important implications for ocean life, as polar oceans support a much higher density of life due to their higher oxygen content. Water polluted with plant nutrients such as nitrates or phosphates may stimulate growth of algae by a process called eutrophication and the decay of these organisms and other biomaterials may reduce the O_2 content in eutrophic water bodies. Scientists assess this aspect of water quality by measuring the water's biochemical oxygen demand, or the amount of O_2 needed to restore it to a normal concentration.

Analysis

Paleoclimatologists measure the ratio of oxygen-18 and oxygen-16 in the shells and skeletons of marine organisms to determine the climate millions of years ago. Seawater molecules that contain the lighter isotope, oxygen-16, evaporate at a slightly faster rate than water molecules containing the 12% heavier oxygen-18, and this disparity increases at lower temperatures. During periods of lower global temperatures, snow and rain from that evaporated water tends to be higher in oxygen-16, and the seawater left behind tends to be higher in oxygen-18. Marine organisms then incorporate more oxygen-18 into their skeletons and shells than they would in a warmer climate. Paleoclimatologists also directly measure this ratio in the water molecules of ice core samples as old as hundreds of thousands of years.

Planetary geologists have measured the relative quantities of oxygen isotopes in samples from the Earth, the Moon, Mars, and meteorites, but were long unable to obtain reference values for the isotope ratios in the Sun, believed to be the same as those of the primordial solar nebula. Analysis of a silicon wafer exposed to the solar wind in space and returned by the crashed Genesis spacecraft has shown that the Sun has a higher proportion of oxygen-16 than does the Earth. The measurement implies that an unknown process depleted oxygen-16 from the Sun's disk of protoplanetary material prior to the coalescence of dust grains that formed the Earth.

Oxygen presents two spectrophotometric absorption bands peaking at the wavelengths 687 and 760 nm. Some remote sensing scientists have proposed using the measurement of the radiance coming from vegetation canopies in those bands to characterize plant health status from a satellite platform. This approach exploits the fact that in those bands it is possible to discriminate the vegetation's reflectance from its fluorescence, which is much weaker. The measurement is technically difficult owing to the low signal-to-noise ratio and the physical structure of vegetation; but it has been proposed as a possible method of monitoring the carbon cycle from satellites on a global scale.

Biological Production and Role of O_2

Photosynthesis and Respiration

In nature, free oxygen is produced by the light-driven splitting of water during oxygenic photosynthesis. According to some estimates, green algae and cyanobacteria in marine environments provide about 70% of the free oxygen produced on Earth, and the rest is produced by terrestrial plants. Other estimates of the oceanic contribution to atmospheric oxygen are higher, while some estimates are lower, suggesting oceans produce ~45% of Earth's atmospheric oxygen each year.

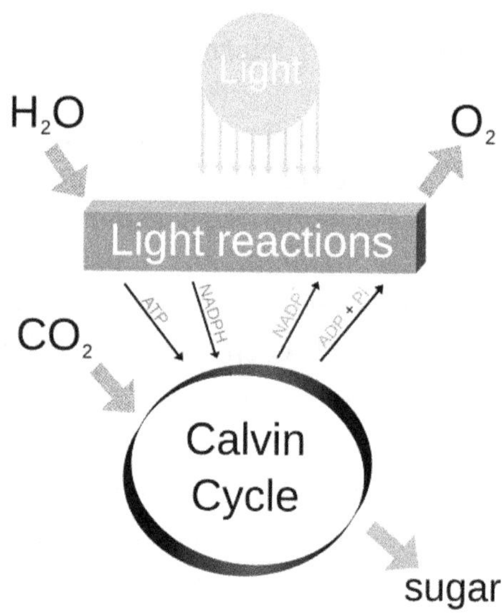

Fig: Photosynthesis Splits Water to Liberate O_2 and Fixes CO_2 into Sugar in what is called a Calvin Cycle

A simplified overall formula for photosynthesis is

$$6CO_2 + 6H_2O + photons \rightarrow C_6H_{12}O_6 + 6O_2$$

Photolytic oxygen evolution occurs in the thylakoid membranes of photosynthetic organisms and requires the energy of four photons. Many steps are involved, but the result is the formation of a proton gradient across the thylakoid membrane, which is used to synthesize adenosine triphosphate (ATP) via photophosphorylation.

The O_2 remaining (after production of the water molecule) is released into the atmosphere.

Oxygen is used in mitochondria in the generation of ATP during oxidative phosphorylation. The reaction for aerobic respiration is essentially the reverse of photosynthesis and is simplified as

$$C_6H_{12}O_6 + 6O_2 \rightarrow 6\,CO_2 + 6H_2O + 2880\ kJ\,/\,mol$$

In vertebrates, O_2 diffuses through membranes in the lungs and into red blood cells. Hemoglobin binds O_2, changing color from bluish red to bright red (CO_2 is released from another part of hemoglobin through the Bohr effect). Other animals use hemocyanin (molluscs and some arthropods) or hemerythrin (spiders and lobsters). A liter of blood can dissolve 200 cm^3 of O_2.

Until the discovery of anaerobic metazoa, oxygen was thought to be a requirement for all complex life.

Reactive oxygen species, such as superoxide ion (O^-_2) and hydrogen peroxide ($H_{20\,2}$), are reactive by-products of oxygen use in organisms. Parts of the immune system of higher organisms create peroxide, superoxide, and singlet oxygen to destroy invading microbes. Reactive oxygen species also play an important role in the hypersensitive response of plants against pathogen attack. Oxygen is damaging to obligately anaerobic organisms, which were the dominant form of early life on Earth until O_2 began to accumulate in the atmosphere about 2.5 billion years ago during the Great Oxygenation Event, about a billion years after the first appearance of these organisms.

An adult human at rest inhales 1.8 to 2.4 grams of oxygen per minute. This amounts to more than 6 billion tonnes of oxygen inhaled by humanity per year.

Living Organisms

The free oxygen partial pressure in the body of a living vertebrate organism is highest in the respiratory system, and decreases along any arterial system, peripheral tissues, and venous system, respectively. Partial pressure is the pressure that oxygen would have if it alone occupied the volume.

Build-Up in the Atmosphere

Free oxygen gas was almost nonexistent in Earth's atmosphere before photosynthetic archaea and bacteria evolved, probably about 3.5 billion years ago. Free oxygen first appeared in significant quantities during the Paleoproterozoic era (between 3.0 and 2.3 billion years ago). Even if there was much dissolved iron in the oceans when oxygenic photosynthesis was getting more common, it appears the banded iron formations were created by anoxyenic or micro-aerophilic iron-oxidizing bacteria which dominated the deeper areas of the photic zone, while oxygen-producing cyanobacteria covered the shallows. Free oxygen began to outgas from the oceans 3–2.7 billion years ago, reaching 10% of its present level around 1.7 billion years ago.

The presence of large amounts of dissolved and free oxygen in the oceans and atmosphere may have driven most of the extant anaerobic organisms to extinction during the Great Oxygenation Event (*oxygen catastrophe*) about 2.4 billion years ago. Cellular respiration using O_2 enables aerobic organisms to produce much more ATP than anaerobic organisms. Cellular respiration of O_2 occurs in all eukaryotes, including all complex multicellular organisms such as plants and animals.

Since the beginning of the Cambrian period 540 million years ago, atmospheric O_2 levels have fluctuated between 15% and 30% by volume. Towards the end of the Carboniferous period (about 300 million years ago) atmospheric O_2 levels reached a maximum of 35% by volume, which may have contributed to the large size of insects and amphibians at this time.

Variations in atmospheric oxygen concentration have shaped past climates. When oxygen declined, atmospheric density dropped, which in turn increased surface evaporation, causing precipitation increases and warmer temperatures.

At the current rate of photosynthesis it would take about 2,000 years to regenerate the entire O_2 in the present atmosphere. It is estimated that oxygen on Earth will last for about one billion years.

Extraterrestrial Free Oxygen

In the field of astrobiology and in the search for extraterrestrial life oxygen is a strong biosignature. That said it might not be a definite biosignature, being possibly produced abiotically on celestial bodies with processes and conditions (such as a peculiar hydrosphere) which allow free oxygen, like with Europa's and Ganymede's thin oxygen atmospheres.

Industrial Production

One hundred million tonnes of O_2 are extracted from air for industrial uses annually by two primary methods. The most common method is fractional distillation of liquefied air, with N_2 distilling as a vapor while O_2 is left as a liquid.

The other primary method of producing O_2 is passing a stream of clean, dry air through one bed of a pair of identical zeolite molecular sieves, which absorbs the nitrogen and delivers a gas stream that is 90% to 93% O_2. Simultaneously, nitrogen gas is released from the other nitrogen-saturated zeolite bed, by reducing the chamber operating pressure and diverting part of the oxygen gas from the producer bed through it, in the reverse direction of flow. After a set cycle time the operation of the two beds is interchanged, thereby allowing for a continuous supply of gaseous oxygen to be pumped through a pipeline. This is known as pressure swing adsorption. Oxygen gas is increasingly obtained by these non-cryogenic technologies.

Oxygen gas can also be produced through electrolysis of water into molecular oxygen and hydrogen. DC electricity must be used: if AC is used, the gases in each limb consist of hydrogen and oxygen in the explosive ratio 2:1. A similar method is the electrocatalytic O_2 evolution from oxides and oxoacids. Chemical catalysts can be used as well, such as in chemical oxygen generators or oxygen candles that are used as part of the life-support equipment on submarines, and are still part of standard equipment on commercial airliners in case of depressurization emergencies. Another air separation method is forcing air to dissolve through ceramic membranes based on zirconium dioxide by either high pressure or an electric current, to produce nearly pure O_2 gas.

Storage

Oxygen storage methods include high-pressure oxygen tanks, cryogenics and chemical compounds. For reasons of economy, oxygen is often transported in bulk as a liquid in specially insulated tankers, since one liter of liquefied oxygen is equivalent to 840 liters of gaseous oxygen at atmospheric pressure and 20°C (68°F). Such tankers are used to refill bulk liquid-oxygen storage containers, which stand outside hospitals and other institutions that need large volumes of pure oxygen gas. Liquid oxygen is passed through heat exchangers, which convert the cryogenic liquid into gas before it enters the building. Oxygen is also stored and shipped in smaller cylinders containing the compressed gas; a form that is useful in certain portable medical applications and oxy-fuel welding and cutting.

Applications

Medical

Uptake of O_2 from the air is the essential purpose of respiration, so oxygen supplementation is used in medicine. Treatment not only increases oxygen levels in the patient's blood, but has the secondary effect of decreasing resistance to blood flow in many types of diseased lungs, easing work load on the heart.

Oxygen therapy is used to treat emphysema, pneumonia, some heart disorders (congestive heart failure), some disorders that cause increased pulmonary artery pressure, and any disease that impairs the body's ability to take up and use gaseous oxygen.

Treatments are flexible enough to be used in hospitals, the patient's home, or increasingly by portable devices. Oxygen tents were once commonly used in oxygen supplementation, but have since been replaced mostly by the use of oxygen masks or nasal cannulas.

Hyperbaric (high-pressure) medicine uses special oxygen chambers to increase the partial pressure of O_2 around the patient and, when needed, the medical staff. Carbon monoxide poisoning, gas gangrene, and decompression sickness (the 'bends') are sometimes addressed with this therapy. Increased O_2 concentration in the lungs helps to displace carbon monoxide from the heme group of hemoglobin. Oxygen gas is poisonous to the anaerobic bacteria that cause gas gangrene, so increasing its partial pressure helps kill them. Decompression sickness occurs in divers who decompress too quickly after a dive, resulting in bubbles of inert gas, mostly nitrogen and helium, forming in the blood. Increasing the pressure of O_2 as soon as possible helps to redissolve the bubbles back into the blood so that these excess gasses can be exhaled naturally through the lungs. Normobaric oxygen administration at the highest available concentration is frequently used as first aid for any diving injury that may involve inert gas bubble formation in the tissues. There is epidemiological support for its use from a statistical study of cases recorded in a long term database.

Life Support and Recreational Use

An application of O_2 as a low-pressure breathing gas is in modern space suits, which surround their occupant's body with the breathing gas. These devices use nearly pure oxygen at about one-third normal pressure, resulting in a normal blood partial pressure of O_2. This trade-off of higher oxygen concentration for lower pressure is needed to maintain suit flexibility.

Scuba and surface-supplied underwater divers and submarines also rely on artificially delivered O_2. Submarines, submersibles and atmospheric diving suits usually operate at normal atmospheric pressure. Breathing air is scrubbed of carbon dioxide by chemical extraction and oxygen is replaced to maintain a constant partial pressure. Ambient pressure divers breathe air or gas mixtures with an oxygen fraction suited to the operating depth. Pure or nearly pure O_2 use in diving at pressures higher than atmospheric is usually limited to rebreathers, or decompression at relatively shallow depths (~6 meters depth, or less), or medical treatment in recompression chambers at pressures up to 2.8 bar, where acute oxygen toxicity can be managed without the risk of drowning. Deeper diving requires significant dilution of O_2 with other gases, such as nitrogen or helium, to prevent oxygen toxicity.

People who climb mountains or fly in non-pressurized fixed-wing aircraft sometimes have supplemental O_2 supplies. Pressurized commercial airplanes have an emergency supply of O_2 automatically supplied to the passengers in case of cabin depressurization. Sudden cabin pressure loss activates chemical oxygen generators above each seat, causing oxygen masks to drop. Pulling on the masks "to start the flow of oxygen" as cabin safety instructions dictate, forces iron filings into the sodium chlorate inside the canister. A steady stream of oxygen gas is then produced by the exothermic reaction.

Oxygen, as a mild euphoric, has a history of recreational use in oxygen bars and in sports. Oxygen bars are establishments found in the United States since the late 1990s that offer higher than normal O_2 exposure for a minimal fee. Professional athletes, especially in American football, sometimes go off-field between plays to don oxygen masks to boost performance. The pharmacological effect is doubted; a placebo effect is a more likely explanation. Available studies support a performance boost from oxygen enriched mixtures only if it is inhaled *during* aerobic exercise.

Other recreational uses that do not involve breathing include pyrotechnic applications, such as George Goble's five-second ignition of barbecue grills.

Industrial

Smelting of iron ore into steel consumes 55% of commercially produced oxygen. In this process, O_2 is injected through a high-pressure lance into molten iron, which removes sulfur impurities and excess carbon as the respective oxides, SO_2 and CO_2. The reactions are exothermic, so the temperature increases to 1,700°C.

Another 25% of commercially produced oxygen is used by the chemical industry. Ethylene is reacted with O_2 to create ethylene oxide, which, in turn, is converted into ethylene glycol; the primary feeder material used to manufacture a host of products, including antifreeze and polyester polymers (the precursors of many plastics and fabrics).

Most of the remaining 20% of commercially produced oxygen is used in medical applications, metal cutting and welding, as an oxidizer in rocket fuel, and in water treatment. Oxygen is used in oxyacetylene welding, burning acetylene with O_2 to produce a very hot flame. In this process, metal up to 60 cm (24 in) thick is first heated with a small oxy-acetylene flame and then quickly cut by a large stream of O_2.

Compounds

The oxidation state of oxygen is -2 in almost all known compounds of oxygen. The oxidation state -1 is found in a few compounds such as peroxides. Compounds containing oxygen in other oxidation states are very uncommon: -1/2 (superoxides), -1/3 (ozonides), 0 (elemental, hypofluorous acid), +1/2 (dioxygenyl), +1 (dioxygen difluoride), and +2 (oxygen difluoride).

Oxides and other Inorganic Compounds

Water (H_{20}) is an oxide of hydrogen and the most familiar oxygen compound. Hydrogen atoms are covalently bonded to oxygen in a water molecule but also have an additional attraction (about 23.3 kJ/mol per hydrogen atom) to an adjacent oxygen atom in a separate molecule. These hydrogen bonds between water molecules hold them approximately 15% closer than what would be expected in a simple liquid with just van der Waals forces.

Due to its electronegativity, oxygen forms chemical bonds with almost all other elements to give corresponding oxides. The surface of most metals, such as aluminium and titanium, are oxidized in the presence of air and become coated with a thin film of oxide that passivates the metal and slows further corrosion. Many oxides of the transition metals are non-stoichiometric compounds, with slightly less metal than the chemical formula would show.

Oxygen is present in the atmosphere in trace quantities in the form of carbon dioxide (CO_2). The Earth's crustal rock is composed in large part of oxides of silicon (silica SiO_2, as found in granite and quartz), aluminium (aluminium oxide Al_2O_3, in bauxite and corundum), iron (iron(III) oxide Fe_2O_3, in hematite and rust), and calcium carbonate (in limestone). The rest of the Earth's crust is also made of oxygen compounds, in particular various complex silicates (in silicate minerals). The Earth's mantle, of much larger mass than the crust, is largely composed of silicates of magnesium and iron.

Water-soluble silicates in the form of Na_4SiO_4, Na_2SiO_3, and $Na_2Si_2O_5$ are used as detergents and adhesives.

Oxygen also acts as a ligand for transition metals, forming transition metal dioxygen complexes, which feature metal–O_2. This class of compounds includes the heme proteins hemoglobin and myoglobin. An exotic and unusual reaction occurs with PtF_6, which oxidizes oxygen to give $O_2^+PtF_6^-$, dioxygenyl hexafluoroplatinate.

Organic Compounds

Among the most important classes of organic compounds that contain oxygen are (where "R" is an organic group): alcohols (R-OH); ethers (R-O-R); ketones (R-CO-R); aldehydes (R-CO-H); carboxylic acids (R-COOH); esters (R-COO-R); acid anhydrides (R-CO-O-CO-R); and amides (R-CO-NR_2). There are many important organic solvents that contain oxygen, including: acetone, methanol, ethanol, isopropanol, furan, THF, diethyl ether, dioxane, ethyl acetate, DMF, DMSO, acetic acid, and formic acid. Acetone ($(CH_3)_2CO$) and phenol (C_6H_5OH) are used as feeder materials in the synthesis of many different substances. Other important organic compounds that contain oxygen are: glycerol, formaldehyde, glutaraldehyde, citric acid, acetic anhydride, and acetamide. Epoxides are ethers in which the oxygen atom is part of a ring of three atoms. The element is similarly found in almost all biomolecules that are important to (or generated by) life.

Oxygen reacts spontaneously with many organic compounds at or below room temperature in a process called autoxidation. Most of the organic compounds that contain oxygen are not made by direct action of O_2. Organic compounds important in industry and commerce that are made by direct oxidation of a precursor include ethylene oxide and peracetic acid.

Safety and Precautions

The NFPA 704 standard rates compressed oxygen gas as nonhazardous to health, nonflammable and nonreactive, but an oxidizer. Refrigerated liquid oxygen (LOX) is given a health hazard rating of 3 (for increased risk of hyperoxia from condensed vapors, and for hazards common to cryogenic liquids such as frostbite), and all other ratings are the same as the compressed gas form.

Toxicity

Oxygen gas (O_2) can be toxic at elevated partial pressures, leading to convulsions and other health problems. Oxygen toxicity usually begins to occur at partial pressures more than 50 kilopascals (kPa), equal to about %50 oxygen composition at standard pressure or 2.5 times the normal sea-level O_2 partial pressure of about 21 kPa. This is not a problem except for patients on mechanical ventilators, since gas supplied through oxygen masks in medical applications is typically composed of only 30–50% O_2 by volume (about 30 kPa at standard pressure).

At one time, premature babies were placed in incubators containing O_2-rich air, but this practice was discontinued after some babies were blinded by the oxygen content being too high.

Breathing pure O_2 in space applications, such as in some modern space suits, or in early spacecraft such as Apollo, causes no damage due to the low total pressures used. In the case of spacesuits, the O_2 partial pressure in the breathing gas is, in general, about 30 kPa (1.4 times normal), and the resulting O_2 partial pressure in the astronaut's arterial blood is only marginally more than normal sea-level O_2 partial pressure.

Oxygen toxicity to the lungs and central nervous system can also occur in deep scuba diving and surface-supplied diving. Prolonged breathing of an air mixture with an O_2 partial pressure more than 60 kPa can eventually lead to permanent pulmonary fibrosis. Exposure to an O_2 partial pressure greater than 160 kPa (about 1.6 atm) may lead to convulsions (normally fatal for divers). Acute oxygen toxicity (causing seizures, its most feared effect for divers) can occur by breathing an air mixture with 21% O_2 at 66 m (217 ft) or more of depth; the same thing can occur by breathing 100% O_2 at only 6 m (20 ft).

Combustion and other Hazards

Highly concentrated sources of oxygen promote rapid combustion. Fire and explosion hazards exist when concentrated oxidants and fuels are brought into close proximity; an ignition event, such as heat or a spark, is needed to trigger combustion. Oxygen is the oxidant, not the fuel.

Concentrated O_2 will allow combustion to proceed rapidly and energetically. Steel pipes and storage vessels used to store and transmit both gaseous and liquid oxygen will act as a fuel; and therefore the design and manufacture of O_2 systems requires special training to ensure that ignition sources are minimized. The fire that killed the Apollo 1 crew in a launch pad test spread so rapidly because the capsule was pressurized with pure O_2 but at slightly more than atmospheric pressure, instead of the 1/3 normal pressure that would be used in a mission.

Liquid oxygen spills, if allowed to soak into organic matter, such as wood, petrochemicals, and asphalt can cause these materials to detonate unpredictably on subsequent mechanical impact.

5.1.2 Oxygen Transport by Haemoglobin and Myoglobin

Oxygen is vital for aerobic respiration in most living organisms, as it plays a key role in energy production. In humans and many other animals, the transport of oxygen from the lungs to tissues is facilitated by two key proteins: hemoglobin and myoglobin. Hemoglobin, found in red blood cells, and myoglobin, present in muscle tissues, are specialized for binding oxygen and releasing it where it's needed.

Hemoglobin and myoglobin are two essential proteins responsible for the transport and storage of oxygen in biological systems. Hemoglobin's cooperative binding and sensitivity to environmental factors make it an efficient transporter of oxygen throughout the body, while myoglobin's high affinity for oxygen ensures that muscle tissues have a constant supply, particularly during periods of increased demand. The delicate balance between oxygen binding, transport, and release is essential for maintaining cellular respiration and energy production. Understanding the roles of hemoglobin and myoglobin in oxygen transport not only provides insight into fundamental physiological processes but also aids in the diagnosis and treatment of disorders related to these critical proteins.

Hemoglobin: Structure and Function

Hemoglobin is a globular protein found in red blood cells that is responsible for the majority of oxygen transport in the blood. It consists of four subunits: two alpha (α) and two beta (β) chains, each containing a heme group capable of binding one oxygen molecule. The heme group consists of a porphyrin ring with an iron (Fe^{2+}) ion at its center, which is essential for oxygen binding.

Hemoglobin's ability to bind oxygen is cooperative, meaning that the binding of one oxygen molecule increases the protein's affinity for binding additional oxygen molecules. This phenomenon is called cooperative binding and is crucial for hemoglobin's efficiency in both picking up oxygen in the lungs and releasing it in tissues that require it. The oxygen dissociation curve for hemoglobin is sigmoidal (S-shaped) due to this cooperativity, allowing it to respond effectively to changes in oxygen demand. When oxygen binds to hemoglobin, it induces a conformational change from the "tense" (T) state, which has a lower affinity for oxygen, to the "relaxed" (R) state, which has a higher affinity. In the lungs, where oxygen levels are high, hemoglobin readily adopts the R state and binds oxygen. In peripheral tissues, where oxygen levels are lower, hemoglobin transitions to the T state, facilitating the release of oxygen to cells.

Myoglobin: Structure and Function

Myoglobin is a single-chain globular protein found primarily in muscle tissues, where it serves as a reservoir for oxygen and aids in its transport to the mitochondria for cellular respiration. Unlike hemoglobin, myoglobin contains only one polypeptide chain and one heme group, allowing it to bind a single oxygen molecule. Myoglobin's structure is simpler than hemoglobin's, but its affinity for oxygen is significantly higher.

This higher affinity for oxygen allows myoglobin to store oxygen within muscle cells, ensuring that there is a readily available supply during periods of intense exercise when oxygen demand exceeds supply. Unlike hemoglobin's cooperative binding, myoglobin exhibits non-cooperative binding, resulting in a hyperbolic oxygen dissociation curve.

This characteristic enables myoglobin to bind oxygen tightly and release it only when the oxygen concentration within the muscle tissue becomes critically low.

Hemoglobin vs. Myoglobin: Key Differences

Though both hemoglobin and myoglobin play essential roles in oxygen transport, their structural differences dictate their distinct physiological functions. Hemoglobin's cooperative binding allows it to efficiently transport oxygen through the bloodstream, releasing oxygen in tissues with varying oxygen demands. In contrast, myoglobin's role is more specialized, ensuring that muscle tissues have a sustained supply of oxygen, particularly during times of oxygen deprivation.

Additionally, hemoglobin's affinity for oxygen is affected by environmental factors such as pH and the presence of carbon dioxide (CO_2).

This allows hemoglobin to respond to metabolic changes and adjust oxygen delivery accordingly. Myoglobin, on the other hand, is not as sensitive to these factors, as its primary role is to store oxygen until it is needed by muscle cells.

Oxygen Binding and Release

The oxygen-binding capacity of hemoglobin is influenced by several factors, including the partial pressure of oxygen (pO_2), pH levels, and the presence of

2,3-bisphosphoglycerate (2,3-BPG). In the lungs, where the pO_2 is high, hemoglobin binds oxygen with high affinity, filling all four of its oxygen-binding sites. As hemoglobin travels to tissues where pO_2 is lower, the oxygen is gradually released.

The Bohr effect describes how changes in pH and CO_2 levels influence hemoglobin's affinity for oxygen. In tissues with high metabolic activity, CO_2 production increases, which lowers pH. This acidic environment causes hemoglobin to release more oxygen, as the lower pH reduces hemoglobin's oxygen-binding affinity. Conversely, in the lungs, where CO_2 levels are lower and pH is higher, hemoglobin's affinity for oxygen increases, promoting oxygen binding.

2,3-BPG, a byproduct of glycolysis in red blood cells, also plays a role in regulating oxygen release. By binding to deoxygenated hemoglobin, 2,3-BPG stabilizes the T state, promoting oxygen release in tissues. This is particularly important under conditions of low oxygen availability, such as high altitudes, where increased levels of 2,3-BPG help hemoglobin release oxygen more readily.

Myoglobin, by contrast, binds oxygen more tightly and releases it only when oxygen levels in the muscle tissues drop to very low levels. This allows myoglobin to maintain an oxygen reservoir for periods of intense muscular activity, such as during exercise or in oxygen-deprived environments.

Role of Hemoglobin in Oxygen Transport

Hemoglobin's primary function is to transport oxygen from the lungs to tissues throughout the body and facilitate the return of carbon dioxide from tissues to the lungs. Oxygen is inhaled into the lungs, where it diffuses into the bloodstream and binds to hemoglobin in red blood cells. Once oxygen is bound, hemoglobin travels through the circulatory system, delivering oxygen to cells and tissues that need it.

In peripheral tissues, hemoglobin releases oxygen, which diffuses into cells and is used for cellular respiration. This process produces energy in the form of adenosine triphosphate (ATP) and generates carbon dioxide as a byproduct. Hemoglobin then picks up CO_2 and returns to the lungs, where CO_2 is expelled during exhalation.

Hemoglobin also plays a role in buffering blood pH by binding to hydrogen ions (H^+), a byproduct of CO_2 conversion in tissues. This buffering capacity helps to maintain acid-base balance in the body, preventing fluctuations in pH that could disrupt cellular processes.

Role of Myoglobin in Muscle Oxygen Supply

Myoglobin is critical in ensuring that muscle tissues have a steady supply of oxygen, particularly during periods of intense activity. When muscles contract and oxygen demand exceeds the rate of oxygen delivery by hemoglobin, myoglobin steps in to release its stored oxygen, ensuring that the mitochondria can continue producing ATP through aerobic respiration.

In diving mammals, such as whales and seals, myoglobin plays an even more crucial role, as these animals are often deprived of external oxygen for extended periods. Their muscle tissues contain significantly higher concentrations of myoglobin compared to land animals, allowing them to store large amounts of oxygen and sustain aerobic respiration during long dives.

Pathological Conditions Related to Hemoglobin and Myoglobin

Abnormalities in hemoglobin and myoglobin function can lead to various medical conditions. For instance, mutations in the genes encoding hemoglobin subunits can result in disorders like sickle cell anemia or thalassemia, both of which affect the protein's oxygen-binding capacity and lead to reduced oxygen transport. In sickle cell anemia, a single mutation in the beta-globin chain causes hemoglobin molecules to polymerize under low oxygen conditions, deforming red blood cells and impairing their ability to transport oxygen.

Myoglobin-related disorders are rarer but can occur in cases of muscle injury or disease. When muscle cells are damaged, myoglobin can be released into the bloodstream in large amounts, potentially leading to myoglobinuria, a condition characterized by the presence of myoglobin in the urine. If myoglobin levels are excessively high, it can lead to kidney damage and acute renal failure.

5.2 OXYGEN TRANSPORT BY HEMERYTHRIN AND HAEMOCYANIN

Oxygen transport is a critical biological function in many organisms, ensuring the efficient delivery of oxygen to tissues for cellular respiration. While hemoglobin plays this role in most vertebrates, other oxygen-binding proteins fulfill similar functions in various invertebrates.

Hemerythrin and hemocyanin are two such proteins, each with distinct characteristics and mechanisms for binding and transporting oxygen. Found in invertebrates such as marine worms, mollusks, and arthropods, these proteins provide insight into the diverse strategies life has evolved to manage oxygen transport under different environmental conditions.

Hemerythrin and hemocyanin represent two distinct oxygen-transport proteins that have evolved in response to the environmental challenges faced by invertebrate species. Hemerythrin, with its non-heme iron active sites, is suited for slow-moving marine invertebrates, while hemocyanin, with its copper-based oxygen binding, is essential for many aquatic arthropods and mollusks. These proteins illustrate the remarkable diversity of biological solutions to the problem of oxygen transport, offering insights into how different species have evolved to survive in a wide range of habitats. Understanding these proteins' structures and functions not only sheds light on invertebrate biology but also highlights the complexity of oxygen transport systems in the natural world.

5.2.1 Hemerythrin: Structure and Oxygen Binding

Hemerythrin is an oxygen-transport protein found primarily in marine invertebrates, such as certain species of marine worms and brachiopods. Unlike hemoglobin, which contains iron in a heme group, hemerythrin contains non-heme iron at its active site. The protein is oligomeric, typically existing as a trimer or octamer, with each subunit containing two iron atoms per active site. These iron atoms are directly responsible for oxygen binding, but the mechanism differs from that of hemoglobin.

In hemerythrin, oxygen binds to the two iron atoms in a bridged fashion, forming a peroxide ion (O_2^{2-}) that links the iron atoms. This binding is reversible, allowing hemerythrin to pick up oxygen in oxygen-rich environments and release it where it is needed, although its oxygen affinity is not regulated by factors like pH or CO_2 as it is in hemoglobin. Hemerythrin turns a characteristic pink or purple color upon oxygenation, a striking difference from the red color associated with oxygenated hemoglobin.

Hemerythrin is less efficient at oxygen transport than hemoglobin, as it lacks the cooperative binding seen in hemoglobin. Each subunit of hemerythrin binds oxygen independently of the others, leading to a more linear oxygen dissociation curve rather than the sigmoidal curve observed in hemoglobin. This makes hemerythrin more suitable for animals that do not require the fine-tuned oxygen delivery mechanisms found in higher vertebrates.

5.2.2 Hemerythrin's Role in Oxygen Transport

Hemerythrin is primarily found in the coelomic fluid of invertebrates, where it transports oxygen from the surrounding environment to tissues. In contrast to hemoglobin, which is present in red blood cells, hemerythrin is typically found dissolved directly in the fluid. This limits its oxygen-carrying capacity, but the presence of multiple iron-binding sites per molecule compensates for this to some extent.

The presence of hemerythrin in marine invertebrates suggests that it is well-adapted to the relatively low-oxygen environments of oceanic habitats. Marine invertebrates tend to have slower metabolisms than terrestrial animals, and hemerythrin's simple oxygen-binding mechanism is sufficient to meet their needs. Additionally, because hemerythrin does not require a heme group, it is less susceptible to damage by carbon monoxide or other competitive inhibitors of oxygen binding, giving it an advantage in certain environments.

5.2.3 Hemocyanin: Structure and Oxygen Binding

Hemocyanin is another oxygen-transport protein, found in arthropods such as spiders, scorpions, and crabs, as well as in mollusks like octopuses and squids. Unlike hemoglobin and hemerythrin, hemocyanin uses copper rather than iron as the oxygen-binding metal. Hemocyanin is a large, extracellular protein, typically existing as a hexamer, dodecamer, or even larger oligomeric structures. Each subunit of hemocyanin contains two copper atoms, which are responsible for binding a single oxygen molecule.

In hemocyanin, the oxygen molecule binds to the copper atoms, causing the protein to turn from colorless (in its deoxygenated state) to blue when oxygenated. This change in color is a hallmark of hemocyanin and is often observed in the blue blood of species such as horseshoe crabs. The oxygen binding in hemocyanin is not cooperative in the same way as hemoglobin, but larger hemocyanin molecules can exhibit a form of allosteric regulation, where the binding of oxygen to one site influences the binding at other sites.

Hemocyanin is highly efficient at binding oxygen in cold and low-oxygen environments, which explains its prevalence in marine and aquatic species. This protein's structure allows it to bind oxygen with high affinity, making it particularly effective in the oxygen-poor conditions found in deep-sea habitats.

5.2.4 Hemocyanin's Role in Oxygen Transport

Hemocyanin is found in the hemolymph, the circulatory fluid of arthropods and mollusks, functioning similarly to how hemoglobin operates in vertebrates. It transports oxygen from the respiratory organs, such as gills, to the tissues where it is needed for cellular respiration. Hemocyanin's ability to transport oxygen in cold environments is critical for the survival of many marine invertebrates, particularly those living in polar or deep-sea ecosystems.

One of the key advantages of hemocyanin is its resistance to changes in pH and carbon dioxide levels. Unlike hemoglobin, which experiences a significant drop in oxygen-binding affinity in response to elevated CO_2 (the Bohr effect), hemocyanin remains relatively stable.

This allows animals that rely on hemocyanin to maintain efficient oxygen transport even in conditions where pH fluctuates, such as during prolonged underwater activity or exposure to varying water temperatures.

Hemocyanin is also notable for its size and structural complexity. While hemoglobin is a relatively small protein, hemocyanin forms massive, multi-subunit complexes that can carry many oxygen molecules simultaneously. This allows for efficient oxygen transport despite the fact that hemocyanin is less soluble in hemolymph than hemoglobin is in blood.

5.2.5 Comparative Analysis: Hemerythrin vs. Hemocyanin

Both hemerythrin and hemocyanin are examples of evolutionary adaptations to specific environmental challenges, demonstrating how invertebrates have evolved alternative oxygen-transport systems that are distinct from hemoglobin. Hemerythrin is adapted for oxygen transport in slow-moving, low-metabolism organisms, such as marine worms, which live in oxygen-poor environments. Its relatively simple oxygen-binding mechanism, along with its reliance on non-heme iron, makes it suitable for these conditions.

Hemocyanin, on the other hand, is better suited for animals that require a more robust oxygen transport system, particularly those living in cold or aquatic environments. The use of copper as the oxygen-binding metal allows hemocyanin to function effectively under low temperatures, and its resistance to pH changes makes it ideal for species that experience fluctuating environmental conditions.

Despite these differences, both hemerythrin and hemocyanin share the common feature of reversible oxygen binding, which is essential for life in environments where oxygen levels can vary significantly. Additionally, both proteins are found in species that lack red blood cells, highlighting a key distinction between vertebrate and invertebrate oxygen-transport systems.

5.2.6 Physiological and Environmental Adaptations

The evolution of hemerythrin and hemocyanin demonstrates how different organisms have adapted to their specific ecological niches. Marine invertebrates, for instance, benefit from hemerythrin's ability to bind oxygen in environments with low oxygen concentrations and from hemocyanin's capacity to function in cold or oxygen-poor water. These proteins also confer advantages in terms of metabolic efficiency and resistance to environmental stresses.

In the case of hemocyanin, its high affinity for oxygen at low temperatures allows marine mollusks and arthropods to thrive in deep-sea environments where oxygen is scarce. Hemerythrin, although less efficient than hemoglobin, provides sufficient oxygen transport for slow-moving species that do not require rapid oxygen delivery to their tissues.

5.3 THE RESPIRATORY CHAIN

The respiratory chain, also known as the electron transport chain (ETC), is a series of biochemical reactions that occur in the mitochondria of cells. It plays a critical role in the process of cellular respiration by converting the energy stored in nutrients into adenosine triphosphate (ATP), the primary energy currency of cells. This chain involves multiple protein complexes and mobile carriers that transport electrons, ultimately leading to the production of ATP. The efficiency of the respiratory chain is crucial for energy production, and its dysfunction can lead to various metabolic disorders. The respiratory chain is a fundamental component of cellular metabolism, responsible for the efficient production of ATP through oxidative phosphorylation. By coupling the transfer of electrons with the creation of a proton gradient, the respiratory chain enables cells to harness the energy stored in nutrients and fuel essential biological processes. However, the chain's complexity and sensitivity to dysfunction also make it a potential source of cellular damage, particularly through the generation of reactive oxygen species.

Understanding the structure, function, and regulation of the respiratory chain is essential for appreciating its role in health and disease. Advances in mitochondrial research continue to shed light on the mechanisms underlying respiratory chain dysfunction and offer potential therapeutic strategies for treating mitochondrial diseases and conditions associated with oxidative stress. The respiratory chain remains a critical area of study in biochemistry, physiology, and medicine, with implications for aging, neurodegenerative diseases, and metabolic disorders.

5.3.1 Components of the Respiratory Chain

The respiratory chain is composed of four main complexes (Complex I, II, III, and IV), along with two mobile electron carriers: ubiquinone (coenzyme Q) and cytochrome c. These components are embedded in the inner mitochondrial membrane, and they work in a coordinated manner to transport electrons from high-energy molecules such as NADH and $FADH_2$ to oxygen, which is reduced to water. Each complex in the chain has a specific role in electron transport and energy production.

Complex I: NADH Dehydrogenase

Complex I, also known as NADH dehydrogenase, is the first and largest protein complex in the respiratory chain. It receives electrons from NADH, a molecule produced during glycolysis and the citric acid cycle. These electrons are transferred to ubiquinone (coenzyme Q), which becomes reduced to ubiquinol. In the process, Complex I pumps protons (H^+ ions) across the inner mitochondrial membrane into the intermembrane space, creating a proton gradient that is essential for ATP production.

Complex II: Succinate Dehydrogenase

Complex II, or succinate dehydrogenase, is unique because it is also a component of the citric acid cycle. It oxidizes succinate to fumarate and, in the process, generates $FADH_2$. The electrons from $FADH_2$ are transferred to ubiquinone, which is reduced to ubiquinol. Unlike Complex I, Complex II does not pump protons across the membrane, but it still contributes to the flow of electrons through the respiratory chain.

Complex III: Cytochrome bc1 Complex

Complex III, also known as the cytochrome bc1 complex, accepts electrons from ubiquinol (the reduced form of ubiquinone) and transfers them to cytochrome c, a small protein that shuttles electrons between Complex III and Complex IV. During this process, Complex III pumps additional protons into the intermembrane space, further contributing to the proton gradient that drives ATP synthesis.

Complex IV: Cytochrome c Oxidase

Complex IV, or cytochrome c oxidase, is the final complex in the respiratory chain. It receives electrons from cytochrome c and transfers them to molecular oxygen (O_2), which is reduced to water (H_2O). This step is crucial because it ensures that the electrons are safely removed from the chain, preventing the accumulation of reactive oxygen species (ROS) that can damage cellular structures. Complex IV also pumps protons across the membrane, maintaining the proton gradient.

5.3.2 The Role of Mobile Electron Carriers

Two mobile electron carriers, ubiquinone (coenzyme Q) and cytochrome c, play key roles in the respiratory chain. Ubiquinone is a lipid-soluble molecule that diffuses through the inner mitochondrial membrane, shuttling electrons from Complex I and II to Complex III. Cytochrome c, on the other hand, is a small, water-soluble protein located in the intermembrane space. It transports electrons from Complex III to Complex IV, ensuring the continuous flow of electrons through the chain.

5.3.3 Proton Gradient and Chemiosmosis

A crucial aspect of the respiratory chain is the creation of a proton gradient across the inner mitochondrial membrane. As electrons pass through Complexes I, III, and IV, protons are pumped from the mitochondrial matrix into the intermembrane space. This creates an electrochemical gradient, often referred to as the proton-motive force, with a higher concentration of protons in the intermembrane space compared to the matrix.

The energy stored in this proton gradient is harnessed by ATP synthase, an enzyme located in the inner mitochondrial membrane. As protons flow back into the matrix through ATP synthase, the enzyme uses this energy to synthesize ATP from adenosine diphosphate (ADP) and inorganic phosphate (Pi). This process, known as chemiosmosis, is the final step of oxidative phosphorylation and is responsible for the majority of ATP production in aerobic cells.

5.3.4 Regulation of the Respiratory Chain

The activity of the respiratory chain is tightly regulated to match the cell's energy demands. Several factors influence its efficiency, including the availability of substrates (NADH and $FADH_2$), the presence of oxygen, and the ADP/ATP ratio. When cellular ATP levels are high, the flow of electrons through the respiratory chain slows down, reducing the rate of oxidative phosphorylation. Conversely, when ATP levels are low and ADP levels are high, the respiratory chain is upregulated to meet the cell's energy needs.

Additionally, the respiratory chain is sensitive to the presence of inhibitors and uncouplers. Inhibitors, such as cyanide or carbon monoxide, block electron flow through specific complexes, leading to a halt in ATP production and, ultimately, cell death. Uncouplers, such as 2,4-dinitrophenol, disrupt the proton gradient by allowing protons to flow back into the matrix without passing through ATP synthase. This results in reduced ATP synthesis and increased heat production.

5.3.5 Reactive Oxygen Species and Mitochondrial Damage

While the respiratory chain is essential for ATP production, it also has the potential to generate harmful byproducts in the form of reactive oxygen species (ROS). Under normal conditions, a small percentage of electrons "leak" from the chain and react with oxygen to form superoxide radicals (O_2^-). These ROS can damage mitochondrial proteins, lipids, and DNA, leading to impaired cellular function and contributing to the aging process and the development of various diseases.

The cell has several defense mechanisms to mitigate the damage caused by ROS, including the enzymes superoxide dismutase, catalase, and glutathione peroxidase, which neutralize ROS and protect mitochondrial integrity. However, excessive ROS production, often associated with mitochondrial dysfunction or oxidative stress, can overwhelm these defenses and lead to cell death through apoptosis or necrosis.

5.3.6 The Respiratory Chain and Disease

Dysfunction in the respiratory chain is implicated in a wide range of human diseases, particularly those involving high-energy tissues such as the brain, heart, and muscles. Mitochondrial diseases, which result from mutations in either nuclear or mitochondrial DNA, can impair the function of the respiratory chain and lead to conditions such as mitochondrial myopathy, Leigh syndrome, and Parkinson's disease.

In addition to genetic disorders, the respiratory chain is also affected by environmental factors, such as toxins and drugs, that inhibit its activity. For example, cyanide poisoning blocks Complex IV, preventing oxygen from being reduced to water and leading to a rapid decline in ATP production. Similarly, certain antibiotics, such as tetracycline, can interfere with mitochondrial protein synthesis and disrupt the respiratory chain.

CHAPTER 6

OXIDOREDUCTASES BASED ON IRON, MANGANESE, AND COPPER

6.1 RIBONUCLEOTIDE REDUCTASES

Ribonucleotide reductase (RNR), also known as ribonucleoside diphosphate reductase, is an enzyme that catalyzes the formation of deoxyribonucleotides from ribonucleotides. It catalyzes this formation by removing the 2'-hydroxyl group of the ribose ring of nucleoside diphosphates (or triphosphates depending on the class of RNR).

This reduction produces deoxyribonucleotides. Deoxyribonucleotides in turn are used in the synthesis of DNA. The reaction catalyzed by RNR is strictly conserved in all living organisms. Furthermore, RNR plays a critical role in regulating the total rate of DNA synthesis so that DNA to cell mass is maintained at a constant ratio during cell division and DNA repair.

A somewhat unusual feature of the RNR enzyme is that it catalyzes a reaction that proceeds via a free radical mechanism of action. The substrates for RNR are ADP, GDP, CDP and UDP. dTDP (deoxythymidine diphosphate) is synthesized by another enzyme (thymidylate kinase) from dTMP (deoxythymidine monophosphate).

6.1.1 Structure

Ribonucleotide reductases are divided into three classes. Class I RNR enzymes are constructed from large alpha subunit and small beta subunits which associate to form an active heterodimeric tetramer.

By reducing NDPs to 2'-dNDPs, the enzyme catalyses the de novo synthesis of deoxyribonucleotides (dNTPs), which are precursors to DNA synthesis and essential for cell proliferation. Class II RNRs produce a 5'-deoxyadenosyl radical by homolytic cleavage of the C-Co bond in adenosylcobalamin. In addition, Class III RNRs contain a stable glycyl radical.

Each Class I alpha monomer consists of three domains:

- One mainly helical domain comprising the 220 N-terminal residues,
- A second large ten-stranded α/β structure comprising 480 residues, and
- A third small five-stranded α/β structure comprising 70 residues.

In Pfam, the second domain has been interpreted as two separate domains:

- A shorter all-alpha N-terminal domain, and
- A longer barrel C-terminal domain.

The Class I beta subunit usually contains a di-metal center and a stable tyrosyl radical. In humans, the beta subunit relies on a di-iron cofactor. In *E. coli*, the tyrosyl radical is located at position 122 (Y122) providing the stable radical for the Class I RNR2 subunits. In *A. aegypti*, this tyrosyl radical is located at position 184 (Y184). The tyrosyl radical is deeply buried inside the protein in a hydrophobic environment, located close to the iron center that is used in the stabilization of a tyrosyl radical. The structure of two μ-oxo-linked irons is dominated by ligands that serve as iron binding sites: four carboxylates [aspartate (D146), glutamate (E177, E240, and E274)] and two histidines (H180 and H277). Association occurs between the C-terminus of RNR2 and the C-terminus of RNR1. Enzymatic activity is dependent on association of the RNR1 and RNR2 subunits. The active site consists of the active dithiol groups from the RNR1 as well as the diferric center and the tyrosyl radical from the RNR2 subunit.

Other residues of RNR2, such as aspartate (D273), tryptophan (W48), and tyrosine (Y356) further stabilize the active-site tyrosyl radical thus allowing electron transfer. These residues help in the transfer of the radical electron from tyrosine (Y122) of RNR2 to cysteine (C439) of RNR1. The electron transfer begins on RNR2 tyrosine (Y122) and continues in RNR2 to tryptophan (W48), which is separated from RNR1 tyrosine (Y731) by 2.5 nanometers. Electron transfer from RNR2 to RNR1 occurs via tyrosine (Y356 to Y731) and continues on through tyrosine (Y730) to cysteine (C439) in the active site. Site-directed mutations of the RNR primary structure indicate that all residues cited above participate in the long distance transfer of the free radical to the active site.

In *A. aegypti* mosquitoes, RNR1 retains most of the crucial amino acid residues, including aspartate (D64) and valine (V292 or V284), that are necessary in allosteric regulation; proline (P210 and P610), leucine (L453 and L473), and methionine (M603) residues that are located in the hydrophobic active site; cysteine (C225, C436 and C451) residues that are involved in removal of a hydrogen atom and transfer of the radical electron at the active site; cysteine (C225 and C436), asparagine (N434), and glutamate (E441) residues that bind the ribonucleotide substrate; tyrosine (Y723 and Y743) residues that dictate the radical transfer; and cysteine (C838 and C841) residues that are used in the regeneration of dithiol groups in the active site.

The yeast *Saccharomyces cerevisiae* possesses a minor isoform of large subunit of ribonucleotide-diphosphate reductase under the designation RNR3 or *YIL066C* on the yeast chromosome IX. It is a paralog of the yeast RNR1, which likely arose from a whole genome duplication event. The RNR complex catalyzes the rate-limiting step in dNTP synthesis, regulated by DNA replication and DNA damage checkpoint pathways via localization of small subunits.

6.1.2 Function

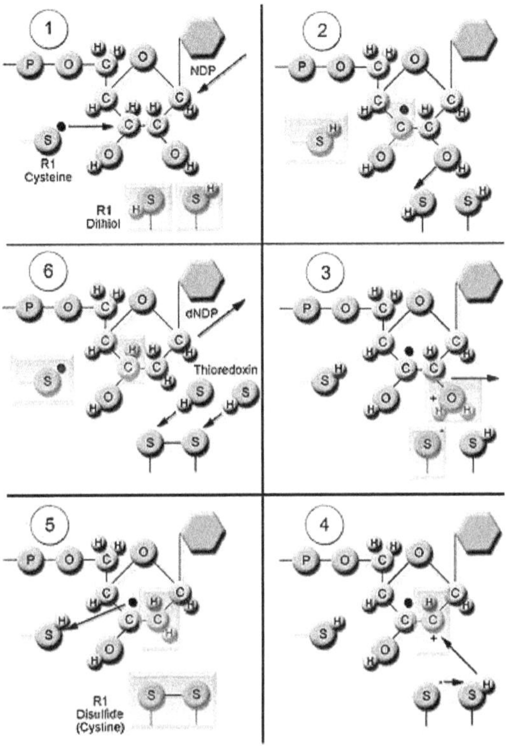

Fig: Mechanism to Catalyze the Conversion of Ribonucleotides to Deoxyribonucleotides. (1) An Electron Transfer on the RNR2 Subunit Activates a RNR1 Cysteine Residue in the Active Site with a Free Radical; (2) The Free Radical Forms a Stable Radical on C-3, and Cysteine Radical Removes a Hydrogen Atom; (3) Cation is Formed on C-2 by Transferring a Hydrogen from a Dithiol Group and is Stabilized by the Radical, Resulting in the Loss of H_2O from C-2; (4) A Hydrogen is Transferred from the Dithiol Group to Reduce the Cation C-2; (5) The C-3 Radical is Reduced by the Hydrogen Removed in Step 2, and the Tyrosyl Radical is Generated; (6) Redoxins Transfer Two Hydrogen to the Disulfide Group that Restores the Original Configuration

The enzyme ribonucleotide reductase (RNR) catalyzes the de novo synthesis of dNDPs. Catalysis of ribonucleoside 5'-diphosphates (NDPs) involves a reduction at the 2'-carbon of ribose 5-phosphate to form the 2'-deoxy derivative-reduced 2'-deoxyribonucleoside 5'-diphosphates (dNDPs). This reduction is initiated with the generation of a free radical. Following a single reduction, RNR requires electrons donated from the dithiol groups of the protein thioredoxin. Regeneration of thioredoxin occurs when nicotinamide adenine dinucleotide phosphate (NADPH) provides two hydrogen atoms that are used to reduce the disulfide groups of thioredoxin.

Three classes of RNR have similar mechanisms for the reduction of NDPs, but differ in the domain that generates the free radical, the specific metal in the metalloprotein structure, and the electron donors. All classes use free-radical chemistry. Class I reductases use an iron center with ferrous to ferric conversion to generate a tyrosyl free radical. Reduction of NDP substrates occurs under aerobic conditions. Class I reductases are divided into IA and IB due to differences in regulation. Class IA reductases are distributed in eukaryotes, eubacteria, bacteriophages, and viruses. Class IB reductases are found in eubacteria. Class IB reductases can also use a radical generated with the stabilization of a binuclear manganese center. Class II reductases generate the free radical 5'-deoxyadenosyl radical from cobalamin (coenzyme B12) and have a simpler structure than class I and class III reductases.

Reduction of NDPs or ribonucleotide 5'-triphosphates (NTPs) occurs under either aerobic or anaerobic conditions. Class II reductases are distributed in archaebacteria, eubacteria, and bacteriophages. Class III reductases use a glycine radical generated with the help of an S-adenosyl methionine and an iron sulphur center. Reduction of NTPs is limited to anaerobic conditions. Class III reductases are distributed in archaebacteria, eubacteria, and bacteriophages. Organisms are not limited to having one class of enzymes. For example, *E. coli* have both class I and class III RNR.

6.1.3 Catalytic Reduction Mechanism

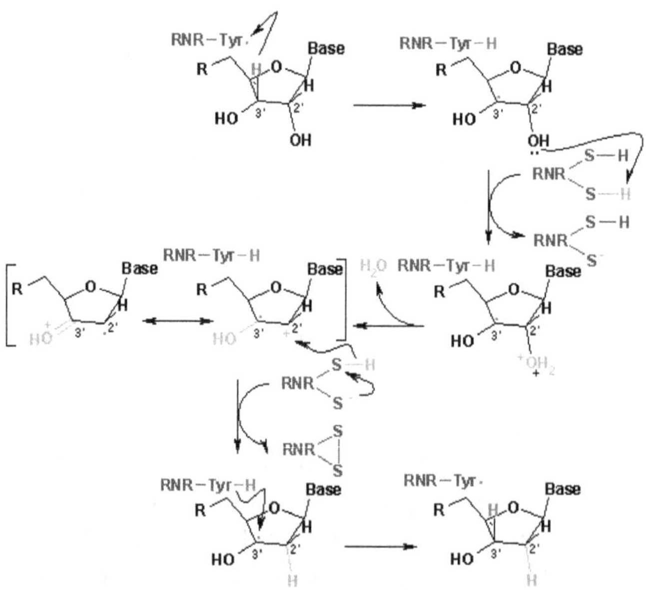

Fig: The Reaction Mechanism of RNR

The mechanism that is currently accepted for the reduction of ribonucleotides to deoxyribonucleotides is depicted in the following scheme. The first step involves the abstraction of the 3'- H of substrate 1 by radical Cys439. Subsequently, the reaction involves the elimination of one water molecule from carbon C-2' of the ribonucleotide, catalyzed by Cys225 and Glu441. In the third step there is a hydrogen atom transfer from Cys225 to carbon C-2' of the 2'-ketyl radical 3, after previous proton transfer from Cys462 to Cys225. At the end of this step, a radical anionic disulfide bridge and the closed-shell ketone intermediate 4 are obtained. This intermediate has been identified during the conversion of several 2'-substituted substrate analogues, as well as with the natural substrate interacting with enzyme mutants. The next step is the oxidation of the anionic disulfide bridge, with concomitant reduction of the substrate, generating 5. The spin density shifts from the sulphur atoms to the C-3' atom of the substrate, with simultaneous proton transfer from Glu441 to carbon C-3'. The last step is the reverse of the first step and involves a hydrogen transfer from Cys439 to C-3', regenerating the initial radical and resulting in the final product 6.

Theoretical models of some steps of these mechanisms using the full model of the R1 protein can be found at the studies performed by Cerqueira *et al.*.

6.1.4 Regulation

Class I RNR comprises RNR1 and RNR2 subunits, which can associate to form a heterodimeric tetramer. RNR1 contains both allosteric sites, mediating regulation of substrate specificity and activity. Depending on the allosteric configuration, one of the four ribonucleotides binds to the active site.

Regulation of RNR is designed to maintain balanced quantities of dNTPs. Binding of effector molecules either increases or decreases RNR activity. When ATP binds to the allosteric activity site, it activates RNR. In contrast, when dATP binds to this site, it deactivates RNR. In addition to controlling activity, the allosteric mechanism also regulates the substrate specificity and ensures the enzyme produces an equal amount of each dNTP for DNA synthesis. In all classes, binding of ATP or dATP to the allosteric site induces reduction of cytidine 5'-diphosphate (CDP) and uridine 5'-diphosphate (UDP); 2'-deoxyguanosine 5'-triphosphate (dGTP) induces reduction of adenosine 5'-diphosphate (ADP); and 2'-deoxythymidine 5'-triphosphate (dTTP) induces reduction of guanosine 5'-diphosphate (GDP).

Class IB reductases are not inhibited by dATP because they lack approximately 50 N-terminal amino acids required for the allosteric activity site. Additionally, it is important that the activity of ribonucleotide reductase be under transcriptional and post-transcriptional control because the synthesis of damage-free DNA relies on a balanced pool of deoxyribonucleotides. Eukaryotic cells with class IA reductases have a mechanism of negative control to turn off synthesis of dNTPs as they accumulate. This mechanism protects the cell from toxic and mutagenic effects that can arise from the overproduction of dNTPs because changes in balanced dNTP pools lead to DNA damage and cell death. Although, the overproduction of dNTPs or an unbalanced supply of them can lead to misincorporation of nucleotides into DNA, the supply of dNTPs supply can allow for DNA repair. p53R2 is a small subunit of ribonucleotide reductase that can induce such repair. Changes within this p53 induced R2 homolog can cause depletion in mitochondrial DNA and consequently p53R2 serves a major factor in dNTP supply. RNR may use the morpheein model of allosteric regulation.

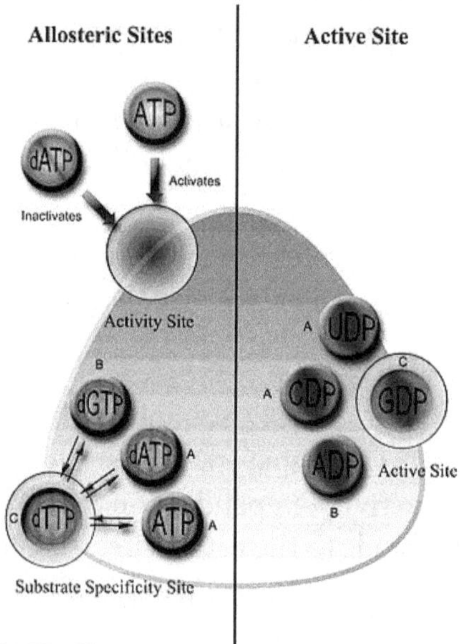

Fig: Regulation of Class I RNR. Class I RNRs are Activated by Binding ATP or Inactivated by Binding dATP to the Activity Site Located on the RNR1 Subunit. When the Enzyme is Activated, Substrates are Reduced if the Corresponding Effectors Bind to the Allosteric Substrate Specificity Site. A = when dATP or ATP is Bound at the Allosteric Site, the Enzyme Accepts UDP and CDP into the Catalytic Site; B = when dGTP is Bound, ADP Enters the Catalytic Site; C = when dTTP is Bound, GDP Enters the Catalytic Site. The Substrates (Ribonucleotides UDP, CDP, ADP, and GDP) are Converted to dNTPs by a Mechanism Involving the Generation of a Free Radical

6.1.5 RNR1 and RNR2 Inhibitors

Generally Class I RNR inhibitors can be divided in three main groups: translation inhibitors, which block the synthesis of the enzyme; dimerization inhibitors that prevent the association of the two RNR subunits (R1 and R2); and catalytic inhibitors that inactivate the subunit R1 and/or subunit R2. Class I RNR can be inhibited by peptides similar to the C-terminus of RNR2. These peptides can compete with

RNR2 for binding to RNR1, and as a result RNR1 does not form an enzymatically active complex with RNR2. Although the C-terminus of RNR2 proteins is different across species, RNR2 can interact with RNR1 across species. When the mouse RNR2 C-terminus was replaced with the *E. coli* RNR2 C-terminal (7 or 33) amino acid residues, the chimeric RNR2 subunit still binds to mouse RNR1 subunits. However, they lack enzymatic activity due probably to the elimination of residues involved in the transfer of the free radical electron from the RNR2 to the RNR1 subunit.

Small peptides can specifically inhibit the RNR2 subunits from binding with RNR1 when they share a significant similarity with the normal RNR2 C-terminus. This inhibition RNR2 binding to RNR1 has been tested successfully in herpes simplex virus (HSV) RNR. When a 7 amino acid oligomer (GAVVNDL) truncated from the C-terminus of the RNR2 subunit was used in competition assays, it prevented the normal RNR2 from forming an enzymatically active complex with RNR1. Other small peptide inhibitors similar to the RNR2 C-terminus have also been used successfully to inhibit HSV RNR enzymatic activity and thus HSV replication. In mice models of stromal keratitis and corneal neovascularization (HSV ocular disease), a small RNR2 C-terminal analog BILD 1263 has been reported to inhibit RNR and is effective in preventing these diseases. In some cases, although treatment with small C-terminal analogs may not stop disease spreading, they can still help in healing. In the acyclovir-resistant HSV (PAAr5), a small peptide inhibitor BILD 1633 has been reported to be 5 to 10 times more potent than BILD 1263 against cutaneous PAAr5 infection. A combination therapy approach (BILD 1633 and acyclovir) is more effective to heal topical lesions in mice. These data suggest that small peptide inhibitors that compete with RNR2 for binding to RNR1 are useful in preventing the spread of HSV.

Gallium inhibits RNR2 by substituting for Fe^{3+} in the active site. Gallium maltolate is an orally bioavailable form of gallium that exploits this inhibitory activity to treat cancer, infections, and other diseases.

The drugs hydroxyurea and Motexafin gadolinium interfere with the action of this enzyme.

6.2 SUPEROXIDE DISMUTASES, SUPEROXIDE REDUCTASES, AND PEROXIDASES

6.2.1 Superoxide Dismutases

Superoxide dismutase (SOD, EC 1.15.1.1) is an enzyme that alternately catalyzes the dismutation (or partitioning) of the superoxide (O^-_2) anion radical into normal molecular oxygen (O_2) and hydrogen peroxide (H_2O_2). Superoxide is produced as a by-product of oxygen metabolism and, if not regulated, causes many types of cell damage. Hydrogen peroxide is also damaging and is degraded by other enzymes such as catalase. Thus, SOD is an important antioxidant defense in nearly all living cells exposed to oxygen.

One exception is *Lactobacillus plantarum* and related lactobacilli, which use intracellular manganese to prevent damage from reactive O^-_2.

Chemical Reaction

SODs catalyze the disproportionation of superoxide:

$$2H^+ + 2O^-_2 \rightarrow O_2 + H_2O_2$$

In this way, O^-_2 is converted into two less damaging species.

The general form, applicable to all the different metal-coordinated forms of SOD, can be written as follows:

$$M_{(n+1)+} - SOD + O^-_2 \rightarrow M_{n+} - SOD + O_2$$

$$M_{n+} - SOD + O^-_2 + 2H^+ \rightarrow M_{(n+1)+} - SOD + H_2O_2$$

The reactions by which SOD-catalyzed dismutation of superoxide for Cu,Zn SOD can be written as follows:

$Cu^{2+} - SOD + O^-_2 \rightarrow Cu^+ - SOD + O_2$ (reduction of copper; oxidation of superoxide)

$Cu^+ - SOD + O^-_2 + 2H^+ \rightarrow Cu^{2+} - SOD + H_2O_2$ (oxidation of copper; reduction of superoxide)

In a series of such reactions, the oxidation state and the charge of the metal cation oscillates between n and n+1: +1 and +2 for Cu, or +2 and +3 for the other metals .

Types

General

Irwin Fridovich and Joe McCord at Duke University discovered the enzymatic activity of superoxide dismutase in 1968. SODs were previously known as a group of metalloproteins with unknown function; for example, CuZnSOD was known as erythrocuprein (or hemocuprein, or cytocuprein) or as the veterinary anti-inflammatory drug "Orgotein".

Likewise, Brewer (1967) identified a protein that later became known as superoxide dismutase as an indophenol oxidase by protein analysis of starch gels using the phenazine-tetrazolium technique.

There are three major families of superoxide dismutase, depending on the protein fold and the metal cofactor: the Cu/Zn type (which binds both copper and zinc), Fe and Mn types (which bind either iron or manganese), and the Ni type (which binds nickel).

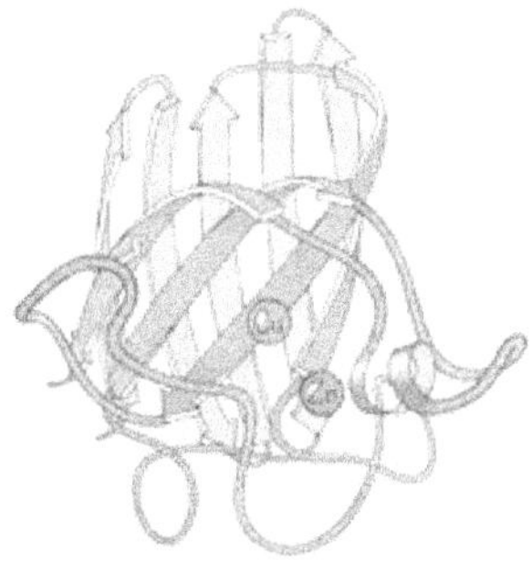

Fig: Ribbon Diagram of Bovine Cu-Zn SOD Subunit

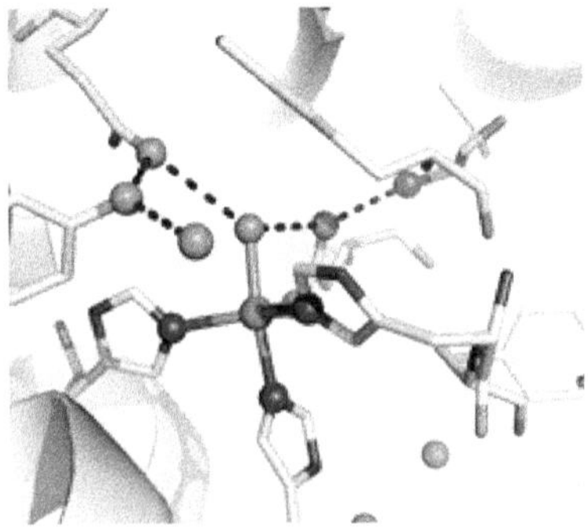

Fig: Active Site of Human Manganese SOD, Manganese Shown in Purple

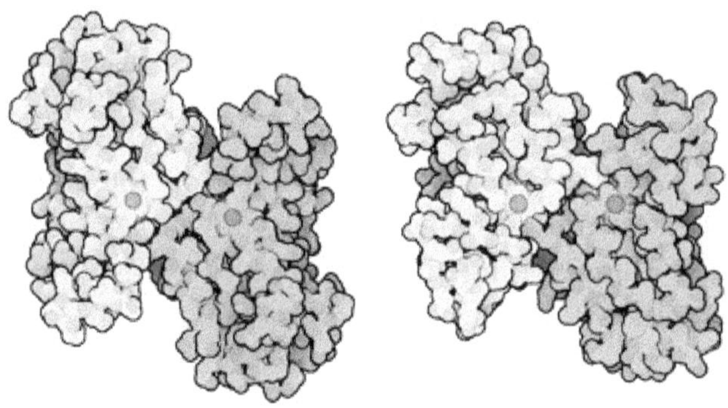

Fig: Mn-SOD vs Fe-SOD Dimers

Copper and zinc – most commonly used by eukaryotes, including humans. The cytosols of virtually all eukaryotic cells contain a SOD enzyme with copper and zinc (Cu-Zn-SOD). For example, Cu-Zn-SOD available commercially is normally purified from bovine red blood cells. The bovine Cu-Zn enzyme is a homodimer of molecular weight 32,500. It was the first SOD whose atomic-detail crystal structure was solved, in 1975.

It is an 8-stranded "Greek key" beta-barrel, with the active site held between the barrel and two surface loops. The two subunits are tightly joined back-to-back, mostly by hydrophobic and some electrostatic interactions. The ligands of the copper and zinc are six histidine and one aspartate side-chains; one histidine is bound between the two metals.

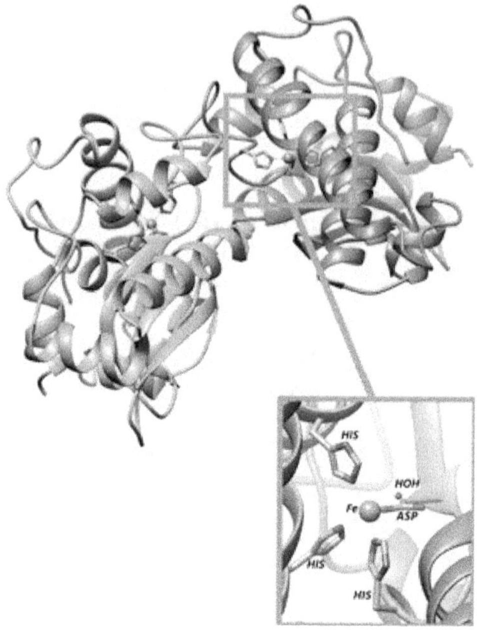

Fig: Active Site for Iron Superoxide DismutaseIron or Manganese—Used by Prokaryotes and Protists, and in Mitochondria and Chloroplasts

- **Iron:** Many bacteria contain a form of the enzyme with iron (Fe-SOD); some bacteria contain Fe-SOD, others Mn-SOD, and some (such as *E. coli*) contain both. Fe-SOD can also be found in the chloroplasts of plants. The 3D structures of the homologous Mn and Fe superoxide dismutases have the same arrangement of alpha-helices, and their active sites contain the same type and arrangement of amino acid side-chains. They are usually dimers, but occasionally tetramers.
- **Manganese:** Nearly all mitochondria, and many bacteria, contain a form with manganese (Mn-SOD): For example, the Mn-SOD found in human mitochondria. The ligands of the manganese ions are 3 histidine side-chains, an aspartate side-chain and a water molecule or hydroxy ligand, depending on the Mn oxidation state (respectively II and III).
- **Nickel–Prokaryotic:** This has a hexameric (6-copy) structure built from right-handed 4-helix bundles, each containing N-terminal hooks that chelate a Ni ion. The Ni-hook contains the motif His-Cys-X-X-Pro-Cys-Gly-X-Tyr; it provides most of the interactions critical for metal binding and catalysis and is, therefore, a likely diagnostic of NiSODs.

In higher plants, SOD isozymes have been localized in different cell compartments. Mn-SOD is present in mitochondria and peroxisomes. Fe-SOD has been found mainly in chloroplasts but has also been detected in peroxisomes, and CuZn-SOD has been localized in cytosol, chloroplasts, peroxisomes, and apoplast.

Human

There are three forms of superoxide dismutase present in humans, in all other mammals, and most chordates. SOD1 is located in the cytoplasm, SOD2 in the mitochondria, and SOD3 is extracellular. The first is a dimer (consists of two units), whereas the others are tetramers (four subunits). SOD1 and SOD3 contain copper and zinc, whereas SOD2, the mitochondrial enzyme, has manganese in its reactive centre. The genes are located on chromosomes 21, 6, and 4, respectively (21q22.1, 6q25.3 and 4p15.3-p15.1).

Plants

In higher plants, superoxide dismutase enzymes (SODs) act as antioxidants and protect cellular components from being oxidized by reactive oxygen species (ROS). ROS can form as a result of drought, injury, herbicides and pesticides, ozone, plant metabolic activity, nutrient deficiencies, photoinhibition, temperature above and below ground, toxic metals, and UV or gamma rays. To be specific, molecular O_2 is reduced to O^-_2 (a ROS called superoxide) when it absorbs an excited electron released from compounds of the electron transport chain. Superoxide is known to denature enzymes, oxidize lipids, and fragment DNA. SODs catalyze the production of O_2 and H_2O_2 from superoxide (O^-_2), which results in less harmful reactants.

When acclimating to increased levels of oxidative stress, SOD concentrations typically increase with the degree of stress conditions. The compartmentalization of different forms of SOD throughout the plant makes them counteract stress very effectively. There are three well-known and -studied classes of SOD metallic coenzymes that exist in plants.

First, Fe SODs consist of two species, one homodimer (containing 1–2 g Fe) and one tetramer (containing 2–4 g Fe). They are thought to be the most ancient SOD metalloenzymes and are found within both prokaryotes and eukaryotes. Fe SODs are most abundantly localized inside plant chloroplasts, where they are indigenous.

Second, Mn SODs consist of a homodimer and homotetramer species each containing a single Mn(III) atom per subunit. They are found predominantly in mitochondrion and peroxisomes. Third, Cu-Zn SODs have electrical properties very different from those of the other two classes.

These are concentrated in the chloroplast, cytosol, and in some cases the extracellular space. Note that Cu-Zn SODs provide less protection than Fe SODs when localized in the chloroplast.

Bacteria

Human white blood cells use enzymes such as NADPH oxidase to generate superoxide and other reactive oxygen species to kill bacteria. During infection, some bacteria (e.g., *Burkholderia pseudomallei*) therefore produce superoxide dismutase to protect themselves from being killed.

Biochemistry

SOD out-competes damaging reactions of superoxide, thus protecting the cell from superoxide toxicity. The reaction of superoxide with non-radicals is spin-forbidden. In biological systems, this means that its main reactions are with itself (dismutation) or with another biological radical such as nitric oxide (NO) or with a transition-series metal.

The superoxide anion radical (O^-_2) spontaneously dismutes to O_2 and hydrogen peroxide (H_2O_2) quite rapidly (~10^5 $M^{-1}s^{-1}$ at pH 7). SOD is necessary because superoxide reacts with sensitive and critical cellular targets. For example, it reacts with the NO radical, and makes toxic peroxynitrite.

Because the uncatalysed dismutation reaction for superoxide requires two superoxide molecules to react with each other, the dismutation rate is second-order with respect to initial superoxide concentration. Thus, the half-life of superoxide, although very short at high concentrations (e.g., 0.05 seconds at 0.1mM) is actually quite long at low concentrations (e.g., 14 hours at 0.1 nM). In contrast, the reaction of superoxide with SOD is first order with respect to superoxide concentration. Moreover, superoxide dismutase has the largest k_{cat}/K_M (an approximation of catalytic efficiency) of any known enzyme (~7 x 10^9 $M^{-1}s^{-1}$), this reaction being limited only by the frequency of collision between itself and superoxide. That is, the reaction rate is "diffusion-limited".

The high efficiency of superoxide dismutase seems necessary: even at the subnanomolar concentrations achieved by the high concentrations of SOD within cells, superoxide inactivates the citric acid cycle enzyme aconitase, can poison energy metabolism, and releases potentially toxic iron. Aconitase is one of several iron-sulfur-containing (de)hydratases in metabolic pathways shown to be inactivated by superoxide.

Stability and Folding Mechanism

SOD1 is an extremely stable protein. In the holo form (both copper and zinc bound) the melting point is > 90 °C. In the apo form (no copper or zinc bound) the melting point is ~60 °C. By differential scanning calorimetry (DSC), holo SOD1 unfolds by a two-state mechanism: from dimer to two unfolded monomers. In chemical denaturation experiments, holo SOD1 unfolds by a three-state mechanism with observation of a folded monomeric intermediate.

Physiology

Superoxide is one of the main reactive oxygen species in the cell. As a consequence, SOD serves a key antioxidant role. The physiological importance of SODs is illustrated by the severe pathologies evident in mice genetically engineered to lack these enzymes. Mice lacking SOD2 die several days after birth, amid massive oxidative stress. Mice lacking SOD1 develop a wide range of pathologies, including hepatocellular carcinoma, an acceleration of age-related muscle mass loss, an earlier incidence of cataracts, and a reduced lifespan. Mice lacking SOD3 do not show any obvious defects and exhibit a normal lifespan, though they are more sensitive to hyperoxic injury. Knockout mice of any SOD enzyme are more sensitive to the lethal effects of superoxide-generating compounds, such as paraquat and diquat (herbicides).

Drosophila lacking SOD1 have a dramatically shortened lifespan, whereas flies lacking SOD2 die before birth. Depletion of SOD1 and SOD2 in the nervous system and muscles of *Drosophila* is associated with reduced lifespan. The accumulation of neuronal and muscular ROS appears to contribute to age-associated impairments. When overexpression of mitochondrial SOD2 is induced, the lifespan of adult *Drosophila* is extended.

Among black garden ants (*Lasius niger*), the lifespan of queens is an order of magnitude greater than of workers despite no systematic nucleotide sequence difference between them. The *SOD3* gene was found to be the most differentially over-expressed in the brains of queen vs worker ants. This finding raises the possibility of an important role of antioxidant function in modulating lifespan.

SOD knockdowns in the worm *C. elegans* do not cause major physiological disruptions. However, the lifespan of *C. elegans* can be extended by superoxide/catalase mimetics suggesting that oxidative stress is a major determinant of the rate of aging.

Knockout or null mutations in SOD1 are highly detrimental to aerobic growth in the budding yeast *Saccharomyces cerevisiae* and result in a dramatic reduction in post-diauxic lifespan. In wild-type *S. cerevisiae*, DNA damage rates increased 3-fold with age, but more than 5-fold in mutants deleted for either the *SOD1* or *SOD2* genes.

Reactive oxygen species levels increase with age in these mutant strains and show a similar pattern to the pattern of DNA damage increase with age. Thus it appears that superoxide dismutase plays a substantial role in preserving genome integrity during aging in *S. cerevisiae*. SOD2 knockout or null mutations cause growth inhibition on respiratory carbon sources in addition to decreased post-diauxic lifespan.

In the fission yeast *Schizosaccharomyces pombe*, deficiency of mitochondrial superoxide dismutase SOD2 accelerates chronological aging.

Several prokaryotic SOD null mutants have been generated, including *E. coli*. The loss of periplasmic CuZnSOD causes loss of virulence and might be an attractive target for new antibiotics.

Role in Disease

Mutations in the first SOD enzyme (SOD1) can cause familial amyotrophic lateral sclerosis (ALS, a form of motor neuron disease). The most common mutation in the U.S. is A4V, while the most intensely studied is G93A. Inactivation of SOD1 causes hepatocellular carcinoma. Diminished SOD3 activity has been linked to lung diseases such as acute respiratory distress syndrome (ARDS) or chronic obstructive pulmonary disease (COPD). Superoxide dismutase is not expressed in neural crest cells in the developing fetus. Hence, high levels of free radicals can cause damage to them and induce dysraphic anomalies (neural tube defects).

Mutations in SOD1 can cause familial ALS (several pieces of evidence also show that wild-type SOD1, under conditions of cellular stress, is implicated in a significant fraction of sporadic ALS cases, which represent 90% of ALS patients.), by a mechanism that is presently not understood, but not due to loss of enzymatic activity or a decrease in the conformational stability of the SOD1 protein. Overexpression of SOD1 has been linked to the neural disorders seen in Down syndrome. In patients with thalassemia, SOD will increase as a form of compensation mechanism. However, in the chronic stage, SOD does not seem to be sufficient and tends to decrease due to the destruction of proteins from the massive reaction of oxidant-antioxidant.

In mice, the extracellular superoxide dismutase (SOD3, ecSOD) contributes to the development of hypertension. Inactivation of SOD2 in mice causes perinatal lethality.

Medical Uses

Supplementary superoxide dimutase has been suggested as a treatment to prevent bronchopulmonary dysplasia in infants who are born preterm, however the effectiveness of his treatment is not clear.

Research

SOD has been used in experimental treatment of chronic inflammation in inflammatory bowel conditions. SOD may ameliorate cis-platinum-induced nephrotoxicity (rodent studies). As "Orgotein" or "ontosein", a pharmacologically-active purified bovine liver SOD, it is also effective in the treatment of urinary tract inflammatory disease in man. For a time, bovine liver SOD even had regulatory approval in several European countries for such use. This was cut short by concerns about prion disease.

An SOD-mimetic agent, TEMPOL, is currently in clinical trials for radioprotection and to prevent radiation-induced dermatitis. TEMPOL and similar SOD-mimetic nitroxides exhibit a multiplicity of actions in diseases involving oxidative stress.

The synthesis of enzymes such as superoxide dismutase, L-ascorbate oxidase, and Delta 1 DNA polymerase is initiated in plants with the activation of genes associated with stress conditions for plants. The most common stress conditions can be injury, drought or soil salinity. Limiting this process initiated by the conditions of strong soil salinity can be achieved by administering exogenous glutamine to plants. The decrease in the level of expression of genes responsible for the synthesis of superoxide dismutase increases with the increase in glutamine concentration.

Cosmetic Uses

SOD may reduce free radical damage to skin—for example, to reduce fibrosis following radiation for breast cancer. Studies of this kind must be regarded as tentative, however, as there were not adequate controls in the study including a lack of randomization, double-blinding, or placebo. Superoxide dismutase is known to reverse fibrosis, possibly through de-differentiation of myofibroblasts back to fibroblasts.

Commercial Sources

SOD is commercially obtained from marine phytoplankton, bovine liver, horseradish, cantaloupe, and certain bacteria. For therapeutic purpose, SOD is usually injected locally. There is no evidence that ingestion of unprotected SOD or SOD-rich foods can have any physiological effects, as all ingested SOD is broken down into amino acids before being absorbed. However, ingestion of SOD bound to wheat proteins could improve its therapeutic activity, at least in theory.

6.2.2 Superoxide Reductases

Superoxide reductase is an enzyme that catalyzes the conversion of highly reactive and toxic superoxide (O_2^-) into less toxic hydrogen peroxide (H_2O_2):

$$\text{reduced rubredoxin} + O_2^- + 2\,H^+ \rightleftharpoons rubredoxin + H_2O_2$$

$$Fe^{2+} + O_2^- + 2\,H^+ \rightleftharpoons Fe^{3+} + H_2O_2$$

Hydrogen peroxide in turn is reduced to water by rubrerythrin. The 3 substrates of this enzyme are reduced rubredoxin, superoxide, and H^+, whereas its two products are rubredoxin and H_2O_2. This enzyme belongs to the family of oxidoreductases, specifically those acting on superoxide as acceptor (only sub-subclass identified to date). The systematic name of this enzyme class is rubredoxin:superoxide oxidoreductase. Other names in common use include neelaredoxin, and desulfoferrodoxin.

6.2.3 Peroxidases

Peroxidases or peroxide reductases are a large group of enzymes which play a role in various biological processes. They are named after the fact that they commonly break up peroxides, and should not be confused with other enzymes that *produce* peroxide, which are often oxidases.

Functionality

Peroxidases typically catalyze a reaction of the form:

$$\mathrm{ROOR'} + \overset{\text{electron donor}}{2\mathrm{e}^-} + 2\mathrm{H}^+ \xrightarrow{\text{Peroxidase}} \mathrm{ROH} + \mathrm{R'OH}$$

Optimal Substrates

For many of these enzymes the optimal substrate is hydrogen peroxide, but others are more active with organic hydroperoxides such as lipid peroxides. Peroxidases can contain a heme cofactor in their active sites, or alternately redox-active cysteine or selenocysteine residues.

The nature of the electron donor is very dependent on the structure of the enzyme.

- For example, horseradish peroxidase can use a variety of organic compounds as electron donors and acceptors. Horseradish peroxidase has an accessible active site, and many compounds can reach the site of the reaction.
- On the other hand, for an enzyme such as cytochrome c peroxidase, the compounds that donate electrons are very specific, due to a very narrow active site.

Classification

Protein families that serve as peroxidases include:

- **Haem-Using:**
  - Haem peroxidase and the related animal heme-dependent peroxidases
  - DyP-type peroxidase family
  - Catalase
  - Some haloperoxidase
  - Di-haem cytochrome c peroxidase
- **Non-Heme:**
  - *Thiol:* glutathione peroxidase, peroxiredoxin
  - Vanadium bromoperoxidase

- Alkyl hydroperoxide reductase
- Manganese peroxidase
- NADH peroxidase

Characterization

The glutathione peroxidase family consists of 8 known human isoforms. Glutathione peroxidases use glutathione as an electron donor and are active with both hydrogen peroxide and organic hydroperoxide substrates. Gpx1, Gpx2, Gpx3, and Gpx4 have been shown to be selenium-containing enzymes, whereas Gpx6 is a selenoprotein in humans with cysteine-containing homologues in rodents.

Amyloid beta, when bound to heme, has been shown to have peroxidase activity. A typical group of peroxidases are the haloperoxidases. This group is able to form reactive halogen species and, as a result, natural organohalogen substances. A majority of peroxidase protein sequences can be found in the PeroxiBase database.

Pathogenic Resistance

While the exact mechanisms have yet to be determined, peroxidases are known to play a part in increasing a plant's defenses against pathogens. Many members of the Solanaceae, notably *Solanum melongena* (eggplant/aubergine) and *Capsicum chinense* (the habanero/Scotch bonnet varieties of chili peppers) use Guaiacol and the enzyme guaiacol peroxidase as a defense against bacterial parasites such as *Ralstonia solanacearum*: the gene expression for this enzyme commences within minutes of bacterial attack.

Applications

Peroxidase can be used for treatment of industrial waste waters. For example, phenols, which are important pollutants, can be removed by enzyme-catalyzed polymerization using horseradish peroxidase. Thus phenols are oxidized to phenoxy radicals, which participate in reactions where polymers and oligomers are produced that are less toxic than phenols. It also can be used to convert toxic materials into less harmful substances. There are many investigations about the use of peroxidase in many manufacturing processes like adhesives, computer chips, car parts, and linings of drums and cans. Other studies have shown that peroxidases may be used successfully to polymerize anilines and phenols in organic solvent matrices. Peroxidases are sometimes used as histological markers. Cytochrome c peroxidase is used as a soluble, easily purified model for cytochrome c oxidase.

6.3 OXYGENASES AND OXIDASES

6.3.1 Oxygenases

An oxygenase is any enzyme that oxidizes a substrate by transferring the oxygen from molecular oxygen O_2 (as in air) to it. The oxygenases form a class of oxidoreductases; their EC number is EC 1.13 or EC 1.14.

Structure

Most oxygenases contain either a metal, usually iron, or an organic cofactor, usually flavin. These cofactors interact with O_2, leading to its transfer to substrate. Oxygenases constitute a major intracellular source of iron and carbon monoxide

Mechanism

Two types of oxygenases are recognized:

- Monooxygenases, or mixed function oxidase, transfer one oxygen atom to the substrate, and reduce the other oxygen atom to water.
- Dioxygenases, or oxygen transferases, incorporate both atoms of molecular oxygen (O_2) into the product(s) of the reaction.

Among the most common monooxygenases are the cytochrome P450 oxidases, responsible for breaking down numerous chemicals in the body.

History

Oxygenases were discovered in 1955 simultaneously by two groups, Osamu Hayaishi from Japan and Howard S. Mason from the US. Hayaishi was awarded the 1986 Wolf Prize in Medicine "for the discovery of the oxygenase enzymes and elucidation of their structure and biological importance."

6.3.2 Oxidases

In biochemistry, an oxidase is an oxidoreductase (any enzyme that catalyzes a redox reaction) that uses dioxygen (O_2) as the electron acceptor. In reactions involving donation of a hydrogen atom, oxygen is reduced to water (H_2O) or hydrogen peroxide (H_2O_2). Some oxidation reactions, such as those involving monoamine oxidase or xanthine oxidase, typically do not involve free molecular oxygen.

The oxidases are a subclass of the oxidoreductases. The use of dioxygen is the only unifying feature; in the EC classification, these enzymes are scattered in many categories.

Examples

An important example is EC 7.1.1.9 cytochrome c oxidase, the key enzyme that allows the body to employ oxygen in the generation of energy and the final component of the electron transfer chain. Other examples are:

- EC 1.1.3.4 Glucose oxidase
- EC 1.4.3.4 Monoamine oxidase

- EC 1.14.-.- Cytochrome P450 oxidase
- EC 1.6.3.1 NADPH oxidase
- EC 1.17.3.2 Xanthine oxidase
- EC 1.1.3.8 L-gulonolactone oxidase
- EC 1.10.3.2 Laccase
- EC 1.4.3.13 Lysyl oxidase
- EC 1.10.3.2 Polyphenol oxidase
- Sulfhydryl oxidase. This enzyme oxidises thiol groups.

Oxidase Test

In microbiology, the oxidase test is used as a phenotypic characteristic for the identification of bacterial strains; it determines whether a given bacterium produces cytochrome oxidases (and therefore utilizes oxygen with an electron transfer chain).

The test is used to determine whether a bacterium is an aerobe or anaerobe. However a bacterium that is Oxidase negative is not necessarily anaerobic, instead showing the bacterium does not possess cytochrome c oxidase.

CHAPTER 7

OXO-TRANSFER PROTEINS BASED ON MOLYBDENUM, TUNGSTEN, AND VANADIUM

7.1 MOLYBDO- AND TUNGSTO-PYRANOPTERINS

Molybdo- and tungsto-pyranopterins are important cofactor molecules that play a crucial role in various biological processes, particularly in the context of enzymatic reactions involving molybdenum and tungsten. These compounds are characterized by their unique chemical structure, which includes a pyranopterin core, and their ability to facilitate redox reactions in different organisms. In summary, molybdo- and tungsto-pyranopterins are essential cofactors that facilitate important enzymatic reactions in various biological systems. Their unique structures and biosynthesis pathways highlight their significance in bioinorganic chemistry. As research continues to unravel the complexities of these compounds, their applications in biotechnology, environmental science, and health will likely expand. Understanding the roles of molybdo- and tungsto-pyranopterins will contribute to advancements in multiple fields, ultimately benefiting society through improved biocatalysts, environmental sustainability, and public health initiatives.

7.1.1 Structure and Composition

Molybdo- and tungsto-pyranopterins contain a pyranopterin backbone, which is a bicyclic structure that features a pyran ring fused to a pteridine-like system. The fundamental difference between the two types of compounds lies in the metal ion associated with the pyranopterin structure. In molybdo-pyranopterins, molybdenum is the central metal, whereas tungsten is the central metal in tungsto-pyranopterins. The metal ions coordinate with the nitrogen atoms in the pyranopterin structure, creating a stable complex that is essential for the catalytic activity of the enzymes that utilize these cofactors.

7.1.2 Biosynthesis

The biosynthesis of molybdo- and tungsto-pyranopterins occurs through a series of enzymatic reactions that begin with the precursor molecule, guanosine triphosphate (GTP). In the case of molybdo-pyranopterin, specific enzymes convert GTP into a series of intermediates, eventually leading to the formation of molybdenum cofactor (MoCo). The process involves the addition of molybdenum ions to the pyranopterin structure, resulting in the formation of molybdo-pyranopterin. Tungsto-pyranopterin biosynthesis follows a similar pathway, with tungsten ions being incorporated into the pyranopterin core.

7.1.3 Biological Significance

Molybdo- and tungsto-pyranopterins serve as essential cofactors for a variety of enzymes, particularly those involved in redox reactions. These enzymes include sulfite oxidase, xanthine oxidase, and aldehyde oxidase, which play critical roles in various metabolic pathways. For instance, sulfite oxidase is responsible for the detoxification of sulfite, a potentially harmful byproduct of sulfur metabolism. Molybdo-pyranopterin is also involved in nitrogen fixation processes, where it aids in the conversion of atmospheric nitrogen into ammonia, a crucial step for plant growth and development.

Tungsto-pyranopterins, while less common than their molybdo counterparts, also exhibit significant biological roles. They are involved in enzymatic reactions that require tungsten as a cofactor, contributing to metabolic processes in certain bacteria and archaea. The unique properties of tungsto-pyranopterins make them suitable for catalyzing specific redox reactions that may not be feasible with other metal cofactors.

7.1.4 Applications in Biotechnology

The unique properties of molybdo- and tungsto-pyranopterins have garnered interest in the field of biotechnology. Their ability to catalyze specific reactions makes them valuable tools for enzyme engineering and the development of biocatalysts. Researchers are exploring the use of these cofactors in bioremediation, where enzymes that utilize molybdo- and tungsto-pyranopterins can break down environmental pollutants and toxic substances. Furthermore, the incorporation of these cofactors into synthetic biology applications holds promise for creating engineered organisms capable of producing valuable compounds.

7.1.5 Research Developments

Recent studies have focused on understanding the mechanisms of action and specificity of molybdo- and tungsto-pyranopterins in various enzymes. Advances in structural biology, including X-ray crystallography and NMR spectroscopy, have provided insights into the binding interactions between these cofactors and their corresponding enzymes. Such research contributes to a deeper understanding of the biochemical pathways in which these cofactors are involved and may lead to novel applications in medicine and environmental science.

7.1.6 Environmental and Health Implications

The significance of molybdo- and tungsto-pyranopterins extends beyond their biological roles; they also have implications for environmental health. The presence of molybdenum and tungsten in the environment can influence soil chemistry and nutrient cycling.

Additionally, understanding the role of these cofactors in biological systems is essential for assessing their potential impact on human health. Molybdenum is an essential trace element for humans, while excessive tungsten exposure can be toxic. Therefore, studying the pathways involving molybdo- and tungsto-pyranopterins can inform public health policies and environmental regulations.

7.1.7 The Xanthine Oxidase Family

The xanthine oxidase family encompasses a group of important enzymes that play critical roles in purine metabolism, as well as in the oxidative stress response within biological systems. These enzymes catalyze the oxidation of hypoxanthine and xanthine to uric acid, a process vital for the proper functioning of various cellular and metabolic processes. The family includes xanthine oxidase (XO) and xanthine dehydrogenase (XDH), which, despite their similarities, exhibit distinct properties and functions.

The xanthine oxidase family, encompassing xanthine oxidase and xanthine dehydrogenase, plays a vital role in purine metabolism and the generation of reactive oxygen species. Their regulatory mechanisms and clinical significance underscore their importance in health and disease. With ongoing research focused on elucidating the intricate functions of these enzymes, we can expect further advancements in our understanding of their roles in various physiological and pathological processes, ultimately leading to improved therapeutic strategies for managing diseases linked to altered purine metabolism and oxidative stress.

Structure and Function

The xanthine oxidase family is characterized by its unique structure, which includes a molybdenum cofactor, iron-sulfur clusters, and flavin adenine dinucleotide (FAD). Xanthine oxidase (XO) and xanthine dehydrogenase (XDH) share a high degree of homology and perform similar enzymatic reactions. Both enzymes catalyze the oxidation of hypoxanthine to xanthine and then from xanthine to uric acid, but the key distinction lies in their electron-donating capabilities.

Xanthine oxidase functions as an oxidase, utilizing molecular oxygen as the electron acceptor, resulting in the production of reactive oxygen species (ROS). In contrast, xanthine dehydrogenase operates as a dehydrogenase, reducing NAD^+ to NADH while converting xanthine to uric acid. This difference in function reflects the physiological context in which these enzymes operate, with XO typically involved in hypoxic conditions where oxidative stress may be heightened.

Enzymatic Mechanism

The enzymatic mechanism of xanthine oxidase and xanthine dehydrogenase involves several key steps. Initially, the substrate binds to the active site of the enzyme, where the molybdenum cofactor facilitates the transfer of electrons. The enzyme first catalyzes the conversion of hypoxanthine to xanthine, followed by the conversion of xanthine to uric acid. During these reactions, the formation of superoxide anions and hydrogen peroxide occurs, particularly in the case of xanthine oxidase. The production of these reactive species is significant as it links xanthine oxidase activity to oxidative stress and inflammation.

Regulation of Xanthine Oxidase Family

The activity of the xanthine oxidase family is tightly regulated at multiple levels. Various factors influence the enzyme's activity, including substrate availability, post-translational modifications, and the redox state of the cell. For instance, xanthine dehydrogenase can be converted to xanthine oxidase through oxidative stress, resulting in increased ROS production. This interconversion is a crucial regulatory mechanism that enables the cell to respond to changing metabolic and oxidative conditions.

Moreover, several endogenous and exogenous factors can modulate xanthine oxidase activity. Hormones, such as insulin, can inhibit the enzyme's activity, while certain drugs, including allopurinol and febuxostat, are specifically designed to inhibit xanthine oxidase, reducing uric acid production in conditions like gout and hyperuricemia.

Clinical Significance

The xanthine oxidase family is of considerable clinical significance due to its role in various diseases. Elevated levels of uric acid, a product of xanthine oxidase activity, can lead to conditions such as gout, characterized by painful joint inflammation due to the deposition of urate crystals. In addition to gout, hyperuricemia has been implicated in cardiovascular diseases, kidney disease, and metabolic syndrome, indicating that the xanthine oxidase family has far-reaching implications for human health.

Moreover, the production of reactive oxygen species by xanthine oxidase links this enzyme family to oxidative stress-related disorders, including cancer and neurodegenerative diseases. In conditions marked by chronic inflammation, the overactivity of xanthine oxidase can exacerbate tissue damage and contribute to disease progression.

Therapeutic Interventions

Given the clinical relevance of the xanthine oxidase family, therapeutic interventions targeting these enzymes have been developed. Allopurinol, a xanthine oxidase inhibitor, is widely used to manage hyperuricemia and prevent gout attacks. By inhibiting xanthine oxidase, allopurinol reduces the production of uric acid, thereby alleviating symptoms and preventing complications associated with high uric acid levels.

In addition to allopurinol, novel xanthine oxidase inhibitors, such as febuxostat, have emerged as effective alternatives for managing gout and related conditions. These inhibitors not only lower uric acid levels but may also exert antioxidant effects, highlighting their potential utility in treating oxidative stress-related diseases.

Research Advances

Recent research has expanded our understanding of the xanthine oxidase family, shedding light on the molecular mechanisms underlying their regulation and function. Advances in structural biology, such as X-ray crystallography and cryo-electron microscopy, have elucidated the three-dimensional structures of xanthine oxidase and xanthine dehydrogenase, providing insights into substrate binding and catalytic mechanisms.

Additionally, studies have investigated the role of xanthine oxidase in various pathophysiological conditions, emphasizing its involvement in inflammation and oxidative stress. Investigations into genetic variations in xanthine oxidase genes may also provide insights into individual susceptibility to diseases associated with altered uric acid metabolism.

Future Directions

The ongoing research into the xanthine oxidase family holds promise for the development of targeted therapies and personalized medicine approaches. Understanding the genetic and environmental factors influencing xanthine oxidase activity may enable clinicians to tailor treatments based on individual patient profiles. Furthermore, exploring the potential of novel xanthine oxidase inhibitors and antioxidants could lead to new strategies for managing conditions associated with oxidative stress and inflammation.

7.1.8 The Sulfite Oxidase Family

The sulfite oxidase family encompasses a group of enzymes that play a critical role in sulfur metabolism, specifically in the oxidation of sulfite to sulfate. This enzymatic process is vital for various biological systems, particularly in the detoxification of sulfite and the regulation of sulfur-containing compounds. The sulfite oxidase family primarily includes the enzyme sulfite oxidase (SO), which is essential for the proper functioning of the sulfur metabolism pathway. This article will explore the structure, function, regulation, clinical significance, and future directions of the sulfite oxidase family.

Structure of Sulfite Oxidase

Sulfite oxidase is a complex enzyme characterized by its unique structure, which includes a molybdenum cofactor, a heme group, and an iron-sulfur cluster. The molybdenum cofactor is crucial for the enzyme's catalytic activity, facilitating the oxidation of sulfite to sulfate. The enzyme typically exists as a homodimer, with each monomer containing active sites where the substrate binds. The heme group, located within the enzyme, acts as an electron acceptor during the oxidation reaction. This intricate structural arrangement allows sulfite oxidase to perform its function efficiently and effectively.

Mechanism of Action

The enzymatic mechanism of sulfite oxidase involves a series of oxidation-reduction reactions. When sulfite binds to the active site of the enzyme, it undergoes oxidation, resulting in the formation of sulfate. During this process, electrons are transferred to the molybdenum cofactor, which subsequently donates these electrons to the heme group. The overall reaction can be summarized as follows: sulfite + O2 → sulfate + H2O. This reaction is crucial for detoxifying sulfite, which can be harmful at elevated concentrations, and converting it into sulfate, a form that can be easily excreted by the body.

Regulation of Sulfite Oxidase Activity

The activity of sulfite oxidase is tightly regulated by various factors, including substrate availability, post-translational modifications, and the cellular redox state. The concentration of sulfite in the cell plays a significant role in regulating enzyme activity. When sulfite levels are high, sulfite oxidase activity increases to facilitate its detoxification. Conversely, when sulfite levels are low, the enzyme's activity may decrease, allowing for the maintenance of proper sulfur balance within the cell.

Post-translational modifications, such as phosphorylation and glycosylation, can also influence sulfite oxidase activity. These modifications can affect the enzyme's stability, localization, and interaction with other cellular components. Additionally, the cellular redox state can modulate sulfite oxidase activity, as oxidative stress may impact the enzyme's function and stability.

Clinical Significance

The sulfite oxidase family is of considerable clinical importance due to its involvement in various metabolic disorders. Deficiencies in sulfite oxidase activity can lead to the accumulation of sulfite in the body, resulting in sulfite oxidase deficiency (SOD), a rare genetic disorder. Individuals with SOD may experience neurological symptoms, such as seizures, developmental delays, and behavioral issues, due to the neurotoxic effects of accumulated sulfite.

Furthermore, sulfite sensitivity is a significant concern for individuals with asthma and other respiratory conditions. Sulfites are commonly used as preservatives in food and beverages, and exposure to these compounds can trigger allergic reactions and exacerbate asthma symptoms. Understanding the role of sulfite oxidase in detoxifying sulfite may provide insights into managing sulfite sensitivity and developing strategies to mitigate its effects.

Role in Sulfur Metabolism

Sulfite oxidase plays a crucial role in sulfur metabolism, a fundamental biochemical process that contributes to the synthesis of various biomolecules, including amino acids, coenzymes, and antioxidants. Sulfur is an essential element for living organisms, and its proper metabolism is vital for maintaining cellular homeostasis. The conversion of sulfite to sulfate by sulfite oxidase is a key step in the sulfur cycle, ensuring the availability of sulfate for various biological functions.

In addition to its role in detoxification, sulfite oxidase is involved in the production of hydrogen sulfide (H2S), a signaling molecule that regulates numerous physiological processes. H2S has been implicated in various biological functions, including vasodilation, neurotransmission, and cellular protection against oxidative stress. The regulation of sulfite oxidase activity, therefore, plays a crucial role in maintaining the balance of sulfur-containing compounds and their physiological effects.

Research Advances

Recent research has provided valuable insights into the structure and function of sulfite oxidase. Advances in structural biology techniques, such as X-ray crystallography and cryo-electron microscopy, have elucidated the three-dimensional structure of the enzyme, revealing critical details about its active site and substrate binding mechanisms. These studies have enhanced our understanding of the enzymatic reaction and provided a foundation for developing potential inhibitors or activators of sulfite oxidase.

Moreover, investigations into the genetic basis of sulfite oxidase deficiency have identified mutations in the gene encoding the enzyme, leading to advancements in genetic screening and diagnosis. Early detection of sulfite oxidase deficiency can enable timely interventions and management strategies, improving the quality of life for affected individuals.

Therapeutic Interventions

Given the clinical significance of sulfite oxidase, therapeutic interventions targeting this enzyme have been explored. In individuals with sulfite oxidase deficiency, enzyme replacement therapy may offer a potential treatment option, providing a source of functional sulfite oxidase to restore normal metabolic activity. Additionally, dietary modifications to limit sulfite intake may be beneficial for individuals with sulfite sensitivity, reducing the risk of allergic reactions and exacerbations.

Furthermore, understanding the regulatory mechanisms governing sulfite oxidase activity may pave the way for the development of pharmacological agents that modulate enzyme function. Such agents could be utilized to enhance detoxification pathways in individuals exposed to elevated sulfite levels, thereby mitigating potential health risks.

Future Directions

Future research on the sulfite oxidase family holds promise for advancing our understanding of sulfur metabolism and its implications for human health. Investigating the roles of sulfite oxidase in various physiological and pathological contexts may unveil new therapeutic targets for managing diseases associated with sulfur metabolism and oxidative stress. Additionally, exploring the interplay between sulfite oxidase and other metabolic pathways could provide insights into the complex regulatory networks governing cellular homeostasis.

The development of novel inhibitors or activators of sulfite oxidase may also lead to innovative therapeutic approaches for managing sulfite-related disorders. Furthermore, expanding our knowledge of sulfite oxidase's role in signaling pathways may enhance our understanding of its broader physiological significance.

7.1.9 The Dimethylsulfoxide (DMSO) Reductase Family

The dimethylsulfoxide (DMSO) reductase family consists of a group of enzymes that play a critical role in the anaerobic respiration of certain microorganisms, particularly in the reduction of DMSO to dimethyl sulfide (DMS) or to methanethiol. These enzymes are crucial in the sulfur cycle and significantly impact both ecological systems and industrial applications.

The dimethylsulfoxide reductase family plays a vital role in sulfur metabolism and the ecological processes associated with the sulfur cycle. These enzymes facilitate the reduction of DMSO, contributing to the production of DMS and other sulfur compounds that significantly impact atmospheric chemistry and microbial metabolism. The structural and functional insights gained from recent research efforts highlight the potential applications of DMSO reductases in industrial processes and environmental sustainability. As our understanding of this enzyme family continues to grow, it is likely that innovative applications and therapeutic interventions will emerge, further emphasizing the importance of DMSO reductases in both biological and industrial contexts.

Structure of DMSO Reductase

DMSO reductases are typically molybdenum-containing enzymes that exhibit a distinctive structural organization. They usually exist as homodimers or heterodimers, with each monomer containing a molybdenum cofactor necessary for catalytic activity. The active site of DMSO reductase contains not only the molybdenum atom but also other metal cofactors, such as iron or tungsten, which assist in the electron transfer processes involved in substrate reduction. The enzyme's structure allows it to bind to DMSO and facilitate its conversion through a series of redox reactions.

The overall three-dimensional arrangement of DMSO reductase contributes to its specificity for DMSO as a substrate, enabling efficient catalysis. Structural studies using techniques such as X-ray crystallography have revealed intricate details about the enzyme's active site and how it interacts with substrates, providing insights into its enzymatic mechanism.

Mechanism of Action

The enzymatic mechanism of DMSO reductase involves the transfer of electrons from electron donors, typically through a series of redox reactions. When DMSO binds to the active site, it undergoes a reduction process facilitated by the molybdenum cofactor. The reaction can be summarized as follows:

$$DMSO + 2el^{-} + 2H^{+} \rightarrow DMS + H2O$$

In this process, electrons are transferred from a suitable electron donor, such as formate or reduced ferredoxin, to DMSO, converting it to DMS or methanethiol, depending on the specific enzyme and reaction conditions. The efficiency of this electron transfer process is critical for the enzyme's overall activity and is influenced by factors such as substrate concentration and the presence of inhibitors.

Biological Significance

DMSO reductases play a vital role in the sulfur cycle within anaerobic environments, such as wetlands, sediments, and the gastrointestinal tracts of certain animals. By facilitating the reduction of DMSO to DMS or methanethiol, these enzymes contribute to the release of sulfur compounds into the atmosphere, influencing both local and global sulfur biogeochemical cycles. DMS is a significant contributor to atmospheric chemistry, as it can be oxidized to sulfate aerosols, which play a role in cloud formation and climate regulation.

Additionally, DMSO reductases are integral to the metabolic processes of various microorganisms, including certain bacteria and archaea. These organisms utilize DMSO as a terminal electron acceptor in anaerobic respiration, allowing them to thrive in environments devoid of oxygen. This metabolic versatility enhances their ecological niche and contributes to the diversity of microbial life.

Regulation of DMSO Reductase Activity

The activity of DMSO reductase is subject to regulation by several factors, including the availability of substrates, the concentration of electron donors, and environmental conditions such as pH and temperature. In microbial systems, the expression of DMSO reductase genes can be induced under anaerobic conditions or when DMSO is present in the environment. This regulatory mechanism ensures that microorganisms can adapt their metabolic pathways according to the availability of electron acceptors and donors.

Moreover, post-translational modifications, such as phosphorylation, can influence the activity and stability of DMSO reductases. These modifications may alter the enzyme's conformation, affecting substrate binding and catalytic efficiency. Understanding these regulatory mechanisms can provide insights into the ecological roles of DMSO reductases and their potential applications.

Industrial Applications

The unique properties of DMSO reductases have prompted interest in their potential industrial applications. DMSO, a widely used solvent in pharmaceuticals and biochemistry, can be reduced to DMS, a compound with various applications, including as a flavoring agent, antifungal agent, and as a precursor for synthesizing other sulfur-containing compounds. DMSO reductases could be harnessed for biotechnological processes to produce DMS in a more environmentally friendly manner compared to traditional chemical methods.

Additionally, the ability of DMSO reductases to catalyze specific reactions involving sulfur compounds makes them attractive candidates for bioremediation efforts. For instance, these enzymes could be used to detoxify environments contaminated with DMSO or other sulfur-containing pollutants, contributing to environmental sustainability.

Research Advances

Recent research has advanced our understanding of the DMSO reductase family, focusing on elucidating their structure, function, and regulation. Investigations using advanced techniques such as cryo-electron microscopy and molecular dynamics simulations have provided deeper insights into the dynamics of enzyme-substrate interactions and the enzymatic mechanisms involved in DMSO reduction.

Moreover, studies examining the genetic basis of DMSO reductase expression in various microorganisms have shed light on the evolutionary adaptations that enable these organisms to thrive in diverse environments. Understanding the evolutionary history of DMSO reductases may also reveal insights into their biochemical properties and potential applications.

Future Directions

The exploration of DMSO reductases holds promise for future research and applications. Investigating the enzymatic mechanisms and substrate specificity of various DMSO reductase family members may unveil novel catalytic properties that can be exploited for industrial applications. Additionally, engineering these enzymes through techniques such as directed evolution could enhance their efficiency and broaden their substrate range, facilitating their use in biotechnological processes.

Furthermore, understanding the ecological roles of DMSO reductases in sulfur cycling can inform strategies for environmental management and bioremediation. Research efforts focused on isolating and characterizing novel DMSO-reducing microorganisms from diverse habitats could lead to discovering new enzymes with unique properties and applications.

7.2 VANADATE-DEPENDENT HALOPEROXIDASES

Vanadate-dependent haloperoxidases (VHPs) are a unique group of enzymes that play a significant role in catalyzing the oxidation of halides, such as chloride, bromide, and iodide, using hydrogen peroxide as the oxidizing agent. These enzymes are distinct due to their dependence on vanadate ions (VO_4^{3-}) for activity, which serves as a crucial cofactor in the catalytic process. VHPs have garnered interest due to their biochemical significance, ecological roles, and potential applications in biocatalysis and environmental remediation.

Vanadate-dependent haloperoxidases are a fascinating group of enzymes that play critical roles in biological processes, ecological interactions, and industrial applications.

Their ability to catalyze the oxidation of halides using hydrogen peroxide has significant implications for environmental sustainability, synthetic chemistry, and food safety. As research continues to uncover the intricacies of VHPs, their potential for innovative applications in biocatalysis and bioremediation will likely expand, underscoring the importance of these enzymes in both natural and industrial contexts.

The ongoing exploration of vanadate-dependent haloperoxidases represents a promising frontier in biochemistry, ecology, and biotechnology, with the potential to contribute to sustainable practices and environmental health.

7.2.1 Structure and Mechanism of Action

Vanadate-dependent haloperoxidases are typically found in certain fungi, bacteria, and some plants. The enzyme structure generally consists of multiple domains that contribute to its catalytic function and stability. The active site of VHPs contains vanadium, which is integrated into a vanadyl ion (VO^{2+}) that mimics the transition state during the oxidation reactions. This unique structure enables VHPs to catalyze the halogenation of organic compounds through an electrophilic aromatic substitution mechanism.

The catalytic mechanism of VHPs involves several steps. Initially, the enzyme binds to halide ions in the presence of hydrogen peroxide. The vanadyl ion then facilitates the transfer of an oxygen atom from hydrogen peroxide to the halide, resulting in the formation of a halogenated product. The reaction can be summarized as follows:

$$\text{Halide} + H_2O_2 \xrightarrow{\text{VHP}} \text{Halogenated Product} + H_2O$$

This catalytic process is crucial for the biosynthesis of halogenated compounds, which are essential for various biological functions.

7.2.2 Biological Significance

VHPs play important roles in various biological processes, particularly in the metabolism of halogenated compounds and the biosynthesis of secondary metabolites. In fungi, for example, these enzymes are involved in the production of halogenated compounds that serve as antifungal agents and contribute to the organism's survival in competitive environments. Additionally, VHPs may participate in the detoxification of halogenated pollutants in contaminated environments, thus contributing to bioremediation efforts.

Moreover, the halogenated products generated by VHPs can have ecological significance. Many of these compounds exhibit antimicrobial and antifungal properties, playing a role in the natural defense mechanisms of plants and microorganisms. The ability to produce halogenated metabolites allows these organisms to interact with their environment more effectively, competing with other species for resources.

7.2.3 Sources and Distribution

Vanadate-dependent haloperoxidases have been identified in various organisms, including fungi, bacteria, and some plants. Among the most studied sources are the halophilic archaea and marine bacteria, where VHPs play a crucial role in the biogeochemical cycling of halogens. Marine environments, in particular, have been found to be rich in haloperoxidases, highlighting the ecological importance of these enzymes in coastal and oceanic ecosystems. Fungal species, such as certain strains of Aspergillus and Penicillium, have also been reported to possess VHPs. These fungi are known for their ability to produce diverse secondary metabolites, including halogenated compounds that exhibit significant bioactivity. The presence of VHPs in these organisms underscores the evolutionary adaptation of fungi to utilize halogens for survival and competition.

7.2.4 Environmental Implications

The environmental implications of vanadate-dependent haloperoxidases are noteworthy, particularly regarding their role in the biodegradation of halogenated pollutants. Halogenated organic compounds are prevalent in various industrial processes and can accumulate in the environment, leading to ecological and health concerns. VHPs can facilitate the breakdown of these compounds, transforming them into less harmful substances.

Research into the bioremediation potential of VHPs is ongoing, with studies focusing on their application in detoxifying contaminated sites. By harnessing the catalytic capabilities of these enzymes, it may be possible to develop biotechnological approaches to mitigate the impact of halogenated pollutants in soil and water systems.

7.2.5 Industrial Applications

The unique catalytic properties of vanadate-dependent haloperoxidases present numerous opportunities for industrial applications. Their ability to catalyze the halogenation of organic compounds makes them attractive for synthetic chemistry, particularly in the production of halogenated pharmaceuticals and agrochemicals. The use of VHPs in biocatalysis offers an environmentally friendly alternative to traditional chemical synthesis methods, potentially reducing the reliance on harsh reagents and conditions.

Additionally, VHPs can be utilized in food preservation and safety. The antimicrobial properties of halogenated compounds produced by VHPs can be applied to enhance the shelf life of food products, contributing to food safety and quality. Research into optimizing the production and activity of VHPs for these applications is ongoing, with the goal of developing sustainable practices in the food industry.

7.2.6 Research Advances and Future Directions

Recent advances in the study of vanadate-dependent haloperoxidases have provided valuable insights into their structure, function, and potential applications. Techniques such as X-ray crystallography and site-directed mutagenesis have facilitated the elucidation of the enzyme's active site and catalytic mechanism. Understanding the interactions between vanadate and the enzyme is crucial for optimizing its activity and stability. Future research directions include exploring the diversity of VHPs in various organisms, as well as their evolutionary adaptations to different environments. Investigating the potential of VHPs as biocatalysts in industrial processes and bioremediation efforts will further contribute to the understanding of their applications. Additionally, studies aimed at engineering VHPs for enhanced performance and specificity could open new avenues in biotechnology and synthetic biology.

7.3 MODEL CHEMISTRY

Model chemistry serves as an essential tool in the field of computational chemistry, providing a framework for understanding the electronic structure and properties of molecules. It involves the development of simplified representations of complex chemical systems, allowing researchers to predict molecular behavior, reactivity, and interactions. This approach is particularly useful in the study of large biomolecules, materials, and reaction mechanisms, where traditional experimental methods may be challenging or impossible to implement. The use of model chemistry not only enhances our understanding of chemical systems but also contributes significantly to advancements in drug design, materials science, and nanotechnology.

Model chemistry is a vital component of modern computational chemistry, providing valuable tools for understanding the electronic structure and properties of molecules. Its historical development, theoretical foundations, and diverse applications underscore its significance in various scientific fields. As computational methods continue to evolve and improve, model chemistry will play an increasingly important role in advancing our knowledge of complex chemical systems, facilitating breakthroughs in drug discovery, materials science, and environmental chemistry. The ongoing exploration of model chemistry promises to yield exciting discoveries and innovations, further enhancing our understanding of the molecular world.

7.3.1 Historical Context

The roots of model chemistry can be traced back to the early days of quantum mechanics and the advent of computational methods for solving the Schrödinger equation. In the 1920s and 1930s, scientists such as Erwin Schrödinger and Werner Heisenberg laid the groundwork for understanding the behavior of electrons in atoms and molecules. However, as the complexity of chemical systems grew, it became clear that exact solutions were often impractical. This realization spurred the development of various approximative methods and models to facilitate the study of larger and more intricate molecules. The introduction of computational techniques, such as Hartree-Fock theory in the 1950s and the subsequent development of density functional theory (DFT) in the 1970s, marked significant milestones in model chemistry. These methods enabled chemists to perform calculations on molecular systems with increasing accuracy and efficiency, paving the way for the modern era of computational chemistry.

7.3.2 Theoretical Foundations of Model Chemistry

Model chemistry is built upon various theoretical frameworks that define how molecular systems can be represented and analyzed. Central to these frameworks is the concept of electronic structure theory, which seeks to describe the arrangement of electrons in atoms and molecules. Different models, including quantum mechanics and molecular mechanics, provide distinct approaches to studying electronic interactions and molecular behavior.

Quantum Mechanics

Quantum mechanics forms the backbone of model chemistry, providing the theoretical basis for understanding electron behavior at the atomic and molecular levels. The Schrödinger equation, which describes the wave function of a quantum system, is fundamental in determining the energy levels and spatial distribution of electrons. Various approximations, such as the Born-Oppenheimer approximation, allow for the simplification of complex molecular systems by treating nuclei and electrons separately.

Molecular Mechanics

Molecular mechanics offers an alternative approach to modeling molecular systems, focusing on classical mechanics principles rather than quantum mechanics. This method uses force fields to describe the potential energy of molecules based on their atomic interactions, enabling the study of large biomolecules and materials over extended periods. Molecular mechanics is particularly useful for simulating conformational changes, protein folding, and dynamics in complex systems.

7.3.3 Types of Model Chemistry

Model chemistry encompasses a variety of computational methods and approaches, each tailored to specific types of molecular systems and research objectives. Some of the most widely used types include:

Ab Initio Methods

Ab initio methods rely on fundamental physical principles to predict molecular properties without empirical parameters. These methods, such as Hartree-Fock and post-Hartree-Fock techniques, provide high levels of accuracy for small to medium-sized molecular systems. However, they can become computationally expensive for larger systems, necessitating the use of more efficient approximations.

Density Functional Theory (DFT)

Density functional theory has become a cornerstone of model chemistry due to its balance between accuracy and computational efficiency. DFT focuses on the electron density rather than wave functions, allowing for the study of larger molecular systems with reasonable computational resources.

This method is particularly effective in predicting molecular geometries, reaction energies, and electronic properties.

Semi-Empirical Methods

Semi-empirical methods combine empirical parameters with quantum mechanical principles to simplify calculations.

These approaches are useful for larger systems where full quantum mechanical treatment may be impractical. Semi-empirical methods, such as PM3 and AM1, offer a compromise between accuracy and computational cost, making them suitable for various applications in materials science and drug discovery.

Molecular Dynamics

Molecular dynamics simulations are essential for studying the time-dependent behavior of molecular systems.

By integrating classical equations of motion, researchers can explore the dynamics of complex biological systems, protein-ligand interactions, and material properties over time. Molecular dynamics provides valuable insights into reaction mechanisms and conformational changes that are difficult to capture through static models.

7.3.4 Applications of Model Chemistry

The applications of model chemistry are vast and diverse, spanning numerous fields of research and industry. Some key areas include:

Drug Design and Discovery

In the pharmaceutical industry, model chemistry plays a pivotal role in drug design and discovery processes. By utilizing computational methods to predict molecular interactions, researchers can identify potential drug candidates, optimize their structures, and evaluate their binding affinities to target proteins. This approach significantly reduces the time and resources required for experimental screening, enhancing the efficiency of drug development.

Materials Science

Model chemistry is instrumental in materials science, where it aids in the design and characterization of novel materials with tailored properties. Researchers use computational methods to explore the relationships between molecular structure and material performance, allowing for the development of advanced materials for applications in electronics, catalysis, and energy storage.

Environmental Chemistry

In environmental chemistry, model chemistry helps assess the behavior and fate of pollutants in various systems. By modeling chemical reactions, transport processes, and interactions between pollutants and environmental matrices, researchers can predict the environmental impact of contaminants and develop effective remediation strategies.

Biochemistry and Biophysics

Model chemistry is crucial in understanding the molecular basis of biological processes. It aids in the study of enzyme mechanisms, protein folding, and the dynamics of biomolecular interactions. By providing insights into these processes, model chemistry contributes to our understanding of fundamental biochemical pathways and disease mechanisms.

7.3.5 Challenges and Limitations

Despite its numerous advantages, model chemistry also faces several challenges and limitations. One major issue is the trade-off between accuracy and computational efficiency. While high-level ab initio methods offer exceptional accuracy, they can be prohibitively resource-intensive for large systems. Conversely, more efficient methods, such as molecular mechanics, may sacrifice accuracy in favor of speed.

Additionally, the choice of model and parameters can significantly influence the results obtained from computational simulations. Careful consideration must be given to the selection of force fields, basis sets, and other computational settings to ensure reliable predictions. Moreover, the validation of computational results against experimental data remains essential for establishing the credibility of model chemistry findings.

7.3.6 Future Directions in Model Chemistry

The future of model chemistry is promising, with ongoing advancements in computational techniques and technology. As computational power continues to grow, researchers can tackle increasingly complex molecular systems, enabling deeper insights into chemical phenomena. The integration of machine learning and artificial intelligence with traditional computational methods holds the potential to revolutionize model chemistry, offering new strategies for predicting molecular properties and behaviors.

Furthermore, interdisciplinary collaborations among chemists, biologists, and materials scientists will enhance the application of model chemistry in addressing real-world challenges. By leveraging computational approaches alongside experimental techniques, researchers can accelerate discoveries in drug design, materials development, and environmental sustainability.

CHAPTER 8
THE SULFUR CYCLE

8.1 ENVIRONMENTAL SULFUR CYCLING

Environmental sulfur cycling is a critical biogeochemical process that plays a fundamental role in maintaining ecosystem balance and influencing climate dynamics. Sulfur is an essential element for life, participating in various biological functions and chemical reactions. The cycling of sulfur involves multiple interconnected processes that include its release into the atmosphere, uptake by organisms, and eventual return to the earth.

Environmental sulfur cycling is a complex and vital process that plays a crucial role in supporting life and maintaining ecosystem balance. Through various biogeochemical transformations, sulfur compounds are recycled within terrestrial and aquatic systems, influencing nutrient dynamics and biological activity. Understanding the intricacies of sulfur cycling is essential for addressing environmental challenges such as acid rain, climate change, and biodiversity loss. Effective monitoring and management strategies can help mitigate the adverse impacts of sulfur cycling and promote healthier ecosystems for future generations. As research advances, a deeper understanding of sulfur cycling will continue to inform environmental policies and sustainable practices, contributing to a more resilient and sustainable planet.

8.1.1 The Role of Sulfur in the Environment

Sulfur is a non-metal element found in various forms in the environment, including sulfide, sulfate, and elemental sulfur. It is a vital component of amino acids, vitamins, and coenzymes, making it essential for the growth and functioning of living organisms. Sulfur is primarily present in the form of sulfate (SO_4^{2-}) in the atmosphere and oceans, while sulfide (S^{2-}) is predominantly found in sediments and soils. The diverse chemical forms of sulfur allow it to participate in a wide range of biological and geological processes.

8.1.2 Sulfur Sources and Inputs

The sulfur cycle begins with various natural and anthropogenic sources that contribute to the sulfur pool in the environment. Natural sources include volcanic eruptions, geothermal activities, and the weathering of rocks, which release sulfur compounds into the atmosphere. For instance, volcanic eruptions emit sulfur dioxide (SO_2), which can oxidize to form sulfate aerosols. These aerosols can be deposited back to the Earth's surface through precipitation, thus contributing to the sulfur pool in soils and water bodies.

Anthropogenic activities, such as fossil fuel combustion, industrial processes, and agricultural practices, significantly contribute to sulfur emissions. The burning of coal and oil releases large amounts of sulfur dioxide into the atmosphere, leading to acid rain formation. This anthropogenic input of sulfur has profound implications for environmental health and ecosystem stability.

8.1.3 The Biological Component of Sulfur Cycling

The biological component of the sulfur cycle involves various microorganisms that play a pivotal role in transforming sulfur compounds through processes such as oxidation, reduction, and assimilation. Key groups of organisms involved in sulfur cycling include sulfur-oxidizing bacteria, sulfur-reducing bacteria, and sulfate-reducing bacteria. These microorganisms facilitate the conversion of sulfur species, making sulfur available to plants and other organisms.

Sulfur Oxidation

Sulfur oxidation is a crucial process in the sulfur cycle that involves the conversion of sulfide (S^{2-}) or elemental sulfur (S) to sulfate (SO_4^{2-}). Sulfur-oxidizing bacteria, such as *Thiobacillus* and *Sulfolobus*, play a vital role in this transformation. These microorganisms utilize sulfur compounds as energy sources, leading to the production of sulfate, which is then available for uptake by plants and other organisms. The oxidation of sulfur compounds also contributes to the acidification of soils and waters, impacting nutrient availability and biological activity.

Sulfur Reduction

Conversely, sulfur reduction involves the conversion of sulfate back to sulfide, primarily through the activity of sulfate-reducing bacteria (SRB) such as *Desulfovibrio* and *Desulfotomaculum*. These bacteria utilize sulfate as an electron acceptor in anaerobic conditions, producing sulfide as a byproduct. Sulfide can precipitate with metal ions to form metal sulfides, influencing the geochemistry of sediments and soils. Sulfate-reducing bacteria are particularly important in anaerobic environments, such as wetlands and oceanic sediments, where they contribute to organic matter degradation and nutrient cycling.

8.1.4 The Role of Plants and Animals

Plants and animals also play significant roles in the sulfur cycle through their interactions with sulfur compounds. Plants take up sulfate from the soil, incorporating it into essential amino acids, proteins, and other biomolecules.

Sulfur is crucial for plant metabolism and growth, influencing crop yields and food security. Once consumed by herbivores, sulfur-containing compounds are transferred through the food web, contributing to the overall cycling of sulfur in ecosystems.

In addition to its nutritional role, sulfur also influences the health of ecosystems. For example, sulfur compounds can act as signaling molecules in plant defense against pathogens. Moreover, animals excrete sulfur in the form of sulfate or sulfide, returning it to the soil and water, where it can be recycled by microorganisms.

8.1.5 Sulfur Cycling in Aquatic Systems

The sulfur cycle is also significantly influenced by aquatic systems, where both freshwater and marine environments facilitate various sulfur transformations. In oceans, sulfate is abundant and plays a crucial role in supporting diverse microbial communities that drive sulfur cycling. Sulfate-reducing bacteria in anoxic marine sediments contribute to the production of hydrogen sulfide, which can be released into the water column.

In freshwater ecosystems, sulfur cycling is influenced by nutrient dynamics and the availability of organic matter. Eutrophication, often driven by nutrient runoff, can enhance microbial activity and sulfate reduction, leading to elevated sulfide levels and potential toxicity to aquatic organisms. Additionally, the interplay between sulfur cycling and carbon cycling in aquatic systems can influence the overall health and stability of these ecosystems.

8.1.6 Environmental Impacts of Sulfur Cycling

The sulfur cycle has significant implications for environmental health and climate dynamics. One of the most well-known impacts of sulfur cycling is the formation of acid rain. When sulfur dioxide is released into the atmosphere, it can undergo oxidation to form sulfuric acid (H_4SO_4), which, when deposited through precipitation, can harm aquatic ecosystems, soil quality, and vegetation. Acid rain can lead to nutrient leaching, reduced soil fertility, and impaired growth of plants and aquatic organisms.

Moreover, sulfur compounds also play a role in climate regulation. Sulfate aerosols, formed from the oxidation of sulfur dioxide, can influence cloud formation and reflect sunlight, potentially cooling the Earth's surface. However, this cooling effect is temporary and does not address the underlying causes of climate change. Additionally, changes in sulfur cycling due to anthropogenic activities can have far-reaching effects on biodiversity and ecosystem resilience.

8.1.7 Monitoring and Management of Sulfur Cycling

Given the environmental implications of sulfur cycling, effective monitoring and management strategies are essential for mitigating adverse impacts. Monitoring sulfur emissions from industrial sources, transportation, and agriculture is crucial for understanding regional sulfur dynamics. Governments and organizations can implement regulations to limit sulfur emissions, promoting cleaner technologies and practices.

Furthermore, sustainable agricultural practices can help maintain healthy sulfur cycling in soils. By incorporating sulfur-rich fertilizers and organic matter, farmers can enhance sulfur availability for crops while minimizing the risk of runoff and pollution. Restoration efforts in aquatic systems can also support natural sulfur cycling processes, promoting the recovery of degraded ecosystems.

8.2 BIOGENIC METABOLISM OF SULFUR

Biogenic metabolism of sulfur refers to the biochemical processes through which living organisms utilize, transform, and recycle sulfur compounds within their cellular systems. Sulfur is a crucial element for all life forms, playing a vital role in the synthesis of amino acids, vitamins, and coenzymes. This metabolism is essential not only for maintaining cellular functions but also for influencing broader biogeochemical cycles.

The biogenic metabolism of sulfur is a complex and essential process that underpins the cycling of this vital element in ecosystems. Through various microbial pathways, sulfur is transformed from one chemical form to another, influencing nutrient dynamics and ecosystem health. Understanding the intricacies of sulfur metabolism provides valuable insights into ecological interactions, nutrient cycling, and environmental challenges. As research continues to uncover the roles of sulfur in biological systems, it will enhance our ability to manage ecosystems sustainably and address the impacts of human activities on natural sulfur cycles. Effective monitoring and management strategies can help mitigate the adverse effects of sulfur metabolism, promoting healthier ecosystems and improving environmental resilience.

8.2.1 Importance of Sulfur in Biological Systems

Sulfur is one of the key biogenic elements, alongside carbon, hydrogen, oxygen, nitrogen, phosphorus, and potassium. It is a vital component of amino acids such as cysteine and methionine, which are necessary for protein synthesis. Furthermore, sulfur plays a critical role in the formation of coenzymes, like coenzyme A (CoA), that are essential for cellular metabolism.

The unique chemical properties of sulfur allow it to exist in multiple oxidation states, facilitating a wide range of redox reactions that are fundamental to energy production and metabolic pathways in living organisms.

8.2.2 Key Microorganisms Involved in Sulfur Metabolism

Various microorganisms, including bacteria and archaea, are pivotal to the biogenic metabolism of sulfur. These organisms possess unique enzymatic capabilities that allow them to convert sulfur compounds through a series of metabolic pathways. Sulfur-oxidizing bacteria, such as *Thiobacillus* and *Acidithiobacillus*, play a critical role in the oxidation of sulfide (S^{2-}) and elemental sulfur (S) to sulfate (SO_4^{2-}), which is then available for uptake by plants and other organisms. Conversely, sulfate-reducing bacteria (SRB), such as *Desulfovibrio* and *Desulfotomaculum*, convert sulfate back to sulfide under anaerobic conditions, contributing to the cycling of sulfur in environments like sediments and anaerobic digesters.

8.2.3 Sulfur Oxidation Pathways

The oxidation of sulfur compounds is a vital metabolic process that enhances the bioavailability of sulfur in ecosystems. Sulfur-oxidizing bacteria utilize reduced sulfur compounds as energy sources, leading to the production of sulfate. For example, *Thiobacillus* species oxidize sulfide or elemental sulfur to sulfuric acid (H_2SO_4), significantly influencing soil and water acidity. This process not only generates energy for the bacteria but also contributes to nutrient cycling by making sulfate available for assimilation by plants.

The oxidation pathways can be categorized into aerobic and anaerobic processes. In aerobic conditions, sulfur oxidation typically occurs in the presence of oxygen, enabling bacteria to use oxygen as the terminal electron acceptor. In contrast, some microorganisms can oxidize sulfur compounds anaerobically using alternative electron acceptors like nitrate or ferric iron. This metabolic versatility allows sulfur-oxidizing bacteria to thrive in diverse environmental conditions, including extreme habitats such as hot springs and acidic environments.

8.2.4 Sulfur Reduction Pathways

Sulfate reduction is the process by which sulfate-reducing bacteria convert sulfate back to sulfide, often in anaerobic environments. This metabolic pathway is critical for the recycling of sulfur within ecosystems. Sulfate-reducing bacteria utilize sulfate as an electron acceptor, leading to the production of hydrogen sulfide (H_2S). This process is particularly significant in marine sediments, wetlands, and anaerobic digesters, where organic matter decomposition creates anoxic conditions conducive to SRB activity.

The reduction of sulfate involves a series of enzymatic steps, starting with the activation of sulfate to adenosine 5'-phosphosulfate (APS) and ultimately leading to the formation of sulfide. The energy released during sulfate reduction supports microbial growth and contributes to the degradation of organic matter. Additionally, the production of sulfide has important implications for nutrient cycling and sediment geochemistry, as it can react with metals to form metal sulfides, influencing the bioavailability of nutrients and heavy metals in the environment.

8.2.5 Role of Sulfur in Plant Nutrition

Sulfur is essential for plant growth and development, influencing various physiological processes. Plants primarily uptake sulfur in the form of sulfate from the soil, which is then assimilated into amino acids, proteins, and other sulfur-containing compounds. Sulfur is vital for the synthesis of glutathione, a key antioxidant that protects plants from oxidative stress. Moreover, sulfur is involved in the synthesis of secondary metabolites, including flavonoids and glucosinolates, which play critical roles in plant defense mechanisms against pathogens and herbivores.

In addition to its nutritional role, sulfur compounds can also act as signaling molecules in plants. For instance, the accumulation of sulfide can trigger defense responses and induce systemic acquired resistance against pathogens. Understanding sulfur metabolism in plants is essential for developing strategies to enhance crop productivity, particularly in sulfur-deficient soils.

8.2.6 Sulfur Cycling in Ecosystems

The biogenic metabolism of sulfur is integral to sulfur cycling in ecosystems. Microbial activities drive the transformation of sulfur compounds, ensuring that sulfur is continuously recycled within the environment. This cycling is interconnected with other biogeochemical cycles, such as carbon and nitrogen, highlighting the complexity of nutrient dynamics in ecosystems.

In aquatic environments, sulfur cycling is influenced by microbial communities that facilitate both sulfur oxidation and reduction. The balance between these processes determines the overall sulfur dynamics in the water column and sediments. For example, in eutrophic conditions where organic matter accumulation occurs, sulfate reduction may dominate, leading to elevated levels of hydrogen sulfide, which can be toxic to aquatic life.

In terrestrial ecosystems, sulfur cycling is affected by soil microorganisms, which play critical roles in sulfur oxidation and reduction. The interactions between plants, microbes, and sulfur compounds influence soil fertility and nutrient availability. Moreover, anthropogenic activities, such as fossil fuel combustion and agricultural practices, can disrupt natural sulfur cycling, leading to environmental issues such as acid rain and nutrient imbalances.

8.2.7 Environmental Implications of Sulfur Metabolism

The biogenic metabolism of sulfur has significant environmental implications. The conversion of sulfur compounds by microorganisms influences soil and water quality, contributing to nutrient availability and ecosystem health. However, human activities have altered natural sulfur cycles, leading to increased sulfur emissions and environmental degradation. The release of sulfur dioxide (SO_2) from industrial processes and vehicle emissions contributes to the formation of acid rain, which can harm aquatic ecosystems, damage forests, and impair soil health.

Moreover, the production of hydrogen sulfide during anaerobic degradation processes can pose risks to human and environmental health. High concentrations of hydrogen sulfide can be toxic to aquatic organisms and may lead to the development of dead zones in water bodies, where oxygen levels are insufficient to support marine life.

To mitigate the environmental impacts of sulfur metabolism, it is essential to implement sustainable practices that promote healthy sulfur cycling. This includes adopting cleaner technologies, reducing sulfur emissions from industrial sources, and improving agricultural practices to minimize nutrient runoff.

CHAPTER 9

NITROGENASE AND NITROGEN CYCLE ENZYMES

9.1 OVERVIEW AND NATIVE NITROGENASE

Nitrogenase is a crucial enzyme complex that facilitates the conversion of atmospheric nitrogen (N_2) into ammonia (NH_3), a process known as nitrogen fixation. This biochemical reaction is fundamental to the nitrogen cycle, allowing for the incorporation of inert atmospheric nitrogen into biologically available forms. Nitrogen is an essential nutrient for all living organisms, forming the backbone of amino acids, proteins, nucleic acids, and other vital biomolecules.

Native nitrogenase is a vital enzyme complex that plays a fundamental role in the nitrogen cycle by facilitating the conversion of atmospheric nitrogen into biologically available ammonia. Its diverse forms and mechanisms reflect the adaptability of nitrogen-fixing organisms in various ecological niches. The ecological and agricultural significance of nitrogenase underscores its potential to enhance soil fertility and support sustainable agricultural practices. As research continues to uncover the intricacies of nitrogenase and its mechanisms, there is a promising opportunity to address global nitrogen needs sustainably and contribute to food security while minimizing environmental impacts. Understanding and harnessing the power of nitrogen-fixing microorganisms will be crucial for creating resilient and sustainable agricultural systems in the face of increasing global food demand.

9.1.1 The Importance of Nitrogen Fixation

Nitrogen fixation is a key ecological process, as atmospheric nitrogen constitutes about 78% of the Earth's atmosphere yet remains largely inaccessible to most living organisms. This limitation necessitates a mechanism to convert atmospheric nitrogen into forms that can be assimilated by plants and subsequently by animals. Nitrogenase plays an essential role in this process, enabling various organisms, including certain bacteria and archaea, to utilize atmospheric nitrogen directly.

The availability of nitrogen in a fixed form is crucial for plant growth and productivity. Most plants rely on nitrogen in the form of nitrates (NO_3^-) or ammonium (NH_4^+), which are taken up from the soil. By facilitating nitrogen fixation, nitrogenase significantly impacts agriculture and ecosystems, enhancing soil fertility and supporting crop yields. Understanding how nitrogenase operates and its various forms in nature can lead to improved agricultural practices and the development of sustainable nitrogen-fixing systems.

9.1.2 Structure and Composition of Nitrogenase

Native nitrogenase is a complex enzyme consisting of two major protein components: the dinitrogenase (also known as the Mo-nitrogenase) and the dinitrogenase reductase. The dinitrogenase is responsible for the actual reduction of atmospheric nitrogen to ammonia, while the dinitrogenase reductase provides the necessary electrons for the reduction reaction.

The dinitrogenase contains a molybdenum (Mo) cofactor, which is essential for its catalytic activity. This cofactor contains a unique metal cluster that plays a pivotal role in the binding and reduction of nitrogen. The dinitrogenase reductase, on the other hand, is involved in the transfer of electrons from reduced ferredoxin or flavodoxin to the dinitrogenase.

Nitrogenase is sensitive to oxygen, as oxygen can inactivate the enzyme by oxidizing its metal centers. Consequently, many nitrogen-fixing bacteria possess specialized adaptations to protect nitrogenase from oxygen exposure, such as anaerobic lifestyles or the formation of specialized structures known as heterocysts.

9.1.3 Mechanism of Nitrogen Fixation

The mechanism by which nitrogenase converts atmospheric nitrogen into ammonia is a complex series of biochemical reactions. The nitrogen fixation process begins with the binding of nitrogen gas to the active site of the dinitrogenase. The enzyme undergoes conformational changes to facilitate the formation of a transition state, where nitrogen is activated for reduction. This process involves the transfer of electrons, facilitated by the dinitrogenase reductase, which donates electrons derived from reduced ferredoxin or flavodoxin.

During the nitrogen reduction process, two molecules of nitrogen are reduced to form two molecules of ammonia and one molecule of nitrogen gas is released. This reaction is highly energetically demanding and requires considerable energy input, typically derived from the hydrolysis of ATP (adenosine triphosphate). The nitrogenase complex is capable of catalyzing this reaction under mild conditions, allowing for the fixation of nitrogen at ambient temperatures and pressures.

9.1.4 Diversity of Nitrogenases

While the molybdenum-containing nitrogenase is the most well-studied form of nitrogenase, other variants exist in nature. One such variant is the vanadium (V) nitrogenase, which contains vanadium instead of molybdenum as a cofactor. Vanadium nitrogenase is found in certain bacteria, particularly those inhabiting nitrogen-limited environments. These organisms have evolved to utilize vanadium in nitrogen fixation, demonstrating the versatility of nitrogenase across different ecological niches.

Additionally, some bacteria possess iron-only nitrogenases, which utilize iron as the primary metal cofactor. Iron-only nitrogenases are particularly intriguing, as they can function under conditions where molybdenum and vanadium are scarce, providing insights into the evolution of nitrogen-fixing mechanisms in varying environments. This diversity highlights the adaptability of nitrogen-fixing organisms and their capacity to thrive in diverse habitats.

9.1.5 Ecological and Agricultural Significance

The ecological and agricultural significance of nitrogenase cannot be overstated. Nitrogen-fixing bacteria play a vital role in maintaining soil fertility and supporting plant growth. Many legumes, for example, form symbiotic relationships with nitrogen-fixing bacteria, such as *Rhizobium*, within their root nodules. This symbiosis allows plants to access a steady supply of fixed nitrogen, which enhances their growth and productivity.

In addition to symbiotic nitrogen fixation, free-living nitrogen-fixing bacteria contribute to the overall nitrogen pool in ecosystems. These bacteria, found in soil and aquatic environments, can fix nitrogen independently, enriching the nutrient content of their surroundings. By enhancing nitrogen availability, these microorganisms play a crucial role in supporting diverse plant communities and sustaining ecosystem health.

The agricultural application of nitrogen-fixing bacteria has gained significant attention as a means to reduce dependence on synthetic nitrogen fertilizers. Incorporating nitrogen-fixing crops into agricultural systems can improve soil fertility, reduce environmental pollution from fertilizer runoff, and enhance sustainable farming practices. Furthermore, research into enhancing nitrogen fixation in crops through genetic engineering and microbial inoculants holds promise for improving global food security.

9.1.6 Challenges and Future Directions

Despite the essential role of nitrogenase in nitrogen fixation, several challenges remain. The efficiency of nitrogen fixation varies significantly among different organisms and environmental conditions. Moreover, the energy requirements for the nitrogen fixation process can be considerable, necessitating further research into optimizing these processes for agricultural applications.

Additionally, the reliance on synthetic nitrogen fertilizers in modern agriculture has led to environmental concerns, including soil degradation, water pollution, and greenhouse gas emissions. Addressing these challenges requires a comprehensive understanding of nitrogenase biology and the development of sustainable nitrogen-fixing strategies. Future research efforts are focusing on elucidating the molecular mechanisms of nitrogenase, exploring its evolutionary history, and identifying new nitrogen-fixing organisms. Advances in biotechnology, such as the use of CRISPR and other genetic engineering techniques, may enable the enhancement of nitrogen fixation in crops, leading to increased agricultural productivity while minimizing environmental impacts.

9.2 NITROGENASE MODELS AND MODEL REACTIONS

Nitrogenase is a complex enzyme that plays a pivotal role in the biological fixation of atmospheric nitrogen into ammonia, a process essential for sustaining life on Earth. Understanding nitrogenase has significant implications for agriculture and ecology, given the reliance of plants and animals on nitrogen in organic forms. Research on nitrogenase models and model reactions has emerged as a vital area of study, allowing scientists to explore the mechanistic intricacies of this enzyme. By employing both natural and synthetic models, researchers can investigate the enzyme's function, structure, and reactivity, providing insight into the processes that underlie nitrogen fixation.

Nitrogenase models and model reactions play a crucial role in elucidating the mechanisms of nitrogen fixation and advancing our understanding of this essential biochemical process. Through a combination of biochemical and synthetic approaches, researchers can explore the intricacies of nitrogenase structure, function, and reactivity. As the world faces increasing challenges related to food security and environmental sustainability, harnessing the potential of nitrogenase and its models will be vital in developing innovative solutions for agricultural productivity and ecosystem health. By continuing to investigate the mechanisms and applications of nitrogenase, scientists can contribute to a more sustainable future that leverages the power of biological nitrogen fixation for the benefit of humanity and the environment.

9.2.1 The Importance of Nitrogen Fixation

Nitrogen is a fundamental element for all living organisms, serving as a building block for amino acids, proteins, nucleic acids, and various other biological molecules. However, the majority of atmospheric nitrogen exists in an inert diatomic form (N_2), making it inaccessible to most organisms. Nitrogen fixation is the process through which atmospheric nitrogen is converted into ammonia (NH_3), allowing it to be assimilated by plants and subsequently into the food chain. Nitrogenase is the key enzyme that facilitates this process in certain bacteria and archaea, making it a critical component of the nitrogen cycle.

Given the vital role nitrogenase plays in agriculture and ecosystem health, understanding its mechanisms through model studies has become increasingly important. Models and model reactions can simulate the function of nitrogenase, enabling researchers to unravel its complex biochemistry and apply this knowledge to improve agricultural practices, enhance crop yields, and address global food security issues.

9.2.2 Overview of Nitrogenase Structure

Nitrogenase consists of two primary components: the dinitrogenase (or Mo-nitrogenase) and the dinitrogenase reductase. The dinitrogenase contains a molybdenum cofactor that is integral to its catalytic activity, while the dinitrogenase reductase provides the necessary electrons for the reduction of nitrogen. Together, these components form a sophisticated enzyme complex capable of converting inert nitrogen gas into biologically available ammonia.

The dinitrogenase is responsible for the reduction of N_2 to NH_3 and operates in a highly regulated manner, involving several electron transfers and ATP hydrolysis. Understanding the structural intricacies of nitrogenase, including the arrangement of its metal centers and the coordination environment around the active sites, is crucial for designing effective model reactions that can mimic its functionality.

9.2.3 Types of Nitrogenase Models

Researchers employ various nitrogenase models to study the enzyme's mechanisms and to explore potential synthetic analogs. The two main categories of models include biochemical models, which utilize natural nitrogenase components, and synthetic models, which are designed to replicate the enzyme's activity using synthetic materials.

Biochemical Models

Biochemical models often involve the use of purified nitrogenase proteins from different nitrogen-fixing organisms. By isolating the dinitrogenase and dinitrogenase reductase, researchers can study their interactions, structural features, and catalytic properties in vitro.

This approach allows for the examination of the enzyme's kinetics, substrate specificity, and electron transfer mechanisms.

Additionally, studies on nitrogen-fixing symbiotic relationships, such as those between legumes and rhizobia, provide insights into the natural context of nitrogenase activity.

These systems reveal how environmental factors, such as soil composition and oxygen availability, influence nitrogen fixation and how plants optimize their symbiotic relationships for enhanced nitrogen uptake.

Synthetic Models

Synthetic nitrogenase models aim to replicate the essential features of nitrogenase using non-biological materials, such as metal complexes and nanoparticles. These models typically incorporate transition metals like molybdenum, iron, or cobalt to mimic the metal centers found in natural nitrogenase. Researchers design these synthetic systems to study specific aspects of nitrogen fixation, including nitrogen activation, bond cleavage, and ammonia synthesis.

The advantages of synthetic models include the ability to precisely control reaction conditions and the potential for high-throughput screening of catalytic activity. This allows scientists to explore new catalysts and optimize reaction parameters to enhance nitrogen fixation processes. By analyzing the reactivity and efficiency of these synthetic models, researchers can gain insights into the fundamental principles governing nitrogenase function.

9.2.4 Model Reactions and Mechanistic Insights

Model reactions designed to mimic nitrogenase activity can provide valuable insights into the enzyme's mechanisms. These reactions often involve the activation of nitrogen gas and its conversion into ammonia under controlled laboratory conditions. By utilizing both biochemical and synthetic models, researchers can investigate key reaction steps, intermediate species, and the influence of various ligands on catalytic activity.

Nitrogen Activation

The first step in nitrogen fixation involves the activation of the N_2 molecule, which is a challenging process due to the strong triple bond between the nitrogen atoms. Model reactions aim to replicate this activation step by utilizing metal complexes that can effectively bind and activate nitrogen. For example, transition metal complexes have been shown to form stable adducts with nitrogen, facilitating the subsequent reduction to ammonia.

Ammonia Production

Once nitrogen has been activated, the next step involves the reduction of the activated nitrogen species to form ammonia. This reaction typically requires the transfer of multiple electrons and protons, a process that can be studied using model reactions. Synthetic models employing metal catalysts have demonstrated the ability to produce ammonia from activated nitrogen species, providing valuable insights into the reduction mechanisms and potential pathways for ammonia synthesis.

Influence of Ligands

The choice of ligands in nitrogenase models can significantly influence the reactivity and selectivity of the catalytic process. Researchers explore various ligands, including phosphines, amines, and other coordination compounds, to determine their impact on nitrogen activation and ammonia production. By systematically varying ligand structures and coordination environments, scientists can identify optimal conditions for enhancing nitrogenase-like activity.

9.2.5 Challenges in Nitrogenase Research

Despite the progress made in understanding nitrogenase through models and model reactions, several challenges remain. One of the primary difficulties is the complexity of the nitrogen fixation process, which involves multiple electron transfers, conformational changes, and substrate interactions. Additionally, the sensitivity of nitrogenase to oxygen can complicate experimental designs, as many reactions must be conducted under anaerobic conditions.

Moreover, the variability of nitrogenase structures among different organisms adds another layer of complexity to model studies. Each nitrogenase variant may exhibit unique properties and mechanisms, necessitating tailored approaches for studying their function and activity.

9.2.6 Future Directions

Research on nitrogenase models and model reactions is an active area of inquiry, with significant implications for agriculture and environmental sustainability. Future studies are likely to focus on several key areas, including:

- **Enhanced Synthetic Models:** Development of more sophisticated synthetic models that can accurately replicate nitrogenase activity and mechanisms. This could involve exploring new metal complexes and ligand systems to optimize nitrogen fixation processes.
- **Genetic Engineering:** Leveraging advances in genetic engineering and synthetic biology to create engineered nitrogen-fixing organisms with enhanced capabilities for nitrogen fixation. This could involve modifying existing nitrogen-fixing bacteria or developing entirely new strains.
- **Environmental Applications:** Investigating the potential for using nitrogenase models in bioremediation and environmental applications. Understanding how nitrogen fixation can be harnessed to improve soil health and fertility could have far-reaching implications for sustainable agriculture.
- **Integrative Approaches:** Combining experimental and computational approaches to gain a comprehensive understanding of nitrogenase mechanisms. Advances in computational chemistry and molecular modeling can provide insights into the dynamics of nitrogenase reactions at the atomic level.

9.3 DENITRIFICATION

Denitrification is a microbially facilitated process where nitrate (NO_3^-) is reduced and ultimately produces molecular nitrogen (N_2) through a series of intermediate gaseous nitrogen oxide products. Facultative anaerobic bacteria perform denitrification as a type of respiration that reduces oxidized forms of nitrogen in response to the oxidation of an electron donor such as organic matter. The preferred nitrogen electron acceptors in order of most to least thermodynamically favorable include nitrate (NO_3^-), nitrite (NO_2^-), nitric oxide (NO), nitrous oxide (N_2O) finally resulting in the production of dinitrogen (N_2) completing the nitrogen cycle. Denitrifying microbes require a very low oxygen concentration of less than 10%, as well as organic C for energy. Since denitrification can remove NO_3^-, reducing its leaching to groundwater, it can be strategically used to treat sewage or animal residues of high nitrogen content. Denitrification can leak N_2O, which is an ozone-depleting substance and a greenhouse gas that can have a considerable influence on global warming.

The process is performed primarily by heterotrophic bacteria (such as *Paracoccus denitrificans* and various pseudomonads), although autotrophic denitrifiers have also been identified (e.g., *Thiobacillus denitrificans*). Denitrifiers are represented in all main phylogenetic groups. Generally several species of bacteria are involved in the complete reduction of nitrate to N_2, and more than one enzymatic pathway has been identified in the reduction process. The denitrification process does not only provide energy to the organism performing nitrate reduction to dinitrogen gas, but also some anaerobic ciliates can use denitrifying endosymbionts to gain energy similar to the use of mitochondria in oxygen respiring organisms.

Direct reduction from nitrate to ammonium, a process known as dissimilatory nitrate reduction to ammonium or DNRA, is also possible for organisms that have the nrf-gene. This is less common than denitrification in most ecosystems as a means of nitrate reduction. Other genes known in microorganisms which denitrify include *nir* (nitrite reductase) and *nos* (nitrous oxide reductase) among others; organisms identified as having these genes include *Alcaligenes faecalis, Alcaligenes xylosoxidans*, many in the genus *Pseudomonas, Bradyrhizobium japonicum*, and *Blastobacter denitrificans*.

9.3.1 Half Reactions

Denitrification generally proceeds through some combination of the following half reactions, with the enzyme catalyzing the reaction in parentheses:

- $NO_3^- + 2\,H^+ + 2\,e^- \rightarrow NO_2^- + H_2O$ (Nitrate reductase)
- $NO_2^- + 2\,H^+ + e^- \rightarrow NO + H_2O$ (Nitrite reductase)
- $2\,NO + 2\,H^+ + 2\,e^- \rightarrow N_{2O} + H_2O$ (Nitric-oxide reductase)

- $N_{2O} + 2 H^+ + 2 e^- \rightarrow N_2 + H_2O$ (Nitrous-oxide reductase)

The complete process can be expressed as a net balanced redox reaction, where nitrate (NO_3^-) gets fully reduced to dinitrogen (N_2):

- $2 NO_3^- + 10 e^- + 12 H^+ \rightarrow N_2 + 6 H_2O$

9.3.2 Conditions of Denitrification

In nature, denitrification can take place in both terrestrial and marine ecosystems. Typically, denitrification occurs in anoxic environments, where the concentration of dissolved and freely available oxygen is depleted. In these areas, nitrate (NO_3^-) or nitrite (NO_2^-) can be used as a substitute terminal electron acceptor instead of oxygen (O_2), a more energetically favourable electron acceptor. Terminal electron acceptor is a compound that gets reduced in the reaction by receiving electrons. Examples of anoxic environments can include soils, groundwater, wetlands, oil reservoirs, poorly ventilated corners of the ocean and seafloor sediments.

Furthermore, denitrification can occur in oxic environments as well. High activity of denitrifiers can be observed in the intertidal zones, where the tidal cycles cause fluctuations of oxygen concentration in sandy coastal sediments. For example, the bacterial species *Paracoccus denitrificans* engages in denitrification under both oxic and anoxic conditions simultaneously. Upon oxygen exposure, the bacteria is able to utilize nitrous oxide reductase, an enzyme that catalyzes the last step of denitrification. Aerobic denitrifiers are mainly Gram-negative bacteria in the phylum Proteobacteria. Enzymes NapAB, NirS, NirK and NosZ are located in the periplasm, a wide space bordered by the cytoplasmic and the outer membrane in Gram-negative bacteria.

A variety of environmental factors can influence the rate of denitrification on an ecosystem-wide scale. For example, temperature and pH have been observed to impact denitrification rates. In the bacterial species, *Pseudomonas mandelii,* expression of denitrifying genes was reduced at temperatures below 30 °C and a pH below 5, while activity was largely unaffected between a pH of 6-8. Organic carbon as an electron donor is a common limiting nutrient for denitrification as observed in benthic sediments and wetlands. Nitrate and oxygen can also be potential limiting factors for denitrification, although the latter only has an observed limiting effect in wet soils. Oxygen likely affects denitrification in multiple ways—because most denitrifiers are facultative, oxygen can inhibit rates, but it can also stimulate denitrification by facilitating nitrification and the production of nitrate. In wetlands as well as deserts, moisture is an environmental limitation to rates of denitrification.

Additionally, environmental factors can also influence whether denitrification proceeds to completion, characterized by the complete reduction of NO_3^- to N_2

rather than releasing N_2O as an end product. Soil pH and texture are both factors that can moderate denitrification, with higher pH levels driving the reaction more to completion. Nutrient composition, particularly the ratio of carbon to nitrogen, is a strong contributor to complete denitrification, with a 2:1 ratio of C:N being able to facilitate full nitrate reduction regardless of temperature or carbon source. Copper, as a co-factor for nitrite reductase and nitrous-oxide reductase, also promoted complete denitrification when added as a supplement. Besides nutrients and terrain, microbial community composition can also affect the ratio of complete denitrification, with prokaryotic phyla Actinomycetota and Thermoproteota being responsible for greater release of N_2 than N_2O compared to other prokaryotes.

Denitrification can lead to a condition called isotopic fractionation in the soil environment. The two stable isotopes of nitrogen, ^{14}N and ^{15}N are both found in the sediment profiles. The lighter isotope of nitrogen, ^{14}N, is preferred during denitrification, leaving the heavier nitrogen isotope, ^{15}N, in the residual matter. This selectivity leads to the enrichment of ^{14}N in the biomass compared to ^{15}N. Moreover, the relative abundance of ^{14}N can be analyzed to distinguish denitrification apart from other processes in nature.

9.3.3 Use in Wastewater Treatment

Denitrification is commonly used to remove nitrogen from sewage and municipal wastewater. It is also an instrumental process in constructed wetlands and riparian zones for the prevention of groundwater pollution with nitrate resulting from excessive agricultural or residential fertilizer usage. Wood chip bioreactors have been studied since the 2000s and are effective in removing nitrate from agricultural run off and even manure.

Reduction under anoxic conditions can also occur through process called anaerobic ammonium oxidation (anammox):

- $NH_4^+ + NO_2^- \rightarrow N_2 + 2\,H_2O$

In some wastewater treatment plants, compounds such as methanol, ethanol, acetate, glycerin, or proprietary products are added to the wastewater to provide a carbon and electron source for denitrifying bacteria. The microbial ecology of such engineered denitrification processes is determined by the nature of the electron donor and the process operating conditions. Denitrification processes are also used in the treatment of industrial wastewater. Many denitrifying bioreactor types and designs are available commercially for the industrial applications, including Electro-Biochemical Reactors (EBRs), membrane bioreactors (MBRs), and moving bed bioreactors (MBBRs).

Aerobic denitrification, conducted by aerobic denitrifiers, may offer the potential to eliminate the need for separate tanks and reduce sludge yield. There are less stringent alkalinity requirements because alkalinity generated during denitrification can partly compensate for the alkalinity consumption in nitrification.

9.3.4 Non-Biological Denitrification

A variety of non-biological methods can remove nitrate. These include methods that can destroy nitrogen compounds, such as chemical and electrochemical methods, and those that selectively transfer nitrate to a concentrated waste stream, such as ion exchange or reverse osmosis. Chemical removal of nitrate can occur through advanced oxidation processes, although it may produce hazardous byproducts. Electrochemical methods can remove nitrate by via a voltage applied across electrodes, with degradation usually occurring at the cathode. Effective cathode materials include transition metals, post transition metals, and semi-conductors like TiO_2. Electrochemical methods can often avoid requiring costly chemical additives, but their effectiveness can be constrained by the pH and ions present. Reverse osmosis is highly effective in removing small charged solutes like nitrate, but it may also remove desirable nutrients, create large volumes of wastewater, and require increased pumping pressures. Ion exchange can selectively remove nitrate from water without large waste streams, but do require regeneration and may face challenges with absorption of undesired ions.

9.4 NITRIC OXIDE

Nitric oxide (nitrogen oxide or nitrogen monoxide) is a colorless gas with the formula NO. It is one of the principal oxides of nitrogen. Nitric oxide is a free radical: it has an unpaired electron, which is sometimes denoted by a dot in its chemical formula (˙N=O or ˙NO). Nitric oxide is also a heteronuclear diatomic molecule, a class of molecules whose study spawned early modern theories of chemical bonding.

An important intermediate in industrial chemistry, nitric oxide forms in combustion systems and can be generated by lightning in thunderstorms. In mammals, including humans, nitric oxide is a signaling molecule in many physiological and pathological processes.

It was proclaimed the "Molecule of the Year" in 1992. The 1998 Nobel Prize in Physiology or Medicine was awarded for discovering nitric oxide's role as a cardiovascular signalling molecule. Its impact extends beyond biology, with applications in medicine, such as the development of sildenafil (Viagra), and in industry, including semiconductor manufacturing.

Nitric oxide should not be confused with nitrogen dioxide (NO_2), a brown gas and major air pollutant, or with nitrous oxide (N_2O), an anesthetic gas.

9.4.1 History

Nitric oxide (NO) was first identified by Joseph Priestley in the late 18th century, originally seen as merely a toxic byproduct of combustion and an environmental pollutant. Its biological significance was later uncovered in the 1980s when researchers Robert F. Furchgott, Louis J. Ignarro, and Ferid Murad discovered its critical role as a vasodilator in the cardiovascular system, a breakthrough that earned them the 1998 Nobel Prize in Physiology or Medicine.

9.4.2 Physical Properties

Electronic Configuration

The ground state electronic configuration of NO is, in united atom notation:

$$(1\sigma)^2(2\sigma)^2(3\sigma)^2(4\sigma^*)^2(5\sigma)^2(1\pi)^4(2\pi^*)^1$$

The first two orbitals are actually pure atomic $1s_O$ and $1s_N$ from oxygen and nitrogen respectively and therefore are usually not noted in the united atom notation. Orbitals noted with an asterisk are antibonding. The ordering of 5σ and 1π according to their binding energies is subject to discussion. Removal of a 1π electron leads to 6 states whose energies span over a range starting at a lower level than a 5σ electron an extending to a higher level. This is due to the different orbital momentum couplings between a 1π and a 2π electron.

The lone electron in the 2π orbital makes NO a doublet ($X\ ^2\Pi$) in its ground state whose degeneracy is split in the fine structure from spin-orbit coupling with a total momentum $J=3/2$ or $J=1/2$.

Dipole

The dipole of NO has been measured experimentally to 0.15740 D and is oriented from O to N ($^-NO^+$) due to the transfer of negative electronic charge from oxygen to nitrogen.

9.4.3 Reactions

With Di- and Triatomic Molecules

Upon condensing to a liquid, nitric oxide dimerizes to dinitrogen dioxide, but the association is weak and reversible. The N–N distance in crystalline NO is 218 pm, nearly twice the N–O distance. Since the heat of formation of ·NO is endothermic, NO can be decomposed to the elements. Catalytic converters in cars exploit this reaction:

$$2\ ^{\cdot}NO \rightarrow O_2 + N_2$$

When exposed to oxygen, nitric oxide converts into nitrogen dioxide:

$$2\ ^{\cdot}NO + O_2 \rightarrow 2\ ^{\cdot}NO_2$$

This reaction is thought to occur via the intermediates ONOO• and the red compound ONOONO.

In water, nitric oxide reacts with oxygen to form nitrous acid (HNO_2). The reaction is thought to proceed via the following stoichiometry:

$$4\ {}^{\bullet}NO + O_2 + 2\ H_2O \rightarrow 4\ HNO_2$$

Nitric oxide reacts with fluorine, chlorine, and bromine to form the nitrosyl halides, such as nitrosyl chloride:

$$2\ {}^{\bullet}NO + Cl_2 \rightarrow 2\ NOCl$$

With NO_2, also a radical, NO combines to form the intensely blue dinitrogen trioxide:

$${}^{\bullet}NO + {}^{\bullet}NO_2 \rightleftharpoons ON\text{-}NO_2$$

Organic Chemistry

The addition of a nitric oxide moiety to another molecule is often referred to as *nitrosylation*. The Traube reaction is the addition of a two equivalents of nitric oxide onto an enolate, giving a diazeniumdiolate (also called a *nitrosohydroxylamine*). The product can undergo a subsequent retro-aldol reaction, giving an overall process similar to the haloform reaction. For example, nitric oxide reacts with acetone and an alkoxide to form a diazeniumdiolate on each α position, with subsequent loss of methyl acetate as a by-product:

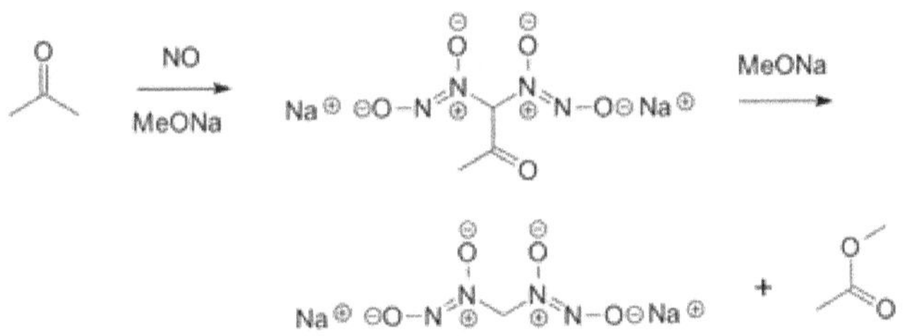

This reaction, which was discovered around 1898, remains of interest in nitric oxide prodrug research. Nitric oxide can also react directly with sodium methoxide, ultimately forming sodium formate and nitrous oxide by way of an *N*-methoxydiazeniumdiolate.

Coordination Complexes

Nitric oxide reacts with transition metals to give complexes called metal nitrosyls. The most common bonding mode of nitric oxide is the terminal linear type (M-NO). Alternatively, nitric oxide can serve as a one-electron pseudohalide. In such complexes, the M-N-O group is characterized by an angle between 120° and 140°. The NO group can also bridge between metal centers through the nitrogen atom in a variety of geometries.

9.4.4 Production and Preparation

In commercial settings, nitric oxide is produced by the oxidation of ammonia at 750–900°C (normally at 850°C) with platinum as catalyst in the Ostwald process:

$$4\ NH_3 + 5\ O_2 \rightarrow 4\ {}^{\cdot}NO + 6\ H_2O$$

The uncatalyzed endothermic reaction of oxygen (O_2) and nitrogen (N_2), which is effected at high temperature (>2000°C) by lightning has not been developed into a practical commercial synthesis:

$$N_2 + O_2 \rightarrow 2\ {}^{\cdot}NO$$

Laboratory Methods

In the laboratory, nitric oxide is conveniently generated by reduction of dilute nitric acid with copper:

$$8\ HNO_3 + 3\ Cu \rightarrow 3\ Cu(NO_3)_2 + 4\ H_2O + 2\ {}^{\cdot}NO$$

An alternative route involves the reduction of nitrous acid in the form of sodium nitrite or potassium nitrite:

$$2\ NaNO_2 + 2\ NaI + 2\ H_2SO_4 \rightarrow I_2 + 2\ Na_2SO_4 + 2\ H_2O + 2\ {}^{\cdot}NO$$

$$2\ NaNO_2 + 2\ FeSO_4 + 3\ H_2SO_4 \rightarrow Fe_2(SO_4)_3 + 2\ NaHSO_4 + 2\ H_2O + 2\ {}^{\cdot}NO$$

$$3\ KNO_2 + KNO_3 + Cr_2O_3 \rightarrow 2\ K_2CrO_4 + 4\ {}^{\cdot}NO$$

The iron(II) sulfate route is simple and has been used in undergraduate laboratory experiments. So-called NONOate compounds are also used for nitric oxide generation.

9.4.5 Detection and Assay

Nitric oxide concentration can be determined using a chemiluminescent reaction involving ozone. A sample containing nitric oxide is mixed with a large quantity of ozone. The nitric oxide reacts with the ozone to produce oxygen and nitrogen dioxide, accompanied with emission of light (chemiluminescence):

$${}^{\cdot}NO + O_3 \rightarrow {}^{\cdot}NO_2 + O_2 + h\nu$$

which can be measured with a photodetector. The amount of light produced is proportional to the amount of nitric oxide in the sample.

Other methods of testing include electroanalysis (amperometric approach), where ·NO reacts with an electrode to induce a current or voltage change. The detection of NO radicals in biological tissues is particularly difficult due to the short lifetime and concentration of these radicals in tissues. One of the few practical methods is spin trapping of nitric oxide with iron-dithiocarbamate complexes and subsequent detection of the mono-nitrosyl-iron complex with electron paramagnetic resonance (EPR).

A group of fluorescent dye indicators that are also available in acetylated form for intracellular measurements exist. The most common compound is 4,5-diaminofluorescein (DAF-2).

9.4.6 Environmental Effects

Acid Rain Deposition

Nitric oxide reacts with the hydroperoxyl radical ($HO^{\bullet}_2$) to form nitrogen dioxide (NO_2), which then can react with a hydroxyl radical ($HO^{\bullet}$) to produce nitric acid (HNO_3):

$$^{\bullet}NO + HO^{\bullet}_2 \rightarrow {}^{\bullet}NO_2 + HO^{\bullet}$$

$$^{\bullet}NO_2 + HO^{\bullet} \rightarrow HNO_3$$

Nitric acid, along with sulfuric acid, contributes to acid rain deposition.

Ozone Depletion

$^{\bullet}NO$ participates in ozone layer depletion. Nitric oxide reacts with stratospheric ozone to form O_2 and nitrogen dioxide:

$$^{\bullet}NO + O_3 \rightarrow {}^{\bullet}NO_2 + O_2$$

This reaction is also utilized to measure concentrations of $^{\bullet}NO$ in control volumes.

Precursor to NO2

Nitric oxide can transform into nitrogen dioxide (this can happen with the hydroperoxy radical, $HO^{\bullet}_2$, or diatomic oxygen, O_2). Symptoms of short-term nitrogen dioxide exposure include nausea, dyspnea and headache. Long-term effects could include impaired immune and respiratory function.

9.4.7 Biological Functions

NO is a gaseous signaling molecule. It is a key vertebrate biological messenger, playing a role in a variety of biological processes. It is a bioproduct in almost all types of organisms, including bacteria, plants, fungi, and animal cells.

Nitric oxide, an endothelium-derived relaxing factor (EDRF), is biosynthesized endogenously from L-arginine, oxygen, and NADPH by various nitric oxide synthase (NOS) enzymes. Reduction of inorganic nitrate may also make nitric oxide. One of the main enzymatic targets of nitric oxide is guanylyl cyclase. The binding of nitric oxide to the heme region of the enzyme leads to activation, in the presence of iron. Nitric oxide is highly reactive (having a lifetime of a few seconds), yet diffuses freely across membranes. These attributes make nitric oxide ideal for a transient paracrine (between adjacent cells) and autocrine (within a single cell) signaling molecule. Once nitric oxide is converted to nitrates and nitrites by oxygen and water, cell signaling is deactivated.

The endothelium (inner lining) of blood vessels uses nitric oxide to signal the surrounding smooth muscle to relax, resulting in vasodilation and increasing blood flow. Sildenafil (Viagra) is a drug that uses the nitric oxide pathway. Sildenafil does not produce nitric oxide, but enhances the signals that are downstream of the nitric oxide pathway by protecting cyclic guanosine monophosphate (cGMP) from degradation by cGMP-specific phosphodiesterase type 5 (PDE5) in the corpus cavernosum, allowing for the signal to be enhanced, and thus vasodilation. Another endogenous gaseous transmitter, hydrogen sulfide (H_2S) works with NO to induce vasodilation and angiogenesis in a cooperative manner.

Nasal breathing produces nitric oxide within the body, while oral breathing does not.

9.4.8 Occupational Safety and Health

In the U.S., the Occupational Safety and Health Administration (OSHA) has set the legal limit (permissible exposure limit) for nitric oxide exposure in the workplace as 25 ppm (30 mg/m^3) over an 8-hour workday. The National Institute for Occupational Safety and Health (NIOSH) has set a recommended exposure limit (REL) of 25 ppm (30 mg/m^3) over an 8-hour workday. At levels of 100 ppm, nitric oxide is immediately dangerous to life and health.

9.4.9 Explosion Hazard

Liquid nitrogen oxide is very sensitive to detonation even in the absence of fuel, and can be initiated as readily as nitroglycerin. Detonation of the endothermic liquid oxide close to its b.p. (-152°C) generated a 100 kbar pulse and fragmented the test equipment. It is the simplest molecule that is capable of detonation in all three phases. The liquid oxide is sensitive and may explode during distillation, and this has been the cause of industrial accidents. Gaseous nitric oxide detonates at about 2300 m/s, but as a solid it can reach a detonation velocity of 6100 m/s.

CHAPTER 10

THE METHANE CYCLE AND NICKEL ENZYMES

10.1 INTRODUCTION

The methane cycle is an essential aspect of global carbon cycling, involving the production, consumption, and transformation of methane (CH_4), a potent greenhouse gas. Methane is a significant contributor to climate change, with a global warming potential much higher than that of carbon dioxide over a 20-year period. However, methane is also a valuable energy source, making its regulation and transformation a key focus in both environmental and industrial fields. Central to the methane cycle are nickel enzymes, which play crucial roles in the biological processes of methane metabolism. In particular, enzymes like methyl-coenzyme M reductase (MCR) in methanogens (methane-producing organisms) and methane monooxygenase (MMO) in methanotrophs (methane-consuming organisms) highlight the unique involvement of nickel in methane biochemistry.

Nickel enzymes play an essential role in the methane cycle, facilitating both the production and consumption of methane through their involvement in methanogenesis and methanotrophy. Enzymes like MCR and CODH exemplify the critical function of nickel in biological methane metabolism, providing insight into the mechanisms that regulate global methane levels. As research continues to uncover the complexities of these enzymes, their potential applications in biotechnology and environmental management will become increasingly important in addressing global energy needs and climate challenges. Understanding and harnessing the power of nickel enzymes in the methane cycle could hold the key to more sustainable and efficient methane utilization and reduction strategies.

10.1.1 The Methane Cycle

The methane cycle consists of two primary processes: methanogenesis, where methane is produced, and methanotrophy, where methane is consumed or oxidized. These processes are critical components of the global carbon cycle, contributing to the regulation of atmospheric methane levels. Methanogenesis occurs primarily in anaerobic environments, such as wetlands, rice paddies, and the digestive tracts of ruminants, where organic matter decomposes without oxygen. On the other hand, methanotrophy typically occurs in aerobic environments, such as soils and sediments, where methanotrophic bacteria oxidize methane as their primary carbon and energy source.

Methanogenesis

Methanogenesis is the biological production of methane, predominantly carried out by archaea known as methanogens. These microorganisms thrive in anaerobic environments and convert carbon compounds such as carbon dioxide (CO_2), acetate, and methyl compounds into methane. Methanogenesis plays a vital role in decomposing organic matter in anaerobic ecosystems and significantly contributes to natural methane emissions.

The process of methanogenesis involves several metabolic pathways, including hydrogenotrophic methanogenesis (CO_2 reduction), acetoclastic methanogenesis (acetate cleavage), and methylotrophic methanogenesis (conversion of methylated compounds). The last step of methanogenesis in all pathways is catalyzed by the nickel-containing enzyme MCR, which reduces methyl-coenzyme M (CH_3-S-CoM) to methane. This enzymatic reaction is fundamental to methane production in nature.

Methanotrophy

Methanotrophy, or methane oxidation, is the biological consumption of methane by methanotrophic bacteria. These organisms use methane as their sole carbon and energy source, transforming it into methanol, formaldehyde, and ultimately carbon dioxide. Methanotrophs play a crucial role in mitigating methane emissions, especially in aerobic environments such as soil and water systems, where they act as natural methane filters. Two types of methanotrophic bacteria are recognized: aerobic methanotrophs and anaerobic methanotrophs. Aerobic methanotrophs oxidize methane in the presence of oxygen, while anaerobic methanotrophs perform methane oxidation in anoxic conditions, often in partnership with sulfate-reducing bacteria. Nickel enzymes are also essential in this oxidation process, playing a key role in the reduction and transformation of methane molecules.

10.1.2 Nickel Enzymes in the Methane Cycle

Nickel is an essential trace element for various biological systems, particularly in enzymes that catalyze key reactions in the methane cycle. Two important classes of nickel enzymes are methyl-coenzyme M reductase (MCR) and carbon monoxide dehydrogenase (CODH). MCR is central to methanogenesis, while CODH plays roles in other carbon cycling processes, including the reverse methanogenesis pathways.

Methyl-Coenzyme M Reductase (MCR)

Methyl-coenzyme M reductase (MCR) is a key enzyme in the methanogenesis process, responsible for catalyzing the final step in methane production. MCR contains a unique nickel-containing tetrapyrrole cofactor known as F430. The nickel ion in F430 undergoes redox changes between Ni(I) and Ni(II) states during the catalysis of methane formation. The structure of MCR consists of large and small subunits arranged around the active site, where the nickel cofactor is located.

MCR catalyzes the reduction of methyl-coenzyme M (CH_3-S-CoM) with coenzyme B (HS-CoB) to form methane and a heterodisulfide (CoM-S-S-CoB). The process is highly regulated and involves several redox steps. The importance of MCR in methane biochemistry has made it a focus for biotechnological applications, particularly in developing bioengineered systems for methane production or mitigation.

Carbon Monoxide Dehydrogenase (CODH)

Carbon monoxide dehydrogenase (CODH) is another nickel-containing enzyme that participates in the methane cycle, particularly in acetogenic and methanogenic organisms. CODH catalyzes the reversible oxidation of carbon monoxide (CO) to carbon dioxide (CO_2). In methanogenic archaea, CODH is involved in the reverse methanogenesis pathway, contributing to carbon cycling and energy conservation in these microorganisms.

The active site of CODH contains a nickel-iron-sulfur cluster, where nickel plays a critical role in the enzyme's redox chemistry. This enzyme is of particular interest because it operates under both aerobic and anaerobic conditions, providing flexibility in various environmental niches.

10.1.3 Nickel Enzymes in Methanotrophy

Methanotrophy involves the oxidation of methane, and while MCR is specific to methanogens, nickel enzymes are also implicated in methanotrophic pathways. While MMO, the primary methane-oxidizing enzyme in aerobic methanotrophs, does not contain nickel, there are growing research areas in anaerobic methanotrophy where nickel-dependent systems are under investigation.

In anaerobic methane oxidation (ANME), methanotrophic archaea often interact with sulfate-reducing bacteria to oxidize methane in the absence of oxygen. The exact role of nickel enzymes in these processes is still being explored, but it is hypothesized that MCR-like systems or novel nickel-dependent proteins could play roles in methane activation and oxidation under anaerobic conditions.

10.1.4 Nickel Enzymes and Biotechnological Applications

Given the central role of nickel enzymes in the methane cycle, they present exciting opportunities for biotechnological innovations. The ability of MCR to catalyze methane production from simple carbon compounds has led to research into bioengineering methanogenic pathways for renewable energy production.

Methane, as a primary component of natural gas, offers a potential source of clean energy when captured and utilized efficiently. Nickel enzymes are also being explored for their potential in mitigating methane emissions. Methane is a potent greenhouse gas, and reducing atmospheric methane levels is a critical component of global climate strategies. By enhancing the activity of methanotrophic bacteria or designing synthetic systems that mimic natural methanotrophy, researchers hope to develop biotechnological solutions for methane capture and conversion.

10.1.5 Environmental and Climate Implications

The methane cycle, driven by the activity of nickel enzymes, has significant implications for global climate systems. Methane emissions from natural and anthropogenic sources, such as agriculture, fossil fuel extraction, and landfills, contribute to global warming. Understanding the role of nickel enzymes in both methane production and consumption is critical for developing strategies to reduce methane emissions and address climate change.

Wetlands, for example, are major natural sources of methane emissions due to the activity of methanogenic archaea in anaerobic soils. On the other hand, soils in upland ecosystems act as methane sinks, where methanotrophic bacteria consume atmospheric methane. Nickel enzymes are integral to both processes, making them a focal point for ecological studies and environmental management practices.

10.1.6 Future Research Directions

Research on nickel enzymes and the methane cycle is a rapidly evolving field with numerous open questions and potential applications. Future studies are likely to focus on several key areas:

- **Enzyme Mechanisms:** A deeper understanding of the mechanisms by which nickel enzymes catalyze methane-related reactions is crucial. Advances in spectroscopy, crystallography, and computational modeling are expected to shed light on the atomic-level details of these enzymes.
- **Anaerobic Methane Oxidation:** The exact role of nickel enzymes in anaerobic methane oxidation (ANME) remains a subject of ongoing research. Discovering new nickel-dependent enzymes in ANME pathways could revolutionize our understanding of methane cycling in anoxic environments.
- **Biotechnological Applications:** As the demand for sustainable energy sources increases, bioengineering nickel enzymes for methane production and capture could have profound impacts on energy and environmental technologies.

- **Environmental Monitoring:** The role of nickel enzymes in natural methane sinks and sources will continue to be a focus for climate change mitigation strategies. Enhancing methanotrophic activity in natural and engineered systems could offer ways to reduce atmospheric methane concentrations.

10.2 METHANOGENESIS

Methanogenesis or biomethanation is the formation of methane coupled to energy conservation by microbes known as methanogens. Organisms capable of producing methane for energy conservation have been identified only from the domain Archaea, a group phylogenetically distinct from both eukaryotes and bacteria, although many live in close association with anaerobic bacteria. The production of methane is an important and widespread form of microbial metabolism. In anoxic environments, it is the final step in the decomposition of biomass. Methanogenesis is responsible for significant amounts of natural gas accumulations, the remainder being thermogenic.

10.2.1 Biochemistry

Methanogenesis in microbes is a form of anaerobic respiration. Methanogens do not use oxygen to respire; in fact, oxygen inhibits the growth of methanogens. The terminal electron acceptor in methanogenesis is not oxygen, but carbon. The two best described pathways involve the use of acetic acid (acetoclastic) or inorganic carbon dioxide (hydrogenotrophic) as terminal electron acceptors:

$$CO_2 + 4\,H_2 \rightarrow CH_4 + 2\,H_2O$$

$$CH_3COOH \rightarrow CH_4 + CO_2$$

During anaerobic respiration of carbohydrates, H_2 and acetate are formed in a ratio of 2:1 or lower, so H_2 contributes only c. 33% to methanogenesis, with acetate contributing the greater proportion. In some circumstances, for instance in the rumen, where acetate is largely absorbed into the bloodstream of the host, the contribution of H_2 to methanogenesis is greater.

However, depending on pH and temperature, methanogenesis has been shown to use carbon from other small organic compounds, such as formic acid (formate), methanol, methylamines, tetramethylammonium, dimethyl sulfide, and methanethiol. The catabolism of the methyl compounds is mediated by methyl transferases to give methyl coenzyme M.

Proposed Mechanism

The biochemistry of methanogenesis involves the following coenzymes and cofactors: F420, coenzyme B, coenzyme M, methanofuran, and methanopterin.

The mechanism for the conversion of CH_3–S bond into methane involves a ternary complex of methyl coenzyme M and coenzyme B fit into a channel terminated by the axial site on nickel of the cofactor F430. One proposed mechanism invokes electron transfer from Ni(I) (to give Ni(II)), which initiates formation of CH_4. Coupling of the coenzyme M thiyl radical (RS·) with HS coenzyme B releases a proton and re-reduces Ni(II) by one-electron, regenerating Ni(I).

Reverse Methanogenesis

Some organisms can oxidize methane, functionally reversing the process of methanogenesis, also referred to as the anaerobic oxidation of methane (AOM). Organisms performing AOM have been found in multiple marine and freshwater environments including methane seeps, hydrothermal vents, coastal sediments and sulfate-methane transition zones. These organisms may accomplish reverse methanogenesis using a nickel-containing protein similar to methyl-coenzyme M reductase used by methanogenic archaea. Reverse methanogenesis occurs according to the reaction:

$$SO^{2-}{}_4 + CH_4 \rightarrow HCO^-{}_3 + HS^- + H_2O$$

Importance in Carbon Cycle

Methanogenesis is the final step in the decay of organic matter. During the decay process, electron acceptors (such as oxygen, ferric iron, sulfate, and nitrate) become depleted, while hydrogen (H_2) and carbon dioxide accumulate. Light organics produced by fermentation also accumulate. During advanced stages of organic decay, all electron acceptors become depleted except carbon dioxide. Carbon dioxide is a product of most catabolic processes, so it is not depleted like other potential electron acceptors.

Only methanogenesis and fermentation can occur in the absence of electron acceptors other than carbon. Fermentation only allows the breakdown of larger organic compounds, and produces small organic compounds. Methanogenesis effectively removes the semi-final products of decay: hydrogen, small organics, and carbon dioxide. Without methanogenesis, a great deal of carbon (in the form of fermentation products) would accumulate in anaerobic environments.

10.2.2 Natural Occurrence

In Ruminants

Enteric fermentation occurs in the gut of some animals, especially ruminants. In the rumen, anaerobic organisms, including methanogens, digest cellulose into forms nutritious to the animal. Without these microorganisms, animals such as cattle would not be able to consume grasses. The useful products of methanogenesis are absorbed by the gut, but methane is released from the animal mainly by belching (eructation). The average cow emits around 250 liters of methane per day. In this way, ruminants contribute about 25% of anthropogenic methane emissions. One method of methane production control in ruminants is by feeding them 3-nitrooxypropanol.

In Humans

Some humans produce flatus that contains methane. In one study of the feces of nine adults, five of the samples contained archaea capable of producing methane. Similar results are found in samples of gas obtained from within the rectum. Even among humans whose flatus does contain methane, the amount is in the range of 10% or less of the total amount of gas.

In Plants

Many experiments have suggested that leaf tissues of living plants emit methane. Other research has indicated that the plants are not actually generating methane; they are just absorbing methane from the soil and then emitting it through their leaf tissues.

In Soils

Methanogens are observed in anoxic soil environments, contributing to the degradation of organic matter. This organic matter may be placed by humans through landfill, buried as sediment on the bottom of lakes or oceans as sediments, and as residual organic matter from sediments that have formed into sedimentary rocks.

In Earth's Crust

Methanogens are a notable part of the microbial communities in continental and marine deep biosphere.

10.2.3 Industry

Methanogenesis can also be beneficially exploited, to treat organic waste, to produce useful compounds, and the methane can be collected and used as biogas, a fuel. It is the primary pathway whereby most organic matter disposed of via landfill is broken down. Some biogas plants use methanogenesis to combine the CO_2 with hydrogen to create more methane.

10.2.4 Role in Global Warming

Atmospheric methane is an important greenhouse gas with a global warming potential 25 times greater than carbon dioxide (averaged over 100 years), and methanogenesis in livestock and the decay of organic material is thus a considerable contributor to global warming. It may not be a net contributor in the sense that it works on organic material which used up atmospheric carbon dioxide when it was created, but its overall effect is to convert the carbon dioxide into methane which is a much more potent greenhouse gas.

Atmospheric Methane

Atmospheric methane is the methane present in Earth's atmosphere. The concentration of atmospheric methane is increasing due to methane emissions, and is causing climate change. Methane is one of the most potent greenhouse gases. Methane's radiative forcing (RF) of climate is direct, and it is the second largest contributor to human-caused climate forcing in the historical period. Methane is a major source of water vapour in the stratosphere through oxidation; and water vapour adds about 15% to methane's radiative forcing effect. The global warming potential (GWP) for methane is about 84 in terms of its impact over a 20-year timeframe, and 28 in terms of its impact over a 100-year timeframe.

Since the beginning of the Industrial Revolution (around 1750) the methane concentration in the atmosphere has increased by about 160%, and this increase is almost entirely caused by human activities. Since 1750 methane has contributed 3% of greenhouse gas (GHG) emissions in terms of mass but is responsible for approximately 23% of radiative or climate forcing. By 2019, global methane concentrations had risen from 722 parts per billion (ppb) in pre-industrial times to 1866 ppb. This is an increase by a factor of 2.6 and the highest value in at least 800,000 years.

Methane increases the amount of ozone (O_3) in the troposphere (4 miles (6 km) to 12 miles (19 km) from the Earth's surface) and also in the stratosphere (from the troposphere to 31 miles (50 km) above the Earth's surface). Both water vapour and ozone are GHGs, which in turn add to climate warming.

Role in Climate Change

Methane (CH_4) in the Earth's atmosphere is a powerful greenhouse gas with a global warming potential (GWP) 84 times greater than CO_2 over a 20-year time frame. Methane is not as persistent as CO_2, and tails off to about 28 times greater than CO_2 over a 100-year time frame.

Radiative or climate forcing is the scientific concept used to measure the human impact on the environment in watts per square meter (W/m^2). It refers to the "difference between solar irradiance absorbed by the Earth and energy radiated back to space" The direct radiative greenhouse gas forcing effect of methane was estimated to be an increase of 0.5 W/m^2 relative to the year 1750 (estimate in 2007).

In their 2021 "Global Methane Assessment" report, the UNEP and CCAC said that their "understanding of methane's effect on radiative forcing" improved with research by teams led by M. Etminan in 2016, and William Collins in 2018. This resulted in an "upward revision" since the 2014 IPCC Fifth Assessment Report (AR5). The "improved understanding" says that prior estimates of the "overall societal impact of methane emissions" were likely underestimated.

Etminan et al. published their new calculations for methane's radiative forcing (RF) in a 2016 Geophysical Research Letters journal article which incorporated the shortwave bands of CH_4 in measuring forcing, not used in previous, simpler IPCC methods. Their new RF calculations which significantly revised those cited in earlier, successive IPCC reports for well mixed greenhouse gases (WMGHG) forcings by including the shortwave forcing component due to CH_4, resulted in estimates that were approximately 20–25% higher. Collins et al. said that CH_4 mitigation that reduces atmospheric methane by the end of the century, could "make a substantial difference to the feasibility of achieving the Paris climate targets", and would provide us with more "allowable carbon emissions to 2100".

In addition to the direct heating effect and the normal feedbacks, the methane breaks down to carbon dioxide and water. This water is often above the tropopause, where little water usually reaches. Ramanathan (1998) notes that both water and ice clouds, when formed at cold lower stratospheric temperatures, are extremely efficient in enhancing the atmospheric greenhouse effect. He also notes that there is a distinct possibility that large increases in methane in future may lead to a surface warming that increases nonlinearly with the methane concentration.

Mitigation efforts to reduce short-lived climate pollutants like methane and black carbon would help combat "near-term climate change" and would support Sustainable Development Goals.

Sources

Any process that results in the production of methane and its release into the atmosphere can be considered a "source". The known sources of methane are predominantly located near the Earth's surface. Two main processes that are responsible for methane production include microorganisms anaerobically converting organic compounds into methane (methanogenesis), which are widespread in aquatic ecosystems, and ruminant animals.

Methane is also released in the Arctic for example from thawing permafrost. Increasing methane emissions are a major contributor to the rising concentration of greenhouse gases in Earth's atmosphere, and are responsible for up to one-third of near-term global heating. During 2019, about 60% (360 million tons) of methane released globally was from human activities, while natural sources contributed about 40% (230 million tons). Reducing methane emissions by capturing and utilizing the gas can produce simultaneous environmental and economic benefits. Since the Industrial Revolution, concentrations of methane in the atmosphere have more than doubled, and about 20 percent of the warming the planet has experienced can be attributed to the gas. About one-third (33%) of anthropogenic emissions are from gas release during the extraction and delivery of fossil fuels; mostly due to gas venting and gas leaks from both active fossil fuel infrastructure and orphan wells. Russia is the world's top methane emitter from oil and gas.

Animal agriculture is a similarly large source (30%); primarily because of enteric fermentation by ruminant livestock such as cattle and sheep. According to the Global Methane Assessment published in 2021, methane emissions from livestock (including cattle) are the largest sources of agricultural emissions worldwide A single cow can make up to 99 kg of methane gas per year. Ruminant livestock can produce 250 to 500 L of methane per day.

Measurement Techniques

Methane was typically measured using gas chromatography. Gas chromatography is a type of chromatography used for separating or analyzing chemical compounds. It is less expensive in general, compared to more advanced methods, but it is more time and labor-intensive. Spectroscopic methods were the preferred method for atmospheric gas measurements due to its sensitivity and precision. Also, spectroscopic methods are the only way of remotely sensing the atmospheric gases. Infrared spectroscopy covers a large spectrum of techniques, one of which detects gases based on absorption spectroscopy. There are various methods for spectroscopic methods, including Differential optical absorption spectroscopy, Laser-induced fluorescence, and Fourier Transform Infrared.

In 2011, cavity ring-down spectroscopy was the most widely used IR absorption technique of detecting methane. It is a form of laser absorption spectroscopy which determines the mole fraction to the order of parts per trillion.

Global Monitoring

CH_4 has been measured directly in the environment since the 1970s. The Earth's atmospheric methane concentration has increased 160% since preindustrial levels in the mid-18th century.

Long term atmospheric measurements of methane by NOAA show that the build up of methane nearly tripled since pre-industrial times since 1750. In 1991 and 1998 there was a sudden growth rate of methane representing a doubling of growth rates in previous years. The June 15, 1991 eruption of Mount Pinatubo, measuring VEI-6—was the second-largest terrestrial eruption of the 20th century. In 2007 it was reported that unprecedented warm temperatures in 1998—the warmest year since surface records were recorded—could have induced elevated methane emissions, along with an increase in wetland and rice field emissions and the amount of biomass burning.

Data from 2007 suggested methane concentrations were beginning to rise again. This was confirmed in 2010 when a study showed methane levels were on the rise for the 3 years 2007 to 2009. After a decade of near-zero growth in methane levels, "globally averaged atmospheric methane increased by [approximately] 7 nmol/mol per year during 2007 and 2008. During the first half of 2009, globally averaged atmospheric CH_4 was [approximately] 7 nmol/mol greater than it was in 2008, suggesting that the increase will continue in 2009." From 2015 to 2019 sharp rises in levels of atmospheric methane have been recorded.

In 2010, methane levels in the Arctic were measured at 1850 nmol/mol which is over twice as high as at any time in the last 400,000 years. According to the IPCC AR5, since 2011 concentrations continued to increase. After 2014, the increase accelerated and by 2017, it reached 1,850 (parts per billion) ppb. The annual average for methane (CH_4) was 1866 ppb in 2019 and scientists reported with "very high confidence" that concentrations of CH_4 were higher than at any time in at least 800,000 years. The largest annual increase occurred in 2021 with current concentrations reaching a record 260% of pre-industrial—with the overwhelming percentage caused by human activity.

In 2013, IPCC scientists said with "very high confidence", that concentrations of atmospheric methane CH_4 "exceeded the pre-industrial levels by about 150% which represented "levels unprecedented in at least the last 800,000 years." The globally averaged concentration of methane in Earth's atmosphere increased by about 150% from 722 ± 25 ppb in 1750 to 1803.1 ± 0.6 ppb in 2011. As of 2016, methane contributed radiative forcing of 0.62 ± 14% Wm^{-2}, or about 20% of the total radiative forcing from all of the long-lived and globally mixed greenhouse gases. The atmospheric methane concentration has continued to increase since 2011 to an average global concentration of 1911.8 ± 0.6 ppb as of 2022. The May 2021 peak was 1891.6 ppb, while the April 2022 peak was 1909.4 ppb, a 0.9% increase.

The Global Carbon Project consortium produces the Global Methane Budget. Working with over fifty international research institutions and 100 stations globally, it updates the methane budget every few years.

In 2013, the balance between sources and sinks of methane was not yet fully understood. Scientists were unable to explain why the atmospheric concentration of methane had temporarily ceased to increase.

The focus on the role of methane in anthropogenic climate change has become more relevant since the mid-2010s.

Natural Sinks or Removal of Atmospheric Methane

The amount of methane in the atmosphere is the result of a balance between the production of methane on the Earth's surface—its source—and the destruction or removal of methane, mainly in the atmosphere—its sink— in an atmospheric chemical process.

Another major natural sink is through oxidation by methanotrophic or methane-consuming bacteria in Earth's soils.

These 2005 NASA computer model simulations—calculated based on data available at that time—illustrate how methane is destroyed as it rises.

As air rises in the tropics, methane is carried upwards through the troposphere—the lowest portion of Earth's atmosphere which is 4 miles (6.4 km) to 12 miles (19 km) from the Earth's surface, into the lower stratosphere—the ozone layer—and then the upper portion of the stratosphere.

This atmospheric chemical process is the most effective methane sink, as it removes 90% of atmospheric methane. This global destruction of atmospheric methane mainly occurs in the troposphere.

Methane molecules react with hydroxyl radicals (OH)—the "major chemical scavenger in the troposphere" that "controls the atmospheric lifetime of most gases in the troposphere". Through this CH_4 oxidation process, atmospheric methane is destroyed and water vapor and carbon dioxide are produced.

While this decreases the concentration of methane in the atmosphere, it is unclear if this leads to a net positive increase in radiative forcing because both water vapor and carbon dioxide are more powerful GHGs factors in terms of affecting the warming of Earth.

This additional water vapor in the stratosphere caused by CH_4 oxidation, adds approximately 15% to methane's radiative forcing effect.

By the 1980s, the global warming problem had been transformed by the inclusion of methane and other non-CO_2 trace-gases—CFCs, N_2O, and O_3— on global warming, instead of focusing primarily on carbon dioxide. Both water and ice clouds, when formed at cold lower stratospheric temperatures, have a significant impact by increasing the atmospheric greenhouse effect. Large increases in future methane could lead to a surface warming that increases nonlinearly with the methane concentration.

Methane also affects the degradation of the ozone layer—the lowest layer of the stratosphere from about 15 to 35 kilometers (9 to 22 mi) above Earth, just above the troposphere. NASA researchers in 2001, had said that this process was enhanced by global warming, because warmer air holds more water vapor than colder air, so the amount of water vapor in the atmosphere increases as it is warmed by the greenhouse effect. Their climate models based on data available at that time, had indicated that carbon dioxide and methane enhanced the transport of water into the stratosphere.

Atmospheric methane could last about 120 years in the stratosphere until it is eventually destroyed through the hydroxyl radicals oxidation process.

There are different ways to quantify the period of time that methane impacts the atmosphere. The average time that a physical methane molecule is in the atmosphere is estimated to be around 9.6 years. However, the average time that the atmosphere will be affected by the emission of that molecule before reaching equilibrium – known as its 'perturbation lifetime' – is approximately twelve years.

The reaction of methane and chlorine atoms acts as a primary sink of Cl atoms and is a primary source of hydrochloric acid (HCl) in the stratosphere.

$$CH_4 + Cl \rightarrow CH_3 + HCl$$

The HCl produced in this reaction leads to catalytic ozone destruction in the stratosphere.

Soils act as a major sink for atmospheric methane through the methanotrophic bacteria that reside within them. This occurs with two different types of bacteria. "High capacity-low affinity" methanotrophic bacteria grow in areas of high methane concentration, such as waterlogged soils in wetlands and other moist environments. And in areas of low methane concentration, "low capacity-high affinity" methanotrophic bacteria make use of the methane in the atmosphere to grow, rather than relying on methane in their immediate environment. Methane oxidation allows methanotrophic bacteria to use methane as a source of energy, reacting methane with oxygen and as a result producing carbon dioxide and water.

$$CH_4 + 2O_2 \rightarrow CO_2 + 2H_2O$$

Forest soils act as good sinks for atmospheric methane because soils are optimally moist for methanotroph activity, and the movement of gases between soil and atmosphere (soil diffusivity) is high. With a lower water table, any methane in the soil has to make it past the methanotrophic bacteria before it can reach the atmosphere. Wetland soils, however, are often sources of atmospheric methane rather than sinks because the water table is much higher, and the methane can be diffused fairly easily into the air without having to compete with the soil's methanotrophs.

Methanotrophic bacteria also occur in the underwater sediments. Their presence can often efficiently limit emissions from sources such as the underwater permafrost in areas like the Laptev Sea.

Atmospheric methane removal is a category of potential approaches being researched to accelerate the breakdown of methane that is in the atmosphere, for the purpose of mitigating some of the impacts of climate change. Atmospheric methane has increased since pre-industrial times from 0.7 ppm to 1.9 ppm. From

2010 to 2019, methane emissions caused 0.5 °C (about 30%) of observed global warming. Global methane emissions approached a record 600 Tg CH_4 per year in 2017.

Methane Concentrations in the Geologic Past

From 1996 to 2004, researchers in the European Project for Ice Coring in Antarctica (EPICA) project were able to drill and analyze gases trapped in the ice cores in Antarctica to reconstruct GHG concentrations in the atmosphere over the past 800,000 years". They found that prior to approximately 900,000 years ago, the cycle of ice ages followed by relatively short warm periods lasted about 40,000 years, but by 800,000 years ago the time interval changed dramatically to cycles that lasted 100,000 years. There were low values of GHG in ice ages, and high values during the warm periods.

This 2016 EPA illustration above is a compilation of paleoclimatology showing methane concentrations over time based on analysis of gas bubbles from EPICA Dome C, Antarctica—approximately 797,446 BCE to 1937 CE, Law Dome, Antarctica—approximately 1008 CE to 1980 CE Cape Grim, Australia—1985 CE to 2015 CE Mauna Loa, Hawaii—1984 CE to 2015 CE and Shetland Islands, Scotland: 1993 CE to 2001 CE

The massive and rapid release of large volumes of methane gas from such sediments into the atmosphere has been suggested as a possible cause for rapid global warming events in the Earth's distant past, such as the Paleocene–Eocene Thermal Maximum, and the Great Dying.

In 2001, NASA's Goddard Institute for Space Studies and Columbia University's Center for Climate Systems Research scientists confirmed that other greenhouse gases apart from carbon dioxide were important factors in climate change in research presented at the annual meeting of the American Geophysical Union (AGU). They offered a theory on the 100,000-year long Paleocene–Eocene Thermal Maximum that occurred approximately 55 million years ago. They posited that there was a vast release of methane that had previously been kept stable through "cold temperatures and high pressure...beneath the ocean floor". This methane release into the atmosphere resulted in the warming of the earth. A 2009 journal article in *Science*, confirmed NASA research that the contribution of methane to global warming had previously been underestimated.

Early in the Earth's history carbon dioxide and methane likely produced a greenhouse effect. The carbon dioxide would have been produced by volcanoes and the methane by early microbes. During this time, Earth's earliest life appeared. According to a 2003 article in the journal *Geology*, these first, ancient bacteria added to the methane concentration by converting hydrogen and carbon dioxide into methane and water. Oxygen did not become a major part of the atmosphere until photosynthetic organisms evolved later in Earth's history. With no oxygen, methane stayed in the atmosphere longer and at higher concentrations than it does today.

Greenhouse Gas

Greenhouse gases (GHGs) are the gases in the atmosphere that raise the surface temperature of planets such as the Earth. What distinguishes them from other gases is that they absorb the wavelengths of radiation that a planet emits, resulting in the greenhouse effect. The Earth is warmed by sunlight, causing its surface to radiate heat, which is then mostly absorbed by greenhouse gases. Without greenhouse gases in the atmosphere, the average temperature of Earth's surface would be about -18°C (0°F), rather than the present average of 15°C (59°F).

The five most abundant greenhouse gases in Earth's atmosphere, listed in decreasing order of average global mole fraction, are: water vapor, carbon dioxide, methane, nitrous oxide, ozone. Other greenhouse gases of concern include chlorofluorocarbons (CFCs and HCFCs), hydrofluorocarbons (HFCs), perfluorocarbons, SF_6, and NF_3. Water vapor causes about half of the greenhouse effect, acting in response to other gases as a climate change feedback.

Human activities since the beginning of the Industrial Revolution (around 1750) have increased carbon dioxide by over 50%, and methane levels by 150%. Carbon dioxide emissions are causing about three-quarters of global warming, while methane emissions cause most of the rest.

The vast majority of carbon dioxide emissions by humans come from the burning of fossil fuels, with remaining contributions from agriculture and industry. Methane emissions originate from agriculture, fossil fuel production, waste, and other sources. The carbon cycle takes thousands of years to fully absorb CO_2 from the atmosphere, while methane lasts in the atmosphere for an average of only 12 years.

Natural flows of carbon happen between the atmosphere, terrestrial ecosystems, the ocean, and sediments. These flows have been fairly balanced over the past 1 million years, although greenhouse gas levels have varied widely in the more distant past. Carbon dioxide levels are now higher than they have been for 3 million years. If current emission rates continue then global warming will surpass 2.0 °C (3.6 °F) sometime between 2040 and 2070. This is a level which the Intergovernmental Panel on Climate Change (IPCC) says is "dangerous".

Properties and Mechanisms

Greenhouse gases are infrared active, meaning that they absorb and emit infrared radiation in the same long wavelength range as what is emitted by the Earth's surface, clouds and atmosphere.

99% of the Earth's dry atmosphere (excluding water vapor) is made up of nitrogen (N_2) (78%) and oxygen (O_2) (21%). Because their molecules contain two atoms of the same element, they have no asymmetry in the distribution of their electrical charges, and so are almost totally unaffected by infrared thermal radiation, with only an extremely minor effect from collision-induced absorption. A further 0.9% of the atmosphere is made up by argon (Ar), which is monatomic, and so completely transparent to thermal radiation. On the other hand, carbon dioxide (0.04%), methane, nitrous oxide and even less abundant trace gases account for less than 0.1% of Earth's atmosphere, but because their molecules contain atoms of different elements, there is an asymmetry in electric charge distribution which allows molecular vibrations to interact with electromagnetic radiation. This makes them infrared active, and so their presence causes greenhouse effect.

Earth absorbs some of the radiant energy received from the sun, reflects some of it as light and reflects or radiates the rest back to space as heat. A planet's surface temperature depends on this balance between incoming and outgoing energy. When Earth's energy balance is shifted, its surface becomes warmer or cooler, leading to a variety of changes in global climate. *Radiative forcing* is a metric calculated in watts per square meter, which characterizes the impact of an external change in a factor that influences climate. It is calculated as the difference in top-of-atmosphere (TOA) energy balance immediately caused by such an external change. A positive forcing, such as from increased concentrations of greenhouse gases, means more energy arriving than leaving at the top-of-atmosphere, which causes additional warming, while negative forcing, like from sulfates forming in the atmosphere from sulfur dioxide, leads to cooling.

Within the lower atmosphere, greenhouse gases exchange thermal radiation with the surface and limit radiative heat flow away from it, which reduces the overall rate of upward radiative heat transfer. The increased concentration of greenhouse gases is also cooling the upper atmosphere, as it is much thinner than the lower layers, and any heat re-emitted from greenhouse gases is more likely to travel further to space than to interact with the fewer gas molecules in the upper layers. The upper atmosphere is also shrinking as the result.

Contributions of Specific Gases to the Greenhouse Effect

Water vapor is the most important greenhouse gas overall, being responsible for 41–67% of the greenhouse effect, but its global concentrations are not directly affected by human activity. While local water vapor concentrations can be affected by developments such as irrigation, it has little impact on the global scale due to its short residence time of about nine days. Indirectly, an increase in global temperatures cause will also increase water vapor concentrations and thus their warming effect, in a process known as water vapor feedback. It occurs because

Clausius–Clapeyron relation establishes that more water vapor will be present per unit volume at elevated temperatures. Thus, local atmospheric concentration of water vapor varies from less than 0.01% in extremely cold regions and up to 3% by mass in saturated air at about 32°C.

Factors Affecting Concentrations

Atmospheric concentrations are determined by the balance between sources (emissions of the gas from human activities and natural systems) and sinks (the removal of the gas from the atmosphere by conversion to a different chemical compound or absorption by bodies of water).

The proportion of an emission remaining in the atmosphere after a specified time is the "airborne fraction" (AF). The *annual airborne fraction* is the ratio of the atmospheric increase in a given year to that year's total emissions. The annual airborne fraction for CO_2 had been stable at 0.45 for the past six decades even as the emissions have been increasing. This means that the other 0.55 of emitted CO_2 is absorbed by the land and atmosphere carbon sinks within the first year of an emission. In the high-emission scenarios, the effectiveness of carbon sinks will be lower, increasing the atmospheric fraction of CO_2 even though the raw amount of emissions absorbed will be higher than in the present.

Major greenhouse gases are well mixed and take many years to leave the atmosphere. The atmospheric lifetime of a greenhouse gas refers to the time required to restore equilibrium following a sudden increase or decrease in its concentration in the atmosphere. Individual atoms or molecules may be lost or deposited to sinks such as the soil, the oceans and other waters, or vegetation and other biological systems, reducing the excess to background concentrations. The average time taken to achieve this is the mean lifetime.

Changes to any of these variables can alter the atmospheric lifetime of a greenhouse gas. For instance, methane's atmospheric lifetime is estimated to have been lower in the 19th century than now, but to have been higher in the second half of the 20th century than after 2000. Carbon dioxide has an even more variable lifetime, which cannot be specified down to a single number. Scientists instead say that while the first 10% of carbon dioxide's airborne fraction (not counting the ~50% absorbed by land and ocean sinks within the emission's first year) is removed "quickly", the vast majority of the airborne fraction – 80% – lasts for "centuries to millennia". The remaining 10% stays for tens of thousands of years. In some models, this longest-lasting fraction is as large as 30%.

Estimates in 2023 found that the current carbon dioxide concentration in the atmosphere may be the highest it has been in the last 14 million years. However the IPCC Sixth Assessment Report estimated similar levels 3 to 3.3 million years ago in the mid-Pliocene warm period. This period can be a proxy for likely climate outcomes with current levels of CO_2.

Carbon dioxide is believed to have played an important effect in regulating Earth's temperature throughout its 4.54 billion year history. Early in the Earth's life, scientists have found evidence of liquid water indicating a warm world even though the Sun's output is believed to have only been 70% of what it is today. Higher carbon dioxide concentrations in the early Earth's atmosphere might help explain this faint young sun paradox. When Earth first formed, Earth's atmosphere may have contained more greenhouse gases and CO_2 concentrations may have been higher, with estimated partial pressure as large as 1,000 kPa (10 bar), because there was no bacterial photosynthesis to reduce the gas to carbon compounds and oxygen. Methane, a very active greenhouse gas, may have been more prevalent as well.

Monitoring

Greenhouse gas monitoring involves the direct measurement of atmospheric concentrations and direct and indirect measurement of greenhouse gas emissions. Indirect methods calculate emissions of greenhouse gases based on related metrics such as fossil fuel extraction.

There are several different methods of measuring carbon dioxide concentrations in the atmosphere, including infrared analyzing and manometry. Methane and nitrous oxide are measured by other instruments, such as the range-resolved infrared differential absorption lidar (DIAL). Greenhouse gases are measured from space such as by the Orbiting Carbon Observatory and through networks of ground stations such as the Integrated Carbon Observation System.

The Annual Greenhouse Gas Index (AGGI) is defined by atmospheric scientists at NOAA as the ratio of total direct radiative forcing due to long-lived and well-mixed greenhouse gases for any year for which adequate global measurements exist, to that present in year 1990. These radiative forcing levels are relative to those present in year 1750 (i.e. prior to the start of the industrial era). 1990 is chosen because it is the baseline year for the Kyoto Protocol, and is the publication year of the first IPCC Scientific Assessment of Climate Change. As such, NOAA states that the AGGI "measures the commitment that (global) society has already made to living in a changing climate. It is based on the highest quality atmospheric observations from sites around the world. Its uncertainty is very low."

There are several surface measurement (including flasks and continuous in situ) networks including NOAA/ERSL, WDCGG, and RAMCES. The NOAA/ESRL Baseline Observatory Network, and the Scripps Institution of Oceanography Network data are hosted at the CDIAC at ORNL. The World Data Centre for Greenhouse Gases

(WDCGG), part of GAW, data are hosted by the JMA. The Reseau Atmospherique de Mesure des Composes an Effet de Serre database (RAMCES) is part of IPSL.

Natural Sources

The natural flows of carbon between the atmosphere, ocean, terrestrial ecosystems, and sediments are fairly balanced; so carbon levels would be roughly stable without human influence. Carbon dioxide is removed from the atmosphere primarily through photosynthesis and enters the terrestrial and oceanic biospheres. Carbon dioxide also dissolves directly from the atmosphere into bodies of water (ocean, lakes, etc.), as well as dissolving in precipitation as raindrops fall through the atmosphere. When dissolved in water, carbon dioxide reacts with water molecules and forms carbonic acid, which contributes to ocean acidity. It can then be absorbed by rocks through weathering. It also can acidify other surfaces it touches or be washed into the ocean.

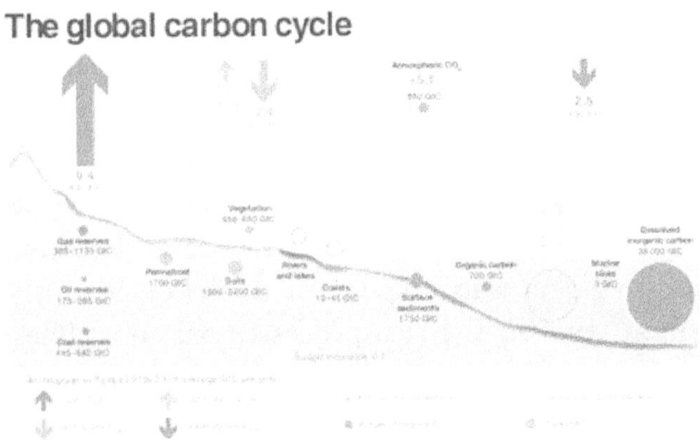

Fig: Schematic Representation of the Overall Perturbation of the Global Carbon Cycle Caused by Anthropogenic Activities, Averaged from 2010 to 2019

The atmospheric carbon cycle accounts for the exchange of gaseous carbon compounds, primarily carbon dioxide (CO_2), between Earth's atmosphere, the oceans, and the terrestrial biosphere. It is one of the faster components of the planet's overall carbon cycle, supporting the exchange of more than 200 billion tons of carbon (i.e. gigatons carbon or GtC) in and out of the atmosphere throughout the course of each year. Atmospheric concentrations of CO_2 remain stable over longer timescales only when there exists a balance between these two flows. Methane (CH_4), Carbon monoxide (CO), and other human-made compounds are present in smaller concentrations and are also part of the atmospheric carbon cycle.

Human-Made Sources

The vast majority of carbon dioxide emissions by humans come from the burning of fossil fuels. Additional contributions come from cement manufacturing, fertilizer production, and changes in land use like deforestation. Methane emissions originate from agriculture, fossil fuel production, waste, and other sources.

If current emission rates continue then temperature rises will surpass 2.0°C (3.6°F) sometime between 2040 and 2070, which is the level the United Nations' Intergovernmental Panel on Climate Change (IPCC) says is "dangerous".

Most greenhouse gases have both natural and human-caused sources. An exception are purely human-produced synthetic halocarbons which have no natural sources. During the pre-industrial Holocene, concentrations of existing gases were roughly constant, because the large natural sources and sinks roughly balanced. In the industrial era, human activities have added greenhouse gases to the atmosphere, mainly through the burning of fossil fuels and clearing of forests.

The major anthropogenic (human origin) sources of greenhouse gases are carbon dioxide (CO_2), nitrous oxide (N_2O), methane and three groups of fluorinated gases (sulfur hexafluoride (SF_6), hydrofluorocarbons (HFCs) and perfluorocarbons (PFCs, sulphur hexafluoride (SF_6), and nitrogen trifluoride (NF_3)). Though the greenhouse effect is heavily driven by water vapor, human emissions of water vapor are not a significant contributor to warming.

Although CFCs are greenhouse gases, they are regulated by the Montreal Protocol which was motivated by CFCs' contribution to ozone depletion rather than by their contribution to global warming. Ozone depletion has only a minor role in greenhouse warming, though the two processes are sometimes confused in the media. In 2016, negotiators from over 170 nations meeting at the summit of the United Nations Environment Programme reached a legally binding accord to phase out hydrofluorocarbons (HFCs) in the Kigali Amendment to the Montreal Protocol. The use of CFC-12 (except some essential uses) has been phased out due to its ozone depleting properties. The phasing-out of less active HCFC-compounds will be completed in 2030.

Reducing Human-Caused Greenhouse Gases

The annual "Emissions Gap Report" by UNEP stated in 2022 that it was necessary to almost halve emissions. "To get on track for limiting global warming to 1.5°C, global annual GHG emissions must be reduced by 45 per cent compared with emissions projections under policies currently in place in just eight years, and they must continue to decline rapidly after 2030, to avoid exhausting the limited remaining atmospheric carbon budget." The report commented that the world should focus on broad-based economy-wide transformations and not incremental change.

In 2022, the Intergovernmental Panel on Climate Change (IPCC) released its Sixth Assessment Report on climate change. It warned that greenhouse gas emissions must peak before 2025 at the latest and decline 43% by 2030 to have a good chance of limiting global warming to 1.5°C (2.7°F). Or in the words of Secretary-General of the United Nations António Guterres: "Main emitters must drastically cut emissions starting this year".

A number of technologies remove greenhouse gases emissions from the atmosphere. Most widely analyzed are those that remove carbon dioxide from the atmosphere, either to geologic formations such as bio-energy with carbon capture and storage and carbon dioxide air capture, or to the soil as in the case with biochar. Many long-term climate scenario models require large-scale human-made negative emissions to avoid serious climate change.

Negative emissions approaches are also being studied for atmospheric methane, called atmospheric methane removal.

Global Warming Potential

Global warming potential (GWP) is an index to measure how much infrared thermal radiation a greenhouse gas would absorb over a given time frame after it has been added to the atmosphere (or *emitted* to the atmosphere). The GWP makes different greenhouse gases comparable with regard to their "effectiveness in causing radiative forcing". It is expressed as a multiple of the radiation that would be absorbed by the same mass of added carbon dioxide (CO_2), which is taken as a reference gas. Therefore, the GWP has a value of 1 for CO_2. For other gases it depends on how strongly the gas absorbs infrared thermal radiation, how quickly the gas leaves the atmosphere, and the time frame being considered.

For example, methane has a GWP over 20 years (GWP-20) of 81.2 meaning that, for example, a leak of a tonne of methane is equivalent to emitting 81.2 tonnes of carbon dioxide measured over 20 years. As methane has a much shorter atmospheric lifetime than carbon dioxide, its GWP is much less over longer time periods, with a GWP-100 of 27.9 and a GWP-500 of 7.95.

The carbon dioxide equivalent (CO_2e or CO_2eq or CO_2-e or CO_2-eq) can be calculated from the GWP. For any gas, it is the mass of CO_2 that would warm the earth as much as the mass of that gas. Thus it provides a common scale for measuring the climate effects of different gases. It is calculated as GWP times mass of the other gas.

Definition

The global warming potential (GWP) is defined as an "index measuring the radiative forcing following an emission of a unit mass of a given substance, accumulated over a chosen time horizon, relative to that of the reference substance, carbon dioxide (CO_2). The GWP thus represents the combined effect of the differing times these substances remain in the atmosphere and their effectiveness in causing radiative forcing."

In turn, *radiative forcing* is a scientific concept used to quantify and compare the external drivers of change to Earth's energy balance. Radiative forcing is the change in energy flux in the atmosphere caused by natural or anthropogenic factors of climate change as measured in watts per meter squared.

As governments develop policies to combat emissions from high-GWP sources, policymakers have chosen to use the 100-year GWP scale as the standard in international agreements. The Kigali Amendment to the Montreal Protocol sets the global phase-down of hydrofluorocarbons (HFCs), a group of high-GWP compounds. It requires countries to use a set of GWP100 values equal to those published in the IPCC's Fourth Assessment Report (AR4). This allows policymakers to have one standard for comparison instead of changing GWP values in new assessment reports. One exception to the GWP100 standard exists: New York state's Climate Leadership and Community Protection Act requires the use of GWP20, despite being a different standard from all other countries participating in phase downs of HFCs.

Calculated Values

A substance's GWP depends on the number of years (denoted by a subscript) over which the potential is calculated. A gas which is quickly removed from the atmosphere may initially have a large effect, but for longer time periods, as it has been removed, it becomes less important. Thus methane has a potential of 25 over 100 years ($GWP_{100} = 25$) but 86 over 20 years ($GWP_{20} = 86$); conversely sulfur hexafluoride has a GWP of 22,800 over 100 years but 16,300 over 20 years (IPCC Third Assessment Report). The GWP value depends on how the gas concentration decays over time in the atmosphere. This is often not precisely known and hence the values should not be considered exact. For this reason when quoting a GWP it is important to give a reference to the calculation.

The GWP for a mixture of gases can be obtained from the mass-fraction-weighted average of the GWPs of the individual gases.

Commonly, a time horizon of 100 years is used by regulators. Water vapour does contribute to anthropogenic global warming, but as the GWP is defined, it is negligible for H_2O: an estimate gives a 100-year GWP between -0.001 and 0.0005.

H_2O can function as a greenhouse gas because it has a profound infrared absorption spectrum with more and broader absorption bands than CO_2. Its concentration in the atmosphere is limited by air temperature, so that radiative forcing by water vapour increases with global warming (positive feedback). But the GWP definition excludes indirect effects. GWP definition is also based on emissions, and anthropogenic emissions of water vapour (cooling towers, irrigation) are removed via precipitation within weeks, so its GWP is negligible.

Applications

Carbon dioxide equivalent (CO_2e or CO_2eq or CO_2-e) of a quantity of gas is calculated from its GWP. For any gas, it is the mass of CO_2 which would warm the earth as much as the mass of that gas. Thus it provides a common scale for measuring the climate effects of different gases. It is calculated as GWP multiplied by mass of the other gas. For example, if a gas has GWP of 100, two tonnes of the gas have CO_2e of 200 tonnes, and 9 tonnes of the gas has CO_2e of 900 tonnes.

On a global scale, the warming effects of one or more greenhouse gases in the atmosphere can also be expressed as an equivalent atmospheric concentration of CO_2. CO_2e can then be the atmospheric concentration of CO_2 which would warm the earth as much as a particular concentration of some other gas or of all gases and aerosols in the atmosphere. For example, CO_2e of 500 parts per million would reflect a mix of atmospheric gases which warm the earth as much as 500 parts per million of CO_2 would warm it. Calculation of the equivalent atmospheric concentration of CO_2 of an atmospheric greenhouse gas or aerosol is more complex and involves the atmospheric concentrations of those gases, their GWPs, and the ratios of their molar masses to the molar mass of CO_2.

CO_2e calculations depend on the time-scale chosen, typically 100 years or 20 years, since gases decay in the atmosphere or are absorbed naturally, at different rates.

The following units are commonly used:

- **By the UN Climate Change Panel (IPCC):** Billion metric tonnes = $n \times 10^9$ tonnes of CO_2 equivalent ($GtCO_2eq$)
- **In Industry:** Million metric tonnes of carbon dioxide equivalents (MMTCDE) and MMT CO_2eq.
- **For Vehicles:** Grams of carbon dioxide equivalent per mile (gCO_2e/mile) or per kilometer (gCO_2e/km)

For example, the table above shows GWP for methane over 20 years at 86 and nitrous oxide at 289, so emissions of 1 million tonnes of methane or nitrous oxide are equivalent to emissions of 86 or 289 million tonnes of carbon dioxide, respectively.

Under the Kyoto Protocol, in 1997 the Conference of the Parties standardized international reporting, by deciding that the values of GWP calculated for the IPCC Second Assessment Report were to be used for converting the various greenhouse gas emissions into comparable CO_2 equivalents.

After some intermediate updates, in 2013 this standard was updated by the Warsaw meeting of the UN Framework Convention on Climate Change (UNFCCC, decision number 24/CP.19) to require using a new set of 100-year GWP values. They published these values in Annex III, and they took them from the IPCC Fourth Assessment

Report, which had been published in 2007. Those 2007 estimates are still used for international comparisons through 2020, although the latest research on warming effects has found other values, as shown in the tables above.

Though recent reports reflect more scientific accuracy, countries and companies continue to use the IPCC Second Assessment Report (SAR) and IPCC Fourth Assessment Report values for reasons of comparison in their emission reports. The IPCC Fifth Assessment Report has skipped the 500-year values but introduced GWP estimations including the climate-carbon feedback (f) with a large amount of uncertainty.

Other Metrics to Compare Greenhouse Gases

The *Global Temperature change Potential* (GTP) is another way to compare gases. While GWP estimates infrared thermal radiation absorbed, GTP estimates the resulting rise in average surface temperature of the world, over the next 20, 50 or 100 years, caused by a greenhouse gas, relative to the temperature rise which the same mass of CO_2 would cause. Calculation of GTP requires modelling how the world, especially the oceans, will absorb heat. GTP is published in the same IPCC tables with GWP.

Another metric called GWP* (pronounced "GWP star") has been proposed to take better account of short-lived climate pollutants (SLCPs) such as methane. A permanent increase in the rate of emission of an SLCP has a similar effect to that of a one-time emission of an amount of carbon dioxide, because both raise the radiative forcing permanently or (in the case of carbon dioxide) practically permanently (since the CO_2 stays in the air for a long time). GWP* therefore assigns an *increase* in emission rate of an SLCP a supposedly equivalent amount (tonnes) of CO_2. However GWP* has been criticised both for its suitability as a metric and for inherent design features which can perpetuate injustices and inequity. Developing countries whose emissions of SLCPs are increasing are "penalized", while developed countries such as Australia or New Zealand which have steady emissions of SLCPs are not penalized in this way, though they may be penalized for their emissions of CO_2.

10.2.5 Extra-Terrestrial Life

The presence of atmospheric methane has a role in the scientific search for extra-terrestrial life. The justification is that on an astronomical timescale, methane in the atmosphere of an Earth-like celestial body will quickly dissipate, and that its presence on such a planet or moon therefore indicates that something is replenishing it. If methane is detected (by using a spectrometer for example) this may indicate that life is, or recently was, present. This was debated when methane was discovered in the Martian atmosphere by M.J. Mumma of NASA's Goddard Flight Center, and verified by the Mars Express Orbiter (2004) and in Titan's atmosphere by the Huygens probe (2005). This debate was furthered with the discovery of 'transient', 'spikes of methane' on Mars by the Curiosity Rover.

It is argued that atmospheric methane can come from volcanoes or other fissures in the planet's crust and that without an isotopic signature, the origin or source may be difficult to identify.

On 13 April 2017, NASA confirmed that the dive of the Cassini orbiter spacecraft on 28 October 2015 discovered an Enceladus plume which has all the ingredients for methanogenesis-based life forms to feed on. Previous results, published in March 2015, suggested hot water is interacting with rock beneath the sea of Enceladus; the new finding supported that conclusion, and add that the rock appears to be reacting chemically. From these observations scientists have determined that nearly 98 percent of the gas in the plume is water, about 1 percent is hydrogen, and the rest is a mixture of other molecules including carbon dioxide, methane and ammonia.

10.3 BIOGENIC OXIDATION OF METHANE

Methane (CH_4) is a significant greenhouse gas, approximately 25 times more potent than carbon dioxide (CO_2) in trapping heat over a 100-year period. While it contributes to global warming, methane is also a key player in the global carbon cycle, serving as an energy source for various microorganisms. Biogenic oxidation of methane refers to the biological processes that consume methane, primarily through microbial activity, converting it to less harmful compounds such as carbon dioxide and water. These processes are crucial in mitigating methane emissions and play an essential role in balancing the methane cycle. The main players in biogenic methane oxidation are aerobic methanotrophs and anaerobic methanotrophic archaea.

Biogenic oxidation of methane is a vital process in regulating methane levels in the atmosphere, with significant implications for climate change and environmental sustainability. Both aerobic and anaerobic methane oxidation play critical roles in the methane cycle, preventing the release of large amounts of methane into the atmosphere. The microbial communities responsible for these processes, particularly methanotrophs and anaerobic methanotrophic archaea, are key players in controlling methane emissions in various ecosystems.

Advances in our understanding of methane oxidation pathways and enzymes, combined with biotechnological innovations, hold promise for developing new strategies to mitigate methane emissions and harness methane as a valuable resource. As global methane emissions continue to rise, biogenic methane oxidation will remain an essential area of research for addressing climate change and environmental challenges.

10.3.1 The Methane Cycle and Its Importance

The methane cycle is a dynamic component of Earth's carbon cycle, involving both methane production (methanogenesis) and methane consumption (methanotrophy). Methane is produced in anaerobic environments such as wetlands, swamps, rice paddies, and the digestive systems of ruminant animals. Methanotrophy, on the other hand, refers to the oxidation of methane by microorganisms, primarily bacteria and archaea, converting it back into carbon dioxide. This process takes place in both aerobic and anaerobic environments and is critical in regulating atmospheric methane levels.

10.3.2 Aerobic Methanotrophy

Aerobic methanotrophy is the oxidation of methane in the presence of oxygen, primarily carried out by methanotrophic bacteria. These bacteria use methane as both a carbon and energy source, converting it through a series of enzymatic reactions into methanol, formaldehyde, and eventually carbon dioxide. Aerobic methanotrophs are typically found in environments where methane and oxygen coexist, such as soils, lakes, and sediments. They are particularly abundant in methane-oxidizing zones, such as those found in the top layers of wetland soils or at the oxic-anoxic interface in water bodies.

Key Enzymes and Pathways

Aerobic methanotrophs utilize a key enzyme called methane monooxygenase (MMO) to catalyze the initial step of methane oxidation, where methane is converted into methanol. There are two forms of MMO: soluble methane monooxygenase (sMMO) and particulate methane monooxygenase (pMMO). pMMO is the more common form, existing as a membrane-bound enzyme, whereas sMMO is soluble in the cytoplasm. Once methane is converted to methanol, the oxidation process proceeds through a series of dehydrogenase reactions, ultimately yielding carbon dioxide.

The ability of methanotrophs to oxidize methane not only helps reduce methane emissions but also allows them to thrive in methane-rich environments. Methanotrophic bacteria are capable of living in a variety of environments, including rice paddies, peat bogs, and the sediments of lakes and rivers, where they play an important role in limiting methane release into the atmosphere.

Environmental Relevance

Aerobic methane oxidation plays a vital role in the methane cycle by reducing methane emissions from aerobic environments such as soil and water. Wetlands, for instance, are among the largest natural sources of methane, yet they are also sites of significant methane oxidation, largely due to the activity of methanotrophic bacteria. Similarly, rice paddies are significant anthropogenic sources of methane, but methanotrophs present in the soil-water interface can substantially mitigate these emissions. Thus, aerobic methanotrophy is an essential natural mechanism for controlling atmospheric methane levels.

10.3.3 Anaerobic Methane Oxidation (AOM)

Anaerobic methane oxidation (AOM) refers to the process where methane is oxidized in the absence of oxygen, typically in environments such as deep-sea sediments, anoxic freshwater systems, and subsurface soils. AOM is primarily mediated by anaerobic methanotrophic archaea (ANME) in cooperation with sulfate-reducing bacteria. Unlike aerobic methanotrophs, anaerobic methanotrophs use alternative electron acceptors, such as sulfate (SO_4^{2-}), nitrate (NO_3^-), and iron (Fe^{3+}), to drive the oxidation of methane. This process is less well understood than aerobic methanotrophy, but it is believed to be responsible for consuming a substantial amount of methane produced in anaerobic environments.

The Role of ANME

The anaerobic methanotrophic archaea (ANME) are a unique group of archaea that mediate AOM. These archaea belong to different phylogenetic groups, including ANME-1, ANME-2, and ANME-3, all of which are involved in different methane oxidation pathways. ANME organisms lack the methane monooxygenase (MMO) enzyme found in aerobic methanotrophs. Instead, they rely on reverse methanogenesis, a process in which methane is oxidized through a pathway similar to methanogenesis but in reverse.

ANME archaea typically form syntrophic associations with sulfate-reducing bacteria, which play a critical role in methane oxidation.

The ANME archaea oxidize methane to carbon dioxide, generating electrons in the process, while the sulfate-reducing bacteria utilize these electrons to reduce sulfate to hydrogen sulfide (H_2S). This syntrophic relationship is essential for AOM, as it enables methane oxidation to proceed under anoxic conditions.

Alternative Electron Acceptors

In addition to sulfate, anaerobic methanotrophs can use other electron acceptors, such as nitrate, iron, and manganese, to oxidize methane.

For instance, nitrate-dependent methane oxidation is carried out by bacteria known as "NC10" bacteria, which perform AOM using nitrate or nitrite as the electron acceptor. Similarly, iron and manganese can serve as electron acceptors for methane oxidation, though these processes are less well understood and are thought to occur in specific environmental niches.

Ecological Importance of AOM

Anaerobic methane oxidation is particularly significant in marine environments, where it prevents large quantities of methane from reaching the atmosphere. Deep-sea sediments, for instance, contain vast methane reserves in the form of methane hydrates. Without the activity of anaerobic methanotrophs, these methane deposits could destabilize and release large amounts of methane into the atmosphere, contributing to global warming. AOM, therefore, acts as a critical methane sink in these environments, preventing catastrophic methane emissions. In freshwater systems, such as lakes and reservoirs, AOM also plays a crucial role in controlling methane emissions. These environments are often rich in organic matter and are prone to methane production via methanogenesis. However, AOM helps mitigate methane release by converting it into carbon dioxide before it can escape into the atmosphere.

10.3.4 Methane Oxidation and Climate Change

Methane is a potent greenhouse gas, and its increasing concentration in the atmosphere is a major contributor to climate change. Both aerobic and anaerobic methane oxidation serve as natural mechanisms for controlling atmospheric methane levels, thus playing a key role in the global climate system. However, human activities, such as fossil fuel extraction, agriculture, and waste management, have significantly increased methane emissions, overwhelming these natural mitigation processes. Efforts to enhance methane oxidation in natural and engineered systems are being explored as potential climate change mitigation strategies. For example, enhancing methanotrophic activity in soils, wetlands, and landfills could help reduce methane emissions. Similarly, promoting AOM in marine environments through nutrient management or microbial enhancement could help prevent methane release from methane hydrates.

10.3.5 Applications of Methane Oxidation in Biotechnology

Methanotrophic bacteria have garnered attention for their potential applications in biotechnology, particularly in areas such as bioenergy, bioremediation, and greenhouse gas mitigation. Methanotrophs can be harnessed to convert methane into useful products, such as biofuels, bioplastics, and chemicals. Additionally, methane oxidation pathways are being explored for their potential in reducing methane emissions from industrial processes, such as oil and gas production. In bioremediation, methanotrophs can be used to degrade methane and other organic pollutants in contaminated environments. These bacteria are capable of metabolizing a wide range of hydrocarbons, making them useful in cleaning up polluted soils and water systems. Moreover, methanotrophs have been studied for their ability to remove volatile organic compounds (VOCs) from the air, providing a natural solution for controlling air pollution.

10.3.6 Future Research Directions

While significant progress has been made in understanding the biogenic oxidation of methane, there are still many unanswered questions and opportunities for future research. One area of interest is the further exploration of anaerobic methane oxidation, particularly the roles of alternative electron acceptors and the metabolic pathways involved. Additionally, understanding how environmental factors, such as temperature, pH, and nutrient availability, influence methanotrophic activity could help improve methane mitigation strategies in natural and engineered systems. Another promising area of research is the development of synthetic biology approaches to enhance methane oxidation. By engineering methanotrophs or developing synthetic enzymes that mimic natural methane oxidation pathways, researchers could create more efficient systems for methane capture and conversion. These efforts could lead to new technologies for reducing methane emissions and transforming methane into valuable resources.

10.4 NICKEL ENZYMES NOT INVOLVED IN METHANE METABOLISM

Nickel enzymes are a unique class of metalloenzymes that utilize nickel as a cofactor to catalyze various biochemical reactions. While many are familiar with the role of nickel enzymes in methane metabolism, particularly in the context of methyl-coenzyme M reductase in methanogens, nickel enzymes also participate in a wide range of other metabolic processes. These processes include hydrogen metabolism, urea degradation, carbon monoxide detoxification, and superoxide reduction, among others. Nickel enzymes are involved in a wide range of metabolic processes beyond methane metabolism. From urease, which plays a key role in nitrogen cycling, to carbon monoxide dehydrogenase, which detoxifies CO, nickel enzymes demonstrate the versatility of this metal in catalysis. Hydrogenases and superoxide dismutase illustrate nickel's essential roles in energy metabolism and protection from oxidative stress, while acetyl-CoA synthase showcases its role in carbon fixation. Understanding these enzymes not only provides insight into the metabolic diversity of nickel but also highlights their importance in global biogeochemical cycles and environmental processes. As research continues, new nickel-dependent enzymes and their functions are likely to be discovered, further expanding our knowledge of the biochemical significance of this metal.

10.4.1 Overview of Nickel as a Cofactor in Enzymes

Nickel is a transition metal that forms complexes with different ligands, making it an ideal cofactor in enzymatic reactions. Its ability to undergo changes in oxidation state (commonly between Ni^{2+} and Ni^{3+}) makes it essential for redox reactions and catalytic processes. Nickel's role as a cofactor is not limited to a particular group of organisms; it is found in bacteria, archaea, plants, and even some animals. The diversity of reactions catalyzed by nickel enzymes speaks to the metal's versatility in biochemical systems. In organisms, nickel is often incorporated into the active sites of enzymes, where it participates directly in catalysis, stabilizes reaction intermediates, or aids in substrate binding.

10.4.2 Urease: A Nickel-Dependent Enzyme in Urea Metabolism

One of the best-studied nickel enzymes not involved in methane metabolism is **urease**, which plays a critical role in nitrogen metabolism by catalyzing the hydrolysis of urea into ammonia and carbon dioxide. Urease is found in a wide range of organisms, including plants, bacteria, fungi, and some invertebrates. This enzyme is vital for nitrogen recycling in the soil and is used by microorganisms to exploit urea as a nitrogen source.

Structure and Function of Urease

Urease contains a nickel ion in its active site, where it facilitates the hydrolysis of urea. The reaction proceeds in two steps: first, the urea is deprotonated, and then water molecules attack the urea, leading to the breakdown into ammonia and carbon dioxide. The nickel ions are essential for this process, as they coordinate the substrate and activate the water molecules involved in the hydrolysis.

Biological and Environmental Relevance

Urease has significant biological and ecological importance. In agriculture, for instance, urease activity in the soil is crucial for the nitrogen cycle, helping to convert urea-based fertilizers into ammonia, which plants can absorb. However, excessive urease activity can lead to ammonia volatilization, which causes nitrogen loss and environmental pollution. Urease inhibitors are often used to reduce this environmental impact. Additionally, urease is a virulence factor for some pathogenic bacteria, such as *Helicobacter pylori*, which causes gastric ulcers by producing ammonia to neutralize stomach acid.

10.4.3 Hydrogenases: Catalyzing Hydrogen Metabolism

Hydrogenases are another class of nickel-dependent enzymes that facilitate the interconversion of hydrogen gas (H_2) into protons and electrons. These enzymes are divided into two main types based on their metal cofactors: nickel-iron hydrogenases (NiFe hydrogenases) and iron-iron hydrogenases (FeFe hydrogenases). While both types catalyze similar reactions, NiFe hydrogenases are particularly important because of their ability to operate in both oxidative and reductive directions, making them key players in microbial energy metabolism.

Nickel in NiFe Hydrogenases

In NiFe hydrogenases, nickel and iron form the catalytic core of the enzyme. The nickel ion plays a direct role in the binding and activation of molecular hydrogen (H_2), while the iron atom participates in electron transfer during the redox reaction. NiFe hydrogenases are found in a variety of microorganisms, including methanogens, cyanobacteria, and purple sulfur bacteria, where they support energy production by catalyzing hydrogen oxidation or production, depending on the organism's metabolic requirements.

Role in Microbial Metabolism

NiFe hydrogenases are essential for the energy metabolism of hydrogen-oxidizing bacteria and archaea. These organisms can use hydrogen gas as a primary energy source, coupling hydrogen oxidation with electron transport chains to generate ATP. Additionally, NiFe hydrogenases are involved in symbiotic relationships, such as those between nitrogen-fixing bacteria and plants, where they help maintain redox balance by utilizing hydrogen produced during nitrogen fixation.

10.4.4 Carbon Monoxide Dehydrogenase: Detoxifying Carbon Monoxide

Carbon monoxide dehydrogenase (CODH) is a nickel-containing enzyme that plays a key role in the oxidation of carbon monoxide (CO) into carbon dioxide (CO_2). Found in a variety of bacteria and archaea, CODH is crucial for both energy production and detoxification in organisms that encounter carbon monoxide as a byproduct of anaerobic fermentation or external environmental sources.

Nickel's Role in Carbon Monoxide Oxidation

In the active site of CODH, nickel participates in the catalysis of CO oxidation. The enzyme contains a nickel-iron-sulfur cluster where CO is bound and subsequently oxidized to CO_2. This reaction is coupled with the reduction of other electron acceptors, such as ferredoxin, which plays a role in energy conservation within the cell. The nickel ion plays a crucial role in stabilizing reaction intermediates during CO oxidation.

Ecological and Environmental Importance

Carbon monoxide dehydrogenase is significant in environments where CO is produced naturally, such as in anaerobic soils, wetlands, and volcanic areas. The ability of CODH to detoxify CO allows organisms to survive in environments where CO would otherwise be toxic. Moreover, some microorganisms can use CO as an energy source, coupling CO oxidation with ATP synthesis, thereby contributing to the carbon cycle.

10.4.5 Superoxide Dismutase: Nickel as a Protective Agent

Superoxide dismutase (SOD) is an essential antioxidant enzyme that protects cells from oxidative stress by catalyzing the dismutation of superoxide radicals (O_2^-) into oxygen and hydrogen peroxide. Most organisms produce multiple forms of SOD, using different metal cofactors, including copper, zinc, manganese, and nickel.

Nickel-Dependent Superoxide Dismutase

Nickel-dependent SOD (Ni-SOD) is found predominantly in prokaryotes, such as cyanobacteria and some proteobacteria. The presence of nickel in the active site allows the enzyme to catalyze the breakdown of superoxide radicals, which are harmful byproducts of cellular respiration. The nickel ion alternates between Ni^{2+} and Ni^{3+} oxidation states during the catalytic cycle, facilitating the conversion of superoxide to oxygen and hydrogen peroxide.

Role in Oxidative Stress Defense

Superoxide radicals are highly reactive molecules that can cause significant damage to cellular components, including DNA, proteins, and lipids. Ni-SOD helps protect cells from oxidative stress, ensuring the survival of microorganisms in environments with high levels of reactive oxygen species. This protective function is critical for organisms living in extreme environments, such as high-altitude lakes or highly irradiated soils, where oxidative stress is prevalent.

10.4.6 Acetyl-CoA Synthase: A Nickel Enzyme in Carbon Fixation

Acetyl-CoA synthase (ACS) is a nickel-dependent enzyme that plays a central role in the Wood-Ljungdahl pathway of carbon fixation. This pathway, found in anaerobic bacteria and archaea, allows these organisms to use carbon dioxide as a carbon source by converting it into acetyl-CoA, a key metabolic intermediate. ACS works in tandem with CODH, with CODH providing carbon monoxide to ACS for the synthesis of acetyl-CoA.

Mechanism of Action

In ACS, nickel is involved in the activation and binding of carbon monoxide, which is then combined with a methyl group and coenzyme A to form acetyl-CoA. The nickel ion stabilizes the carbonyl group of CO, facilitating its incorporation into acetyl-CoA. This process is crucial for autotrophic growth in organisms that rely on CO_2 as a carbon source.

Importance in Carbon Fixation

The Wood-Ljungdahl pathway is one of the oldest carbon fixation pathways and is used by various anaerobic microorganisms, including acetogens and methanogens. ACS, as a key enzyme in this pathway, plays a critical role in enabling these organisms to grow autotrophically by fixing carbon dioxide. This pathway is particularly important in anoxic environments, such as deep-sea hydrothermal vents and anaerobic sediments, where it contributes to the global carbon cycle.

CHAPTER 11
PHOTOSYNTHESIS

11.1 OVERVIEW

Photosynthesis is a system of biological processes by which photosynthetic organisms, such as most plants, algae, and cyanobacteria, convert light energy, typically from sunlight, into the chemical energy necessary to fuel their metabolism. *Photosynthesis* usually refers to oxygenic photosynthesis, a process that produces oxygen. Photosynthetic organisms store the chemical energy so produced within intracellular organic compounds (compounds containing carbon) like sugars, glycogen, cellulose and starches. To use this stored chemical energy, an organism's cells metabolize the organic compounds through cellular respiration. Photosynthesis plays a critical role in producing and maintaining the oxygen content of the Earth's atmosphere, and it supplies most of the biological energy necessary for complex life on Earth.

Some bacteria also perform anoxygenic photosynthesis, which uses bacteriochlorophyll to split hydrogen sulfide as a reductant instead of water, producing sulfur instead of oxygen. Archaea such as *Halobacterium* also perform a type of non-carbon-fixing anoxygenic photosynthesis, where the simpler photopigment retinal and its microbial rhodopsin derivatives are used to absorb green light and power proton pumps to directly synthesize adenosine triphosphate (ATP), the "energy currency" of cells. Such archaeal photosynthesis might have been the earliest form of photosynthesis that evolved on Earth, as far back as the Paleoarchean, preceding that of cyanobacteria.

While the details may differ between species, the process always begins when light energy is absorbed by the reaction centers, proteins that contain photosynthetic pigments or chromophores. In plants, these proteins are chlorophylls (a porphyrin derivative that absorbs the red and blue spectrums of light, thus reflecting green) held inside chloroplasts, abundant in leaf cells.

In bacteria they are embedded in the plasma membrane. In these light-dependent reactions, some energy is used to strip electrons from suitable substances, such as water, producing oxygen gas. The hydrogen freed by the splitting of water is used in the creation of two important molecules that participate in energetic processes: reduced nicotinamide adenine dinucleotide phosphate (NADPH) and ATP.

In plants, algae, and cyanobacteria, sugars are synthesized by a subsequent sequence of light-independent reactions called the Calvin cycle. In this process, atmospheric carbon dioxide is incorporated into already existing organic compounds, such as ribulose bisphosphate (RuBP). Using the ATP and NADPH produced by the light-dependent reactions, the resulting compounds are then reduced and removed to form further carbohydrates, such as glucose. In other bacteria, different mechanisms like the reverse Krebs cycle are used to achieve the same end.

The first photosynthetic organisms probably evolved early in the evolutionary history of life using reducing agents such as hydrogen or hydrogen sulfide, rather than water, as sources of electrons. Cyanobacteria appeared later; the excess oxygen they produced contributed directly to the oxygenation of the Earth, which rendered the evolution of complex life possible. The average rate of energy captured by global photosynthesis is approximately 130 terawatts, which is about eight times the total power consumption of human civilization. Photosynthetic organisms also convert around 100–115 billion tons (91–104 Pg petagrams, or a billion metric tons), of carbon into biomass per year. Photosynthesis was discovered in 1779 by Jan Ingenhousz. He showed that plants need light, not just air, soil, and water.

Photosynthesis is vital for climate processes, as it captures carbon dioxide from the air and binds it into plants, harvested produce and soil. Cereals alone are estimated to bind 3,825 Tg or 3.825 Pg of carbon dioxide every year, i.e. 3.825 billion metric tons.

Most photosynthetic organisms are photoautotrophs, which means that they are able to synthesize food directly from carbon dioxide and water using energy from light. However, not all organisms use carbon dioxide as a source of carbon atoms to carry out photosynthesis; photoheterotrophs use organic compounds, rather than carbon dioxide, as a source of carbon.

In plants, algae, and cyanobacteria, photosynthesis releases oxygen. This oxygenic photosynthesis is by far the most common type of photosynthesis used by living organisms. Some shade-loving plants (sciophytes) produce such low levels of oxygen during photosynthesis that they use all of it themselves instead of releasing it to the atmosphere.

Although there are some differences between oxygenic photosynthesis in plants, algae, and cyanobacteria, the overall process is quite similar in these organisms. There are also many varieties of anoxygenic photosynthesis, used mostly by bacteria, which consume carbon dioxide but do not release oxygen.

Carbon dioxide is converted into sugars in a process called carbon fixation; photosynthesis captures energy from sunlight to convert carbon dioxide into carbohydrates. Carbon fixation is an endothermic redox reaction. In general outline, photosynthesis is the opposite of cellular respiration: while photosynthesis is a process of reduction of carbon dioxide to carbohydrates, cellular respiration is the oxidation of carbohydrates or other nutrients to carbon dioxide. Nutrients used in cellular respiration include carbohydrates, amino acids and fatty acids. These nutrients are oxidized to produce carbon dioxide and water, and to release chemical energy to drive the organism's metabolism.

Photosynthesis and cellular respiration are distinct processes, as they take place through different sequences of chemical reactions and in different cellular compartments (cellular respiration in mitochondria).

The general equation for photosynthesis as first proposed by Cornelis van Niel is:

$$\underset{\text{carbon dioxide}}{CO_2} + \underset{\text{electron donor}}{2H_2A} + \underset{\text{light energy}}{\text{photons}} \rightarrow \underset{\text{carbohydrate}}{[CH_2O]} + \underset{\text{oxidized electron donor}}{2A} + \underset{\text{water}}{H_2O}$$

Since water is used as the electron donor in oxygenic photosynthesis, the equation for this process is:

$$\underset{\text{carbon dioxide}}{CO_2} + \underset{\text{water}}{2H_2O} + \underset{\text{light energy}}{\text{photons}} \rightarrow \underset{\text{carbohydrate}}{[CH_2O]} + \underset{\text{oxygen}}{O_2} + \underset{\text{water}}{H_2O}$$

This equation emphasizes that water is both a reactant in the light-dependent reaction and a product of the light-independent reaction, but canceling *n* water molecules from each side gives the net equation:

$$\underset{\text{carbon dioxide}}{CO_2} + \underset{\text{water}}{H_2O} + \underset{\text{light energy}}{\text{photons}} \rightarrow \underset{\text{carbohydrate}}{[CH_2O]} + \underset{\text{oxygen}}{O_2}$$

Other processes substitute other compounds (such as arsenite) for water in the electron-supply role; for example some microbes use sunlight to oxidize arsenite to arsenate: The equation for this reaction is:

$$\underset{\text{carbon dioxide}}{CO_2} + \underset{\text{arsenite}}{(AsO_3^{3-})} + \underset{\text{light energy}}{\text{photons}} \rightarrow \underset{\text{arsenate}}{(AsO_4^{3-})} + \underset{\text{carbon monoxide}}{CO}$$

Photosynthesis occurs in two stages. In the first stage, *light-dependent reactions* or *light reactions* capture the energy of light and use it to make the hydrogen carrier NADPH and the energy-storage molecule ATP. During the second stage, the *light-independent reactions* use these products to capture and reduce carbon dioxide.

Most organisms that use oxygenic photosynthesis use visible light for the light-dependent reactions, although at least three use shortwave infrared or, more specifically, far-red radiation.

Some organisms employ even more radical variants of photosynthesis. Some archaea use a simpler method that employs a pigment similar to those used for vision in animals. The bacteriorhodopsin changes its configuration in response to sunlight, acting as a proton pump. This produces a proton gradient more directly, which is then converted to chemical energy. The process does not involve carbon dioxide fixation and does not release oxygen, and seems to have evolved separately from the more common types of photosynthesis.

11.1.1 Photosynthetic Membranes and Organelles

In photosynthetic bacteria, the proteins that gather light for photosynthesis are embedded in cell membranes. In its simplest form, this involves the membrane surrounding the cell itself. However, the membrane may be tightly folded into cylindrical sheets called thylakoids, or bunched up into round vesicles called *intracytoplasmic membranes*. These structures can fill most of the interior of a cell, giving the membrane a very large surface area and therefore increasing the amount of light that the bacteria can absorb.

In plants and algae, photosynthesis takes place in organelles called chloroplasts. A typical plant cell contains about 10 to 100 chloroplasts. The chloroplast is enclosed by a membrane. This membrane is composed of a phospholipid inner membrane, a phospholipid outer membrane, and an intermembrane space. Enclosed by the membrane is an aqueous fluid called the stroma. Embedded within the stroma are stacks of thylakoids (grana), which are the site of photosynthesis. The thylakoids appear as flattened disks. The thylakoid itself is enclosed by the thylakoid membrane, and within the enclosed volume is a lumen or thylakoid space. Embedded in the thylakoid membrane are integral and peripheral membrane protein complexes of the photosynthetic system.

Plants absorb light primarily using the pigment chlorophyll. The green part of the light spectrum is not absorbed but is reflected which is the reason that most plants have a green color. Besides chlorophyll, plants also use pigments such as carotenes and xanthophylls. Algae also use chlorophyll, but various other pigments are present, such as phycocyanin, carotenes, and xanthophylls in green algae, phycoerythrin in red algae (rhodophytes) and fucoxanthin in brown algae and diatoms resulting in a wide variety of colors.

These pigments are embedded in plants and algae in complexes called antenna proteins. In such proteins, the pigments are arranged to work together. Such a combination of proteins is also called a light-harvesting complex.

Although all cells in the green parts of a plant have chloroplasts, the majority of those are found in specially adapted structures called leaves. Certain species adapted to conditions of strong sunlight and aridity, such as many *Euphorbia* and cactus species, have their main photosynthetic organs in their stems. The cells in the interior tissues of a leaf, called the mesophyll, can contain between 450,000 and 800,000 chloroplasts for every square millimeter of leaf. The surface of the leaf is coated with a water-resistant waxy cuticle that protects the leaf from excessive evaporation of water and decreases the absorption of ultraviolet or blue light to minimize heating. The transparent epidermis layer allows light to pass through to the palisade mesophyll cells where most of the photosynthesis takes place.

11.1.2 Light-Dependent Reactions

In the light-dependent reactions, one molecule of the pigment chlorophyll absorbs one photon and loses one electron. This electron is taken up by a modified form of chlorophyll called pheophytin, which passes the electron to a quinone molecule, starting the flow of electrons down an electron transport chain that leads to the ultimate reduction of NADP to NADPH. In addition, this creates a proton gradient (energy gradient) across the chloroplast membrane, which is used by ATP synthase in the synthesis of ATP. The chlorophyll molecule ultimately regains the electron it lost when a water molecule is split in a process called photolysis, which releases oxygen.

The overall equation for the light-dependent reactions under the conditions of non-cyclic electron flow in green plants is:

$$2\,H_2O + 2\,NADP^+ + 3\,ADP + 3\,P_i + light \rightarrow 2\,NADPH + 2\,H^+ + 3\,ATP + O_2$$

Not all wavelengths of light can support photosynthesis. The photosynthetic action spectrum depends on the type of accessory pigments present. For example, in green plants, the action spectrum resembles the absorption spectrum for chlorophylls and carotenoids with absorption peaks in violet-blue and red light. In red algae, the action spectrum is blue-green light, which allows these algae to use the blue end of the spectrum to grow in the deeper waters that filter out the longer wavelengths (red light) used by above-ground green plants. The non-absorbed part of the light spectrum is what gives photosynthetic organisms their color (e.g., green plants, red algae, purple bacteria) and is the least effective for photosynthesis in the respective organisms.

Z Scheme

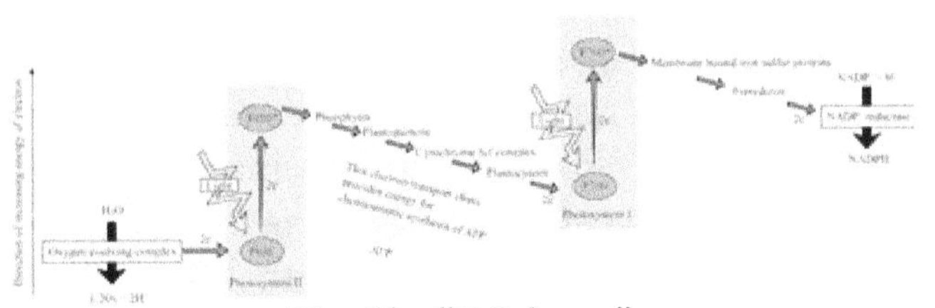

Fig: The "Z Scheme"

In plants, light-dependent reactions occur in the thylakoid membranes of the chloroplasts where they drive the synthesis of ATP and NADPH. The light-dependent reactions are of two forms: cyclic and non-cyclic.

In the non-cyclic reaction, the photons are captured in the light-harvesting antenna complexes of photosystem II by chlorophyll and other accessory pigments. The absorption of a photon by the antenna complex loosens an electron by a process called photoinduced charge separation. The antenna system is at the core of the chlorophyll molecule of the photosystem II reaction center. That loosened electron is taken up by the primary electron-acceptor molecule, pheophytin. As the electrons are shuttled through an electron transport chain, a chemiosmotic potential is generated by pumping proton cations (H^+) across the membrane and into the thylakoid space. An ATP synthase enzyme uses that chemiosmotic potential to make ATP during photophosphorylation, whereas NADPH is a product of the terminal redox reaction in the *Z-scheme.* The electron enters a chlorophyll molecule in Photosystem I. There it is further excited by the light absorbed by that photosystem. The electron is then passed along a chain of electron acceptors to which it transfers some of its energy. The energy delivered to the electron acceptors is used to move hydrogen ions across the thylakoid membrane into the lumen. The electron is eventually used to reduce the coenzyme NADP with an H^+ to NADPH (which has functions in the light-independent reaction); at that point, the path of that electron ends.

The cyclic reaction is similar to that of the non-cyclic but differs in that it generates only ATP, and no reduced NADP (NADPH) is created. The cyclic reaction takes place only at photosystem I. Once the electron is displaced from the photosystem, the electron is passed down the electron acceptor molecules and returns to photosystem I, from where it was emitted, hence the name *cyclic reaction.*

Water Photolysis

Linear electron transport through a photosystem will leave the reaction center of that photosystem oxidized. Elevating another electron will first require re-reduction of the reaction center. The excited electrons lost from the reaction center (P700) of photosystem I are replaced by transfer from plastocyanin, whose electrons come from electron transport through photosystem II. Photosystem II, as the first step of the *Z-scheme*, requires an external source of electrons to reduce its oxidized chlorophyll *a* reaction center. The source of electrons for photosynthesis in green plants and cyanobacteria is water. Two water molecules are oxidized by the energy of four successive charge-separation reactions of photosystem II to yield a molecule of diatomic oxygen and four hydrogen ions. The electrons yielded are transferred to a redox-active tyrosine residue that is oxidized by the energy of $P680^+$. This resets the ability of P680 to absorb another photon and release another photo-dissociated electron. The oxidation of water is catalyzed in photosystem II by a redox-active structure that contains four manganese ions and a calcium ion; this oxygen-evolving complex binds two water molecules and contains the four oxidizing equivalents that are used to drive the water-oxidizing reaction. The hydrogen ions are released in the thylakoid lumen and therefore contribute to the transmembrane chemiosmotic potential that leads to ATP synthesis. Oxygen is a waste product of light-dependent reactions, but the majority of organisms on Earth use oxygen and its energy for cellular respiration, including photosynthetic organisms.

11.1.3 Light-Independent Reactions

Calvin Cycle

In the light-independent (or "dark") reactions, the enzyme RuBisCO captures CO_2 from the atmosphere and, in a process called the Calvin cycle, uses the newly formed NADPH and releases three-carbon sugars, which are later combined to form sucrose and starch. The overall equation for the light-independent reactions in green plants is

$$3\ CO_2 + 9\ ATP + 6\ NADPH + 6\ H^+ \rightarrow C_3H_6O_3\text{-phosphate} + 9\ ADP + 8\ P_i + 6\ NADP^+ + 3\ H_2O$$

Carbon fixation produces the three-carbon sugar intermediate, which is then converted into the final carbohydrate products. The simple carbon sugars photosynthesis produces are then used to form other organic compounds, such as the building material cellulose, the precursors for lipid and amino acid biosynthesis, or as a fuel in cellular respiration. The latter occurs not only in plants but also in animals when the carbon and energy from plants is passed through a food chain.

The fixation or reduction of carbon dioxide is a process in which carbon dioxide combines with a five-carbon sugar, ribulose 1,5-bisphosphate, to yield two molecules of a three-carbon compound, glycerate 3-phosphate, also known as 3-phosphoglycerate. Glycerate 3-phosphate, in the presence of ATP and NADPH produced during the light-dependent stages, is reduced to glyceraldehyde 3-phosphate. This product is also referred to as 3-phosphoglyceraldehyde (PGAL) or, more generically, as triose phosphate.

Most (five out of six molecules) of the glyceraldehyde 3-phosphate produced are used to regenerate ribulose 1,5-bisphosphate so the process can continue. The triose phosphates not thus "recycled" often condense to form hexose phosphates, which ultimately yield sucrose, starch, and cellulose, as well as glucose and fructose.

The sugars produced during carbon metabolism yield carbon skeletons that can be used for other metabolic reactions like the production of amino acids and lipids.

Carbon Concentrating Mechanisms

On Land

In hot and dry conditions, plants close their stomata to prevent water loss. Under these conditions, CO_2 will decrease and oxygen gas, produced by the light reactions of photosynthesis, will increase, causing an increase of photorespiration by the oxygenase activity of ribulose-1,5-bisphosphate carboxylase/oxygenase (RuBisCO) and decrease in carbon fixation. Some plants have evolved mechanisms to increase the CO_2 concentration in the leaves under these conditions.

Plants that use the C_4 carbon fixation process chemically fix carbon dioxide in the cells of the mesophyll by adding it to the three-carbon molecule phosphoenolpyruvate (PEP), a reaction catalyzed by an enzyme called PEP carboxylase, creating the four-carbon organic acid oxaloacetic acid. Oxaloacetic acid or malate synthesized by this process is then translocated to specialized bundle sheath cells where the enzyme RuBisCO and other Calvin cycle enzymes are located, and where CO_2 released by decarboxylation of the four-carbon acids is then fixed by RuBisCO activity to the three-carbon 3-phosphoglyceric acids. The physical separation of RuBisCO from the oxygen-generating light reactions reduces photorespiration and increases CO_2 fixation and, thus, the photosynthetic capacity of the leaf. C_4 plants can produce more sugar than C_3 plants in conditions of high light and temperature. Many important crop plants are C_4 plants, including maize, sorghum, sugarcane, and millet. Plants that do not use PEP-carboxylase in carbon fixation are called C_3 plants because the primary carboxylation reaction, catalyzed by RuBisCO, produces the three-carbon 3-phosphoglyceric acids directly in the Calvin-Benson cycle. Over 90% of plants use C_3 carbon fixation, compared to 3% that use C_4 carbon fixation; however, the evolution of C_4 in over sixty plant lineages makes it a striking example of convergent evolution. C_2 photosynthesis, which involves carbon-concentration by selective breakdown of photorespiratory glycine, is both an evolutionary precursor to C_4 and a useful carbon-concentrating mechanism in its own right.

Xerophytes, such as cacti and most succulents, also use PEP carboxylase to capture carbon dioxide in a process called Crassulacean acid metabolism (CAM). In contrast to C_4 metabolism, which *spatially* separates the CO_2 fixation to PEP from the Calvin cycle, CAM *temporally* separates these two processes. CAM plants have a different leaf anatomy from C_3 plants, and fix the CO_2 at night, when their stomata are open. CAM plants store the CO_2 mostly in the form of malic acid via carboxylation of phosphoenolpyruvate to oxaloacetate, which is then reduced to malate. Decarboxylation of malate during the day releases CO_2 inside the leaves, thus allowing carbon fixation to 3-phosphoglycerate by RuBisCO. CAM is used by 16,000 species of plants.

Calcium-oxalate-accumulating plants, such as *Amaranthus hybridus* and *Colobanthus quitensis,* show a variation of photosynthesis where calcium oxalate crystals function as dynamic carbon pools, supplying carbon dioxide (CO_2) to photosynthetic cells when stomata are partially or totally closed. This process was named alarm photosynthesis. Under stress conditions (e.g., water deficit), oxalate released from calcium oxalate crystals is converted to CO_2 by an oxalate oxidase enzyme, and the produced CO_2 can support the Calvin cycle reactions. Reactive hydrogen peroxide

(H_2O_2), the byproduct of oxalate oxidase reaction, can be neutralized by catalase. Alarm photosynthesis represents a photosynthetic variant to be added to the well-known C4 and CAM pathways. However, alarm photosynthesis, in contrast to these pathways, operates as a biochemical pump that collects carbon from the organ interior (or from the soil) and not from the atmosphere.

In Water

Cyanobacteria possess carboxysomes, which increase the concentration of CO_2 around RuBisCO to increase the rate of photosynthesis. An enzyme, carbonic anhydrase, located within the carboxysome, releases CO_2 from dissolved hydrocarbonate ions (HCO^-_3). Before the CO_2 can diffuse out, RuBisCO concentrated within the carboxysome quickly sponges it up. HCO^-_3 ions are made from CO_2 outside the cell by another carbonic anhydrase and are actively pumped into the cell by a membrane protein. They cannot cross the membrane as they are charged, and within the cytosol they turn back into CO_2 very slowly without the help of carbonic anhydrase. This causes the HCO^-_3 ions to accumulate within the cell from where they diffuse into the carboxysomes. Pyrenoids in algae and hornworts also act to concentrate CO_2 around RuBisCO.

11.1.4 Efficiency

Plants usually convert light into chemical energy with a photosynthetic efficiency of 3–6%. Absorbed light that is unconverted is dissipated primarily as heat, with a small fraction (1–2%) reemitted as chlorophyll fluorescence at longer (redder) wavelengths. This fact allows measurement of the light reaction of photosynthesis by using chlorophyll fluorometers.

Actual plants' photosynthetic efficiency varies with the frequency of the light being converted, light intensity, temperature, and proportion of carbon dioxide in the atmosphere, and can vary from 0.1% to 8%. By comparison, solar panels convert light into electric energy at an efficiency of approximately 6–20% for mass-produced panels, and above 40% in laboratory devices. Scientists are studying photosynthesis in hopes of developing plants with increased yield.

The efficiency of both light and dark reactions can be measured, but the relationship between the two can be complex. For example, the light reaction creates ATP and NADPH energy molecules, which C_3 plants can use for carbon fixation or photorespiration. Electrons may also flow to other electron sinks. For this reason, it is not uncommon for authors to differentiate between work done under non-photorespiratory conditions and under photorespiratory conditions.

Chlorophyll fluorescence of photosystem II can measure the light reaction, and infrared gas analyzers can measure the dark reaction. An integrated chlorophyll fluorometer and gas exchange system can investigate both light and dark reactions when researchers use the two separate systems together. Infrared gas analyzers and some moisture sensors are sensitive enough to measure the photosynthetic assimilation of CO_2 and of ΔH_2O using reliable methods. CO_2 is commonly measured in μmols/(m^2/s), parts per million, or volume per million; and H_2O is commonly measured in mmols/(m^2/s) or in mbars. By measuring CO_2 assimilation, ΔH_2O, leaf temperature, barometric pressure, leaf area, and photosynthetically active radiation (PAR), it becomes possible to estimate, "A" or carbon assimilation, "E" or transpiration, "gs" or stomatal conductance, and "Ci" or intracellular CO_2. However, it is more common to use chlorophyll fluorescence for plant stress measurement, where appropriate, because the most commonly used parameters FV/FM and Y(II) or F/FM' can be measured in a few seconds, allowing the investigation of larger plant populations.

Gas exchange systems that offer control of CO_2 levels, above and below ambient, allow the common practice of measurement of A/Ci curves, at different CO_2 levels, to characterize a plant's photosynthetic response.

Integrated chlorophyll fluorometer – gas exchange systems allow a more precise measure of photosynthetic response and mechanisms. While standard gas exchange photosynthesis systems can measure Ci, or substomatal CO_2 levels, the addition of integrated chlorophyll fluorescence measurements allows a more precise measurement of $C_{C,}$ the estimation of CO_2 concentration at the site of carboxylation in the chloroplast, to replace Ci. CO_2 concentration in the chloroplast becomes possible to estimate with the measurement of mesophyll conductance or g_m using an integrated system.

Photosynthesis measurement systems are not designed to directly measure the amount of light the leaf absorbs, but analysis of chlorophyll fluorescence, P700- and P515-absorbance, and gas exchange measurements reveal detailed information about, e.g., the photosystems, quantum efficiency and the CO_2 assimilation rates. With some instruments, even wavelength dependency of the photosynthetic efficiency can be analyzed.

A phenomenon known as quantum walk increases the efficiency of the energy transport of light significantly. In the photosynthetic cell of an alga, bacterium, or plant, there are light-sensitive molecules called chromophores arranged in an antenna-shaped structure called a photocomplex.

When a photon is absorbed by a chromophore, it is converted into a quasiparticle referred to as an exciton, which jumps from chromophore to chromophore towards the reaction center of the photocomplex, a collection of molecules that traps its energy in a chemical form accessible to the cell's metabolism. The exciton's wave properties enable it to cover a wider area and try out several possible paths simultaneously, allowing it to instantaneously "choose" the most efficient route, where it will have the highest probability of arriving at its destination in the minimum possible time.

Because that quantum walking takes place at temperatures far higher than quantum phenomena usually occur, it is only possible over very short distances. Obstacles in the form of destructive interference cause the particle to lose its wave properties for an instant before it regains them once again after it is freed from its locked position through a classic "hop". The movement of the electron towards the photo center is therefore covered in a series of conventional hops and quantum walks.

11.1.5 Evolution

Fossils of what are thought to be filamentous photosynthetic organisms have been dated at 3.4 billion years old. More recent studies also suggest that photosynthesis may have begun about 3.4 billion years ago, though the first direct evidence of photosynthesis comes from thylakoid membranes preserved in 1.75-billion-year-old cherts.

Oxygenic photosynthesis is the main source of oxygen in the Earth's atmosphere, and its earliest appearance is sometimes referred to as the oxygen catastrophe. Geological evidence suggests that oxygenic photosynthesis, such as that in cyanobacteria, became important during the Paleoproterozoic era around two billion years ago. Modern photosynthesis in plants and most photosynthetic prokaryotes is oxygenic, using water as an electron donor, which is oxidized to molecular oxygen in the photosynthetic reaction center.

Symbiosis and the Origin of Chloroplasts

Several groups of animals have formed symbiotic relationships with photosynthetic algae. These are most common in corals, sponges, and sea anemones. Scientists presume that this is due to the particularly simple body plans and large surface areas of these animals compared to their volumes. In addition, a few marine mollusks, such as *Elysia viridis* and *Elysia chlorotica,* also maintain a symbiotic relationship with chloroplasts they capture from the algae in their diet and then store in their bodies. This allows the mollusks to survive solely by photosynthesis for several months at a time. Some of the genes from the plant cell nucleus have even been transferred to the slugs, so that the chloroplasts can be supplied with proteins they need to survive.

An even closer form of symbiosis may explain the origin of chloroplasts. Chloroplasts have many similarities with photosynthetic bacteria, including a circular chromosome, prokaryotic-type ribosome, and similar proteins in the photosynthetic reaction center. The endosymbiotic theory suggests that photosynthetic bacteria were acquired (by endocytosis) by early eukaryotic cells to form the first plant cells. Therefore, chloroplasts may be photosynthetic bacteria that adapted to life inside plant cells. Like mitochondria, chloroplasts possess their own DNA, separate from the nuclear DNA of their plant host cells and the genes in this chloroplast DNA resemble those found in cyanobacteria. DNA in chloroplasts codes for redox proteins such as those found in the photosynthetic reaction centers. The CoRR Hypothesis proposes that this co-location of genes with their gene products is required for redox regulation of gene expression, and accounts for the persistence of DNA in bioenergetic organelles.

Photosynthetic Eukaryotic Lineages

Symbiotic and kleptoplastic organisms excluded:

- The glaucophytes and the red and green algae—clade Archaeplastida (uni- and multicellular)
- The cryptophytes—clade Cryptista (unicellular)
- The haptophytes—clade Haptista (unicellular)
- The dinoflagellates and chromerids in the superphylum Myzozoa, and Pseudoblepharisma in the phylum Ciliophora—clade Alveolata (unicellular)
- The ochrophytes—clade Stramenopila (uni- and multicellular)
- The chlorarachniophytes and three species of Paulinella in the phylum Cercozoa—clade Rhizaria (unicellular)
- The euglenids—clade Excavata (unicellular)

Except for the euglenids, which are found within the Excavata, all of these belong to the Diaphoretickes.

Archaeplastida and the photosynthetic Paulinella got their plastids, which are surrounded by two membranes, through primary endosymbiosis in two separate events, by engulfing a cyanobacterium. The plastids in all the other groups have either a red or green algal origin, and are referred to as the "red lineages" and the "green lineages".

The only known exception is the ciliate Pseudoblepharisma tenue, which in addition to its plastids that originated from green algae also has a purple sulfur bacterium as symbiont. In dinoflagellates and euglenids the plastids are surrounded by three membranes, and in the remaining lines by four. A nucleomorph, remnants of the original algal nucleus located between the inner and outer membranes of the plastid, is present in the cryptophytes (from a red alga) and chlorarachniophytes (from a green alga). Some dinoflagellates that lost their photosynthetic ability later regained it again through new endosymbiotic events with different algae. While able to perform photosynthesis, many of these eukaryotic groups are mixotrophs and practice heterotrophy to various degrees.

Photosynthetic Prokaryotic Lineages

Early photosynthetic systems, such as those in green and purple sulfur and green and purple nonsulfur bacteria, are thought to have been anoxygenic, and used various other molecules than water as electron donors. Green and purple sulfur bacteria are thought to have used hydrogen and sulfur as electron donors. Green nonsulfur bacteria used various amino and other organic acids as electron donors. Purple nonsulfur bacteria used a variety of nonspecific organic molecules. The use of these molecules is consistent with the geological evidence that Earth's early atmosphere was highly reducing at that time.

With a possible exception of Heimdallarchaeota, photosynthesis is not found in archaea. Haloarchaea are phototrophic and can absorb energy from the sun, but do not harvest carbon from the atmosphere and are therefore not photosynthetic. Instead of chlorophyll they use rhodopsins, which convert light-energy to ion gradients but cannot mediate electron transfer reactions.

In bacteria eight photosynthetic lineages are currently known:

- Cyanobacteria, the only prokaryotes performing oxygenic photosynthesis and the only prokaryotes that contain two types of photosystems (type I (RCI), also known as Fe-S type, and type II (RCII), also known as quinone type). The seven remaining prokaryotes have anoxygenic photosynthesis and use versions of either type I or type II.
- Chlorobi (green sulfur bacteria) Type I
- Heliobacteria Type I
- Chloracidobacterium Type I
- Proteobacteria (purple sulfur bacteria and purple non-sulfur bacteria) Type II
- Chloroflexota (green non-sulfur bacteria) Type II
- Gemmatimonadota Type II
- Eremiobacterota Type II

Cyanobacteria and the Evolution of Photosynthesis

The biochemical capacity to use water as the source for electrons in photosynthesis evolved once, in a common ancestor of extant cyanobacteria (formerly called blue-green algae). The geological record indicates that this transforming event took place early in Earth's history, at least 2450–2320 million years ago (Ma), and, it is speculated, much earlier. Because the Earth's atmosphere contained almost no oxygen during the estimated development of photosynthesis, it is believed that the first photosynthetic cyanobacteria did not generate oxygen. Available evidence from geobiological studies of Archean (>2500 Ma) sedimentary rocks indicates that life existed 3500 Ma, but the question of when oxygenic photosynthesis evolved is still unanswered.

A clear paleontological window on cyanobacterial evolution opened about 2000 Ma, revealing an already-diverse biota of cyanobacteria. Cyanobacteria remained the principal primary producers of oxygen throughout the Proterozoic Eon (2500–543 Ma), in part because the redox structure of the oceans favored photoautotrophs capable of nitrogen fixation. Green algae joined cyanobacteria as the major primary producers of oxygen on continental shelves near the end of the Proterozoic, but only with the Mesozoic (251–66 Ma) radiations of dinoflagellates, coccolithophorids, and diatoms did the primary production of oxygen in marine shelf waters take modern form. Cyanobacteria remain critical to marine ecosystems as primary producers of oxygen in oceanic gyres, as agents of biological nitrogen fixation, and, in modified form, as the plastids of marine algae.

11.1.6 Experimental History

Discovery

Although some of the steps in photosynthesis are still not completely understood, the overall photosynthetic equation has been known since the 19th century.

Jan van Helmont began the research of the process in the mid-17th century when he carefully measured the mass of the soil a plant was using and the mass of the plant as it grew. After noticing that the soil mass changed very little, he hypothesized that the mass of the growing plant must come from the water, the only substance he added to the potted plant. His hypothesis was partially accurate – much of the gained mass comes from carbon dioxide as well as water. However, this was a signaling point to the idea that the bulk of a plant's biomass comes from the inputs of photosynthesis, not the soil itself.

Joseph Priestley, a chemist and minister, discovered that when he isolated a volume of air under an inverted jar and burned a candle in it (which gave off CO_2), the candle would burn out very quickly, much before it ran out of wax. He further discovered that a mouse could similarly "injure" air. He then showed that a plant could restore the air the candle and the mouse had "injured."

In 1779, Jan Ingenhousz repeated Priestley's experiments. He discovered that it was the influence of sunlight on the plant that could cause it to revive a mouse in a matter of hours.

In 1796, Jean Senebier, a Swiss pastor, botanist, and naturalist, demonstrated that green plants consume carbon dioxide and release oxygen under the influence of light. Soon afterward, Nicolas-Théodore de Saussure showed that the increase in mass of the plant as it grows could not be due only to uptake of CO_2 but also to the incorporation of water. Thus, the basic reaction by which organisms use photosynthesis to produce food (such as glucose) was outlined.

Refinements

Cornelis Van Niel made key discoveries explaining the chemistry of photosynthesis. By studying purple sulfur bacteria and green bacteria, he was the first to demonstrate that photosynthesis is a light-dependent redox reaction in which hydrogen reduces (donates its atoms as electrons and protons to) carbon dioxide.

Robert Emerson discovered two light reactions by testing plant productivity using different wavelengths of light. With the red alone, the light reactions were suppressed. When blue and red were combined, the output was much more substantial. Thus, there were two photosystems, one absorbing up to 600 nm wavelengths, the other up to 700 nm. The former is known as PSII, the latter is PSI. PSI contains only chlorophyll "a", PSII contains primarily chlorophyll "a" with most of the available chlorophyll "b", among other pigments. These include phycobilins, which are the red and blue pigments of red and blue algae, respectively, and fucoxanthol for brown algae and diatoms. The process is most productive when the absorption of quanta is equal in both PSII and PSI, assuring that input energy from the antenna complex is divided between the PSI and PSII systems, which in turn powers the photochemistry.

Robert Hill thought that a complex of reactions consisted of an intermediate to cytochrome b_6 (now a plastoquinone), and that another was from cytochrome f to a step in the carbohydrate-generating mechanisms. These are linked by plastoquinone, which does require energy to reduce cytochrome f. Further experiments to prove that the oxygen developed during the photosynthesis of green plants came from water were performed by Hill in 1937 and 1939. He showed that isolated chloroplasts give off oxygen in the presence of unnatural reducing agents like iron oxalate, ferricyanide or benzoquinone after exposure to light. In the Hill reaction:

$$2\,H_2O + 2\,A + (\text{light, chloroplasts}) \rightarrow 2\,AH_2 + O_2$$

A is the electron acceptor. Therefore, in light, the electron acceptor is reduced and oxygen is evolved. Samuel Ruben and Martin Kamen used radioactive isotopes to determine that the oxygen liberated in photosynthesis came from the water.

Melvin Calvin and Andrew Benson, along with James Bassham, elucidated the path of carbon assimilation (the photosynthetic carbon reduction cycle) in plants. The carbon reduction cycle is known as the Calvin cycle, but many scientists refer to it as the Calvin-Benson, Benson-Calvin, or even Calvin-Benson-Bassham (or CBB) Cycle.

Nobel Prize–winning scientist Rudolph A. Marcus was later able to discover the function and significance of the electron transport chain.

Otto Heinrich Warburg and Dean Burk discovered the I-quantum photosynthesis reaction that splits CO_2, activated by the respiration.

In 1950, first experimental evidence for the existence of photophosphorylation *in vivo* was presented by Otto Kandler using intact *Chlorella* cells and interpreting his findings as light-dependent ATP formation. In 1954, Daniel I. Arnon et al. discovered photophosphorylation *in vitro* in isolated chloroplasts with the help of P^{32}.

Louis N. M. Duysens and Jan Amesz discovered that chlorophyll "a" will absorb one light, oxidize cytochrome f, while chlorophyll "a" (and other pigments) will absorb another light but will reduce this same oxidized cytochrome, stating the two light reactions are in series.

Development of the Concept

In 1893, the American botanist Charles Reid Barnes proposed two terms, *photosyntax* and *photosynthesis*, for the biological process of *synthesis of complex carbon compounds out of carbonic acid, in the presence of chlorophyll, under the influence of light*. The term *photosynthesis* is derived from the Greek *phōs* and *sýnthesis*, while another word that he designated was *photosyntax*, from *sýntaxis*. Over time, the term *photosynthesis* came into common usage. Later discovery of anoxygenic photosynthetic bacteria and photophosphorylation necessitated redefinition of the term.

C3 : C4 Photosynthesis Research

In the late 1940s at the University of California, Berkeley, the details of photosynthetic carbon metabolism were sorted out by the chemists Melvin Calvin, Andrew Benson, James Bassham and a score of students and researchers utilizing the carbon-14 isotope and paper chromatography techniques. The pathway of CO_2 fixation by the algae *Chlorella* in a fraction of a second in light resulted in a three carbon molecule called phosphoglyceric acid (PGA). For that original and ground-breaking work, a Nobel Prize in Chemistry was awarded to Melvin Calvin in 1961. In parallel, plant physiologists studied leaf gas exchanges using the new method of infrared gas analysis and a leaf chamber where the net photosynthetic rates ranged from 10 to 13 $\mu mol\ CO_2 \cdot m^{-2} \cdot s^{-1}$, with the conclusion that all terrestrial plants have the same photosynthetic capacities, that are light saturated at less than 50% of sunlight.

Later in 1958–1963 at Cornell University, field grown maize was reported to have much greater leaf photosynthetic rates of 40 $\mu mol\ CO_2 \cdot m^{-2} \cdot s^{-1}$ and not be saturated at near full sunlight. This higher rate in maize was almost double of those observed in other species such as wheat and soybean, indicating that large differences in photosynthesis exist among higher plants. At the University of Arizona, detailed gas exchange research on more than 15 species of monocots and dicots uncovered for the first time that differences in leaf anatomy are crucial factors in differentiating photosynthetic capacities among species. In tropical grasses, including maize, sorghum, sugarcane, Bermuda grass and in the dicot amaranthus, leaf photosynthetic rates were around 38-40 μmol $CO_2 \cdot m^{-2} \cdot s^{-1}$, and the leaves have two types of green cells, i.e. outer layer of mesophyll cells surrounding a tightly packed cholorophyllous vascular bundle sheath cells. This type of anatomy was termed Kranz anatomy in the 19th century by the botanist Gottlieb Haberlandt while studying leaf anatomy of sugarcane. Plant species with the greatest photosynthetic rates and Kranz anatomy showed no apparent photorespiration, very low CO_2 compensation point, high optimum temperature, high stomatal resistances and lower mesophyll resistances for gas diffusion and rates never saturated at full sun light. The research at Arizona was designated a Citation Classic in 1986. These species were later termed C4 plants as the first stable compound of CO_2 fixation in light has four carbons as malate and aspartate. Other species that lack Kranz anatomy were termed C3 type such as cotton and sunflower, as the first stable carbon compound is the three-carbon PGA. At 1000 ppm CO_2 in measuring air, both the C3 and C4 plants had similar leaf photosynthetic rates around 60 μmol $CO_2 \cdot m^{-2} \cdot s^{-1}$ indicating the suppression of photorespiration in C3 plants.

11.1.7 Factors

There are four main factors influencing photosynthesis and several corollary factors. The four main are:

- Light irradiance and wavelength.
- Water absorption.
- Carbon dioxide concentration.
- Temperature.

Total photosynthesis is limited by a range of environmental factors. These include the amount of light available, the amount of leaf area a plant has to capture light (shading by other plants is a major limitation of photosynthesis), the rate at which carbon dioxide can be supplied to the chloroplasts to support photosynthesis, the availability of water, and the availability of suitable temperatures for carrying out photosynthesis.

Light Intensity (Irradiance), Wavelength and Temperature

The process of photosynthesis provides the main input of free energy into the biosphere, and is one of four main ways in which radiation is important for plant life.

The radiation climate within plant communities is extremely variable, in both time and space.

In the early 20th century, Frederick Blackman and Gabrielle Matthaei investigated the effects of light intensity (irradiance) and temperature on the rate of carbon assimilation.

- At constant temperature, the rate of carbon assimilation varies with irradiance, increasing as the irradiance increases, but reaching a plateau at higher irradiance.
- At low irradiance, increasing the temperature has little influence on the rate of carbon assimilation. At constant high irradiance, the rate of carbon assimilation increases as the temperature is increased.

These two experiments illustrate several important points: First, it is known that, in general, photochemical reactions are not affected by temperature. However, these experiments clearly show that temperature affects the rate of carbon assimilation, so there must be two sets of reactions in the full process of carbon assimilation. These are the light-dependent 'photochemical' temperature-independent stage, and the light-independent, temperature-dependent stage. Second, Blackman's experiments illustrate the concept of limiting factors. Another limiting factor is the wavelength of light. Cyanobacteria, which reside several meters underwater, cannot receive the correct wavelengths required to cause photoinduced charge separation in conventional photosynthetic pigments. To combat this problem, Cyanobacteria have a light-harvesting complex called Phycobilisome. This complex is made up of a series of proteins with different pigments which surround the reaction center.

Carbon Dioxide Levels and Photorespiration

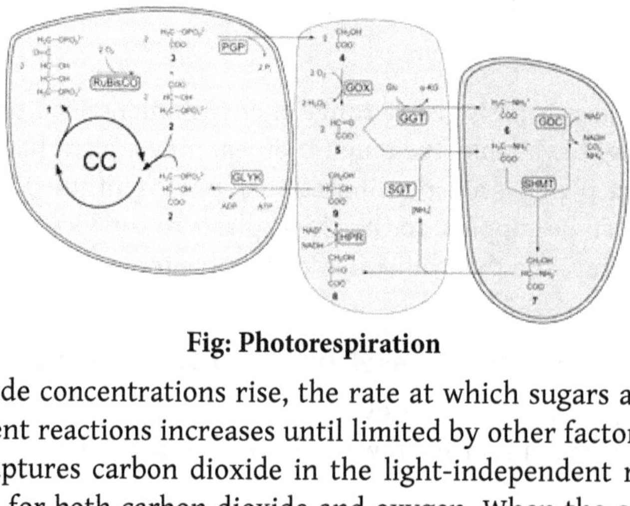

Fig: Photorespiration

As carbon dioxide concentrations rise, the rate at which sugars are made by the light-independent reactions increases until limited by other factors. RuBisCO, the enzyme that captures carbon dioxide in the light-independent reactions, has a binding affinity for both carbon dioxide and oxygen. When the concentration of carbon dioxide is high, RuBisCO will fix carbon dioxide. However, if the carbon dioxide concentration is low, RuBisCO will bind oxygen instead of carbon dioxide. This process, called photorespiration, uses energy, but does not produce sugars.

RuBisCO oxygenase activity is disadvantageous to plants for several reasons:

- One product of oxygenase activity is phosphoglycolate (2 carbon) instead of 3-phosphoglycerate (3 carbon). Phosphoglycolate cannot be metabolized by the Calvin-Benson cycle and represents carbon lost from the cycle. A high oxygenase activity, therefore, drains the sugars that are required to recycle ribulose 5-bisphosphate and for the continuation of the Calvin-Benson cycle.
- Phosphoglycolate is quickly metabolized to glycolate that is toxic to a plant at a high concentration; it inhibits photosynthesis.
- Salvaging glycolate is an energetically expensive process that uses the glycolate pathway, and only 75% of the carbon is returned to the Calvin-Benson cycle as 3-phosphoglycerate. The reactions also produce ammonia (NH_3), which is able to diffuse out of the plant, leading to a loss of nitrogen.

A highly simplified summary is:

$$2 \text{ glycolate} + \text{ATP} \rightarrow \text{3-phosphoglycerate} + \text{carbon dioxide} + \text{ADP} + NH_3$$

The salvaging pathway for the products of RuBisCO oxygenase activity is more commonly known as photorespiration, since it is characterized by light-dependent oxygen consumption and the release of carbon dioxide.

11.2 THE REACTION PATHWAY

Photosynthesis is a fundamental biochemical process by which plants, algae, and certain bacteria convert light energy into chemical energy, primarily in the form of glucose. This process not only sustains the organisms that perform it but also provides energy and organic compounds for nearly all life on Earth. The photosynthetic pathway consists of a series of complex reactions, categorized broadly into two phases: the light-dependent reactions and the light-independent reactions (commonly known as the Calvin Cycle). These reactions take place within the chloroplasts in plants, specifically in the thylakoid membranes and the stroma.

The reaction pathway of photosynthesis is a complex but vital process that sustains life on Earth. Through the light-dependent reactions and the Calvin Cycle, plants, algae, and some bacteria capture and convert solar energy into chemical energy in the form of glucose. The oxygen generated during photosynthesis is essential for aerobic organisms, while the carbon fixation process helps regulate atmospheric CO_2 levels. Understanding the intricacies of this pathway allows scientists to explore new ways to enhance crop productivity, develop sustainable energy sources, and address environmental challenges such as climate change.

11.2.1 Light-Dependent Reactions: The Energy-Harvesting Phase

The light-dependent reactions are the first phase of photosynthesis and occur in the thylakoid membranes of chloroplasts. These reactions require direct sunlight and involve the absorption of light by photosynthetic pigments like chlorophyll. The primary goal of the light-dependent reactions is to convert solar energy into chemical energy in the form of ATP (adenosine triphosphate) and NADPH (nicotinamide adenine dinucleotide phosphate), which are essential for the subsequent light-independent reactions.

Absorption of Light Energy

The process begins when chlorophyll molecules in photosystem II (PSII) absorb photons, or particles of light. This energy excites electrons within the chlorophyll, raising them to a higher energy state. These high-energy electrons are passed through a series of protein complexes known as the electron transport chain (ETC). As electrons move through the ETC, they release energy, which is harnessed to pump protons (H^+ ions) into the thylakoid lumen, creating a proton gradient.

Photolysis of Water

In addition to exciting electrons, photosystem II is also responsible for splitting water molecules, a process known as photolysis. Water molecules are broken down into oxygen, protons, and electrons. The oxygen is released as a byproduct into the atmosphere, while the protons contribute to the proton gradient across the thylakoid membrane. The electrons released from water replace those lost by chlorophyll in PSII, maintaining the flow of electrons through the electron transport chain.

Electron Transport and ATP Synthesis

As electrons continue to move through the electron transport chain, they reach photosystem I (PSI), where they are re-energized by another input of light energy. These electrons are then used to reduce $NADP^+$ into NADPH, an important electron carrier that will be used in the Calvin Cycle. Meanwhile, the proton gradient created by the movement of protons into the thylakoid lumen drives the synthesis of ATP through a process known as chemiosmosis. The enzyme ATP synthase facilitates the production of ATP as protons flow back into the stroma from the thylakoid lumen.

The Role of ATP and NADPH

The ATP and NADPH generated during the light-dependent reactions are crucial for powering the light-independent reactions in the Calvin Cycle. ATP provides the necessary energy, while NADPH supplies the reducing power needed to convert carbon dioxide into glucose. Without the light-dependent phase, the energy required for the Calvin Cycle would not be available, halting the process of photosynthesis.

11.2.2 The Calvin Cycle: Carbon Fixation and Sugar Synthesis

The second phase of photosynthesis, known as the Calvin Cycle or the light-independent reactions, takes place in the stroma of the chloroplast. This phase does not directly depend on sunlight but relies on the ATP and NADPH produced during the light-dependent reactions. The Calvin Cycle is responsible for fixing carbon dioxide (CO_2) from the atmosphere and converting it into organic compounds, specifically glucose, which serves as an energy source for the plant and other organisms.

Carbon Fixation

The Calvin Cycle begins with carbon fixation, a process by which carbon dioxide molecules are incorporated into an organic molecule. This step is catalyzed by the enzyme ribulose-1,5-bisphosphate carboxylase/oxygenase (commonly known as RuBisCO). RuBisCO attaches a molecule of CO_2 to ribulose-1,5-bisphosphate (RuBP), a five-carbon sugar. This results in an unstable six-carbon compound, which quickly splits into two molecules of 3-phosphoglycerate (3-PGA), a three-carbon compound.

Reduction Phase

In the reduction phase, ATP and NADPH generated in the light-dependent reactions are used to convert 3-PGA into glyceraldehyde-3-phosphate (G3P). First, ATP donates a phosphate group to 3-PGA, creating 1,3-bisphosphoglycerate. Next, NADPH donates electrons to 1,3-bisphosphoglycerate, reducing it to G3P. This molecule is an important intermediate in the production of glucose and other carbohydrates. For every six molecules of G3P produced, only one is used to synthesize glucose and other carbohydrates. The remaining five molecules of G3P are recycled to regenerate RuBP, allowing the Calvin Cycle to continue.

Regeneration of RuBP

The final phase of the Calvin Cycle involves the regeneration of RuBP, the molecule that accepts CO_2 during carbon fixation. Five molecules of G3P are rearranged, using ATP, to form three molecules of RuBP. This regeneration is essential for the Calvin Cycle to repeat, enabling continuous carbon fixation and sugar synthesis. The cycle must turn three times to produce one molecule of G3P, and six times to produce one glucose molecule.

11.2.3 The Importance of Photosynthesis for Life on Earth

Photosynthesis is the foundation of life on Earth. It provides energy not only to plants but also to herbivores and the carnivores that consume them. The oxygen released as a byproduct of the light-dependent reactions is essential for the respiration of most living organisms. Additionally, photosynthesis plays a significant role in regulating the levels of carbon dioxide in the atmosphere, which has implications for the global climate and the carbon cycle.

Photosynthesis and the Carbon Cycle

Photosynthesis is a critical component of the carbon cycle, which involves the exchange of carbon between the atmosphere, oceans, and living organisms. By converting atmospheric CO_2 into organic compounds, photosynthesis helps regulate the concentration of greenhouse gases, mitigating the impact of human activities on global warming. Plants act as carbon sinks, absorbing CO_2 from the atmosphere and storing it in the form of biomass.

Human Dependence on Photosynthesis

Humans rely on photosynthesis for food, oxygen, and energy. Agriculture, the foundation of human civilization, is based on the cultivation of photosynthetic organisms such as crops. These plants not only provide food for humans but also feed livestock and other animals that humans depend on for protein. Furthermore, fossil fuels, which power much of the modern world, are derived from ancient photosynthetic organisms that stored solar energy in the form of organic matter millions of years ago.

11.2.4 Factors Affecting Photosynthesis

Photosynthesis is influenced by several environmental factors, including light intensity, carbon dioxide concentration, temperature, and water availability. Understanding these factors is crucial for improving crop yields, addressing food security, and mitigating the effects of climate change.

Light Intensity

The rate of photosynthesis increases with light intensity up to a certain point, beyond which it levels off. This is because chlorophyll molecules can only absorb a limited amount of light at a time. In shaded environments or during cloudy days, photosynthetic rates may decrease, affecting plant growth.

Carbon Dioxide Concentration

Carbon dioxide is a substrate for the Calvin Cycle, and its availability can limit the rate of photosynthesis. In environments with low CO_2 levels, plants may struggle to fix sufficient carbon, leading to reduced growth and productivity. Conversely, elevated CO_2 concentrations can enhance photosynthesis in some plants, a phenomenon known as the CO_2 fertilization effect.

Temperature

Photosynthesis is temperature-dependent, with each plant species having an optimal temperature range for photosynthetic activity. Extreme temperatures can denature enzymes like RuBisCO, impairing the plant's ability to fix carbon and produce energy. Climate change, which alters global temperature patterns, may have significant consequences for photosynthesis and agricultural productivity.

Water Availability

Water is essential for the light-dependent reactions, particularly in the process of photolysis, where water molecules are split to provide electrons and protons. Drought conditions can reduce the availability of water, leading to stomatal closure in plants. This limits CO_2 uptake and reduces photosynthetic rates, ultimately affecting plant growth and crop yields.

11.3 MODELING PHOTOSYNTHESIS

Photosynthesis is one of the most important biological processes on Earth, responsible for converting light energy into chemical energy in plants, algae, and certain bacteria. By transforming carbon dioxide (CO_2) and water (H_2O) into glucose and oxygen (O_2), photosynthesis plays a crucial role in the global carbon cycle, the production of oxygen, and the sustainability of life.

Due to its complexity, photosynthesis has been the subject of various theoretical models, which help scientists better understand and simulate its mechanisms. These models are essential for studying plant behavior, enhancing crop productivity, addressing climate change, and optimizing artificial photosynthetic systems.

Modeling photosynthesis is a critical area of research that bridges biology, chemistry, physics, and environmental science. By simulating the complex reactions and interactions involved in photosynthesis, models help researchers understand how plants convert light into energy, respond to environmental changes, and contribute to global carbon cycling. These models have wide-ranging applications, from improving agricultural productivity and studying climate change to developing artificial photosynthesis systems. While challenges remain, continued advancements in biochemical, gas exchange, canopy, and dynamic models, as well as the integration of genetic data and machine learning, hold great promise for the future of photosynthesis research.

11.3.1 Importance of Modeling Photosynthesis

The complexity of photosynthesis necessitates the development of models that capture the intricate interactions between light, carbon, water, and biochemical processes. These models are invaluable for several reasons:

Enhancing Agricultural Productivity

Photosynthesis models help improve agricultural practices by providing insights into how plants respond to varying environmental conditions. By simulating different light intensities, temperatures, and CO_2 concentrations, researchers can identify strategies to optimize crop yields. This is especially important in the context of global food security, where rising populations and changing climates demand more efficient agricultural systems.

Climate Change Research

Photosynthesis is a major player in the global carbon cycle, removing CO_2 from the atmosphere and storing it in plant biomass. Understanding how photosynthesis is influenced by climate factors such as temperature and atmospheric CO_2 levels is critical for predicting the effects of climate change. Modeling photosynthesis enables scientists to assess the impact of rising CO_2 on plant growth and atmospheric carbon sequestration, helping to inform policy decisions related to climate mitigation.

Development of Artificial Photosynthesis

Research into artificial photosynthesis seeks to mimic natural photosynthesis in order to create sustainable energy sources. Models of photosynthetic processes can guide the design of systems that efficiently convert solar energy into fuels, reducing our dependence on fossil fuels and mitigating greenhouse gas emissions.

Exploring Ecosystem Dynamics

At a broader scale, photosynthesis models are used to study ecosystem dynamics. By integrating photosynthesis into models of forest growth, carbon flux, and plant competition, scientists can predict how ecosystems will respond to environmental changes, such as deforestation, pollution, or global warming.

11.3.2 Types of Photosynthesis Models

Various types of models have been developed to simulate different aspects of photosynthesis. These models range from simple representations of individual reactions to more comprehensive models that account for the entire photosynthetic pathway, as well as the interactions between plants and their environment.

Biochemical Models

Biochemical models focus on the individual reactions involved in photosynthesis, particularly the Calvin Cycle and the electron transport chain. These models describe the kinetics of enzyme-catalyzed reactions, such as RuBisCO's role in carbon fixation, and the transfer of electrons during the light-dependent reactions. By modeling these processes at the molecular level, researchers can investigate how changes in enzyme activity, substrate availability, or temperature affect overall photosynthetic efficiency.

One of the most well-known biochemical models is the Farquhar, von Caemmerer, and Berry (FvCB) model, which simulates the Calvin Cycle based on three limiting factors: the rate of RuBisCO carboxylation, the rate of electron transport, and the regeneration of RuBP (ribulose bisphosphate). This model is widely used in plant physiology to predict photosynthetic rates under varying environmental conditions.

Gas Exchange Models

Gas exchange models simulate the movement of CO_2 and O_2 between plants and the atmosphere. These models are particularly useful for studying how stomatal conductance—the rate at which gases enter and exit the leaf—affects photosynthesis. Stomata are small pores on the surface of leaves that regulate gas exchange, and their opening and closing can significantly influence the rate of carbon fixation.

The Ball-Berry model is a commonly used gas exchange model that relates stomatal conductance to photosynthesis, humidity, and CO_2 concentration. Such models help researchers explore the trade-offs plants face between maximizing carbon intake and minimizing water loss, which is critical for understanding plant responses to drought or increased CO_2 levels.

Canopy and Ecosystem Models

While biochemical and gas exchange models focus on individual leaves or plants, canopy and ecosystem models take a larger-scale approach. These models simulate photosynthesis at the level of plant communities or entire ecosystems, accounting for factors such as light distribution within a canopy, plant competition, and nutrient availability. Canopy models are particularly useful for predicting how forest growth and crop productivity will respond to environmental changes.

The Soil-Plant-Atmosphere (SPA) model is one example of an ecosystem model that integrates photosynthesis, water uptake, and transpiration processes. It simulates how energy and carbon flow through ecosystems, helping scientists predict how land-use changes, climate variations, and agricultural practices impact carbon sequestration and ecosystem health.

Dynamic Models

Dynamic models simulate the temporal changes in photosynthesis in response to fluctuating environmental conditions. Unlike static models, which provide a snapshot of photosynthesis at a given moment, dynamic models capture the diurnal (day-night) cycles of light and temperature that plants experience in nature. This type of modeling is essential for understanding how plants acclimate to changing environments over time.

Dynamic models can be used to explore how photosynthesis is regulated by circadian rhythms, the internal biological clocks that synchronize plant metabolism with the external environment. These models also help identify potential bottlenecks in photosynthetic efficiency, such as periods of low light or high temperatures, which can reduce carbon fixation.

11.3.3 Challenges and Future Directions in Photosynthesis Modeling

While significant progress has been made in developing models of photosynthesis, several challenges remain. One of the key difficulties is the high level of variability in photosynthetic efficiency across different species, environmental conditions, and developmental stages. Models that work well for one plant species may not be directly applicable to others, especially when comparing C3, C4, and CAM (crassulacean acid metabolism) plants, which use different mechanisms for carbon fixation.

Additionally, most current models do not fully account for the effects of nutrient limitations, such as nitrogen and phosphorus availability, on photosynthesis. Incorporating these factors into models will be crucial for accurately predicting plant responses to nutrient-deficient soils or changing environmental conditions.

Another area of ongoing research is the integration of genetic information into photosynthesis models. With the advent of high-throughput sequencing technologies, scientists can now study how genetic variation affects photosynthetic traits. By incorporating genetic data into models, researchers hope to identify genes that enhance photosynthetic efficiency, opening up new possibilities for crop improvement through genetic engineering.

Finally, advancements in computational power and machine learning offer exciting opportunities for improving the accuracy and scalability of photosynthesis models. Machine learning algorithms can analyze large datasets and identify patterns that are difficult to capture with traditional models. This could lead to the development of more robust and versatile models that can simulate photosynthesis across different species, ecosystems, and environmental conditions.

CHAPTER 12

THE BIOCHEMISTRY OF ZINC

12.1 AN OVERVIEW OF ZINC

Zinc is a chemical element with the symbol Zn and atomic number 30. It is a slightly brittle metal at room temperature and has a shiny-greyish appearance when oxidation is removed. It is the first element in group 12 (IIB) of the periodic table. In some respects, zinc is chemically similar to magnesium: both elements exhibit only one normal oxidation state (+2), and the Zn^{2+} and Mg^{2+} ions are of similar size. Zinc is the 24th most abundant element in Earth's crust and has five stable isotopes. The most common zinc ore is sphalerite (zinc blende), a zinc sulfide mineral. The largest workable lodes are in Australia, Asia, and the United States. Zinc is refined by froth flotation of the ore, roasting, and final extraction using electricity (electrowinning).

Zinc is an essential trace element for humans, animals, plants and for microorganisms and is necessary for prenatal and postnatal development. It is the second most abundant trace metal in humans after iron and it is the only metal which appears in all enzyme classes. Zinc is also an essential nutrient element for coral growth as it is an important cofactor for many enzymes.

Zinc deficiency affects about two billion people in the developing world and is associated with many diseases. In children, deficiency causes growth retardation, delayed sexual maturation, infection susceptibility, and diarrhea. Enzymes with a zinc atom in the reactive center are widespread in biochemistry, such as alcohol dehydrogenase in humans. Consumption of excess zinc may cause ataxia, lethargy, and copper deficiency. In marine biomes, notably within polar regions, a deficit of zinc can compromise the vitality of primary algal communities, potentially destabilizing the intricate marine trophic structures and consequently impacting biodiversity.

Brass, an alloy of copper and zinc in various proportions, was used as early as the third millennium BC in the Aegean area and the region which currently includes Iraq, the United Arab Emirates, Kalmykia, Turkmenistan and Georgia. In the second millennium BC it was used in the regions currently including West India, Uzbekistan, Iran, Syria, Iraq, and Israel. Zinc metal was not produced on a large scale until the 12th century in India, though it was known to the ancient Romans and Greeks. The mines of Rajasthan have given definite evidence of zinc production going back to the 6th century BC. The oldest evidence of pure zinc comes from Zawar, in Rajasthan, as early as the 9th century AD when a distillation process was employed to make pure zinc. Alchemists burned zinc in air to form what they called "philosopher's wool" or "white snow".

The element was probably named by the alchemist Paracelsus after the German word *Zinke* (prong, tooth). German chemist Andreas Sigismund Marggraf is credited with discovering pure metallic zinc in 1746. Work by Luigi Galvani and Alessandro Volta uncovered the electrochemical properties of zinc by 1800. Corrosion-resistant zinc plating of iron (hot-dip galvanizing) is the major application for zinc. Other applications are in electrical batteries, small non-structural castings, and alloys such as brass. A variety of zinc compounds are commonly used, such as zinc carbonate and zinc gluconate (as dietary supplements), zinc chloride (in deodorants), zinc pyrithione (anti-dandruff shampoos), zinc sulfide (in luminescent paints), and dimethylzinc or diethylzinc in the organic laboratory.

12.1.1 Characteristics

Physical Properties

Zinc is a bluish-white, lustrous, diamagnetic metal, though most common commercial grades of the metal have a dull finish. It is somewhat less dense than iron and has a hexagonal crystal structure, with a distorted form of hexagonal close packing, in which each atom has six nearest neighbors (at 265.9 pm) in its own plane and six others at a greater distance of 290.6 pm. The metal is hard and brittle at most temperatures but becomes malleable between 100 and 150 °C. Above 210 °C, the metal becomes brittle again and can be pulverized by beating. Zinc is a fair conductor of electricity. For a metal, zinc has relatively low melting (419.5 °C) and boiling point (907 °C). The melting point is the lowest of all the d-block metals aside from mercury and cadmium; for this reason among others, zinc, cadmium, and mercury are often not considered to be transition metals like the rest of the d-block metals.

Many alloys contain zinc, including brass. Other metals long known to form binary alloys with zinc are aluminium, antimony, bismuth, gold, iron, lead, mercury, silver, tin, magnesium, cobalt, nickel, tellurium, and sodium. Although neither zinc nor zirconium is ferromagnetic, their alloy, $ZrZn_2$, exhibits ferromagnetism below 35 K.

Occurrence

Zinc makes up about 75 ppm (%0.0075) of Earth's crust, making it the 24th most abundant element. It also makes up 312 ppm of the solar system, where it is the

22nd most abundant element. Typical background concentrations of zinc do not exceed 1 μg/m^3 in the atmosphere; 300 mg/kg in soil; 100 mg/kg in vegetation; 20 μg/L in freshwater and 5 μg/L in seawater. The element is normally found in association with other base metals such as copper and lead in ores. Zinc is a chalcophile, meaning the element is more likely to be found in minerals together with sulfur and other heavy chalcogens, rather than with the light chalcogen oxygen or with non-chalcogen electronegative elements such as the halogens. Sulfides formed as the crust solidified under the reducing conditions of the early Earth's atmosphere. Sphalerite, which is a form of zinc sulfide, is the most heavily mined zinc-containing ore because its concentrate contains 60–62% zinc.

Other source minerals for zinc include smithsonite (zinc carbonate), hemimorphite (zinc silicate), wurtzite (another zinc sulfide), and sometimes hydrozincite (basic zinc carbonate). With the exception of wurtzite, all these other minerals were formed by weathering of the primordial zinc sulfides.

Identified world zinc resources total about 1.9–2.8 billion tonnes. Large deposits are in Australia, Canada and the United States, with the largest reserves in Iran. The most recent estimate of reserve base for zinc (meets specified minimum physical criteria related to current mining and production practices) was made in 2009 and calculated to be roughly 480 Mt. Zinc reserves, on the other hand, are geologically identified ore bodies whose suitability for recovery is economically based (location, grade, quality, and quantity) at the time of determination. Since exploration and mine development is an ongoing process, the amount of zinc reserves is not a fixed number and sustainability of zinc ore supplies cannot be judged by simply extrapolating the combined mine life of today's zinc mines. This concept is well supported by data from the United States Geological Survey (USGS), which illustrates that although refined zinc production increased 80% between 1990 and 2010, the reserve lifetime for zinc has remained unchanged. About 346 million tonnes have been extracted throughout history to 2002, and scholars have estimated that about 109–305 million tonnes are in use.

Isotopes

Five stable isotopes of zinc occur in nature, with ^{64}Zn being the most abundant isotope (49.17% natural abundance). The other isotopes found in nature are ^{66}Zn (27.73%), ^{67}Zn (4.04%), ^{68}Zn (18.45%), and ^{70}Zn (0.61%).

Several dozen radioisotopes have been characterized. ^{65}Zn, which has a half-life of 243.66 days, is the least active radioisotope, followed by ^{72}Zn with a half-life of 46.5 hours. Zinc has 10 nuclear isomers, of which ^{69m}Zn has the longest half-life, 13.76 h. The superscript *m* indicates a metastable isotope. The nucleus of a metastable isotope is in an excited state and will return to the ground state by emitting a photon in the form of a gamma ray. ^{61}Zn has three excited metastable states and ^{73} Zn has two. The isotopes ^{65}Zn, ^{71}Zn, ^{77}Zn and ^{78}Zn each have only one excited metastable state.

The most common decay mode of a radioisotope of zinc with a mass number lower than 66 is electron capture. The decay product resulting from electron capture is an isotope of copper.

$$^{n}_{30}Zn + e^{-} \rightarrow {}^{n}_{29}Cu$$

The most common decay mode of a radioisotope of zinc with mass number higher than 66 is beta decay (β^{-}), which produces an isotope of gallium.

$$^{n}_{30}Zn \rightarrow {}^{n}_{31}Ga + e^{-} + \nu_{e}$$

12.1.2 Compounds and Chemistry

Reactivity

Zinc has an electron configuration of $[Ar]3d^{10}4s^{2}$ and is a member of the group 12 of the periodic table. It is a moderately reactive metal and strong reducing agent. The surface of the pure metal tarnishes quickly, eventually forming a protective passivating layer of the basic zinc carbonate, $Zn_5(OH)_6(CO3)_2$, by reaction with atmospheric carbon dioxide.

Zinc burns in air with a bright bluish-green flame, giving off fumes of zinc oxide. Zinc reacts readily with acids, alkalis and other non-metals. Extremely pure zinc reacts only slowly at room temperature with acids. Strong acids, such as hydrochloric or sulfuric acid, can remove the passivating layer and the subsequent reaction with the acid releases hydrogen gas.

The chemistry of zinc is dominated by the +2 oxidation state. When compounds in this oxidation state are formed, the outer shell *s* electrons are lost, yielding a bare zinc ion with the electronic configuration $[Ar]3d^{10}$. In aqueous solution an octahedral complex, $[Zn(H_2O)_6]^{2+}$ is the predominant species. The volatilization of zinc in combination with zinc chloride at temperatures above 285 °C indicates the formation of Zn_2Cl_2, a zinc compound with a +1 oxidation state. No compounds of zinc in positive oxidation states other than +1 or +2 are known. Calculations indicate that a zinc compound with the oxidation state of +4 is unlikely to exist. Zn(III) is predicted to exist in the presence of strongly electronegative trianions; however, there exists some doubt around this possibility. But in 2021 another compound was reported with more evidence that had the oxidation state of +3 with the formula $ZnBeB_{11}(CN)_{12}$. Zinc chemistry is similar to the chemistry of the late first-row transition metals, nickel and copper, though it has a filled d-shell and compounds are diamagnetic and mostly colorless. The ionic radii of zinc and magnesium happen to be nearly identical. Because of this some of the equivalent salts have the same crystal structure, and in other circumstances where ionic radius is a determining factor, the chemistry of zinc has much in common with that of magnesium. In other respects, there is little similarity with the late first-row transition metals. Zinc tends to form bonds with a greater degree of covalency and much more stable complexes with N- and S- donors. Complexes of zinc are mostly 4- or 6- coordinate, although 5-coordinate complexes are known.

Zinc(I) Compounds

Zinc(I) compounds are very rare. The $[Zn_2]^{2+}$ ion is implicated by the formation of a yellow diamagnetic glass by dissolving metallic zinc in molten $ZnCl_2$. The $[Zn_2]^{2+}$ core would be analogous to the $[Hg_2]^{2+}$ cation present in mercury(I) compounds. The diamagnetic nature of the ion confirms its dimeric structure. The first zinc(I) compound containing the Zn–Zn bond, $(\eta^5\text{-}C_5Me_5)_2Zn_2$.

Zinc(II) Compounds

Binary compounds of zinc are known for most of the metalloids and all the nonmetals except the noble gases. The oxide ZnO is a white powder that is nearly insoluble in neutral aqueous solutions, but is amphoteric, dissolving in both strong basic and acidic solutions. The other chalcogenides (ZnS, ZnSe, and ZnTe) have varied applications in electronics and optics. Pnictogenides (Zn_3N_2, Zn_3P_2, Zn_3As_2 and Zn_3Sb_2), the peroxide (ZnO_2), the hydride (ZnH_2), and the carbide (ZnC_2) are also known. Of the four halides, ZnF_2 has the most ionic character, while the others ($ZnCl_2$, $ZnBr_2$, and ZnI_2) have relatively low melting points and are considered to have more covalent character.

In weak basic solutions containing Zn^{2+} ions, the hydroxide $Zn(OH)_2$ forms as a white precipitate. In stronger alkaline solutions, this hydroxide is dissolved to form zincates ($[Zn(OH)_4]^{2-}$).

The nitrate $Zn(NO_3)_2$, chlorate $Zn(ClO_3)_2$, sulfate $ZnSO_4$, phosphate $Zn_3(PO_4)_2$, molybdate $ZnMoO_4$, cyanide $Zn(CN)_2$, arsenite $Zn(AsO_2)_2$, arsenate $Zn(AsO_4)_2{\cdot}8H_2O$ and the chromate $ZnCrO_4$ (one of the few colored zinc compounds) are a few examples of other common inorganic compounds of zinc.

Organozinc compounds are those that contain zinc–carbon covalent bonds. Diethylzinc ($(C_2H_5)_2Z_n$) is a reagent in synthetic chemistry. It was first reported in 1848 from the reaction of zinc and ethyl iodide, and was the first compound known to contain a metal–carbon sigma bond.

Test for Zinc

Cobalticyanide paper (Rinnmann's test for Zn) can be used as a chemical indicator for zinc. 4g of $K_3Co(CN)_6$ and 1g of $KClO_3$ is dissolved on 100 ml of water. Paper is dipped in the solution and dried at 100°C. One drop of the sample is dropped onto the dry paper and heated. A green disc indicates the presence of zinc.

12.1.3 History

Ancient Use

Various isolated examples of the use of impure zinc in ancient times have been discovered. Zinc ores were used to make the zinc–copper alloy brass thousands of years prior to the discovery of zinc as a separate element. Judean brass from the 14th to 10th centuries BC contains 23% zinc.

Knowledge of how to produce brass spread to Ancient Greece by the 7th century BC, but few varieties were made. Ornaments made of alloys containing 80–90% zinc, with lead, iron, antimony, and other metals making up the remainder, have been found that are 2,500 years old. A possibly prehistoric statuette containing 87.5% zinc was found in a Dacian archaeological site.

Strabo writing in the 1st century BC (but quoting a now lost work of the 4th century BC historian Theopompus) mentions "drops of false silver" which when mixed with copper make brass. This may refer to small quantities of zinc that is a by-product of smelting sulfide ores. Zinc in such remnants in smelting ovens was usually discarded as it was thought to be worthless.

The manufacture of brass was known to the Romans by about 30 BC. They made brass by heating powdered calamine (zinc silicate or carbonate), charcoal and copper together in a crucible. The resulting calamine brass was then either cast or hammered into shape for use in weaponry. Some coins struck by Romans in the Christian era are made of what is probably calamine brass.

The oldest known pills were made of the zinc carbonates hydrozincite and smithsonite. The pills were used for sore eyes and were found aboard the Roman ship Relitto del Pozzino, wrecked in 140 BC.

The Berne zinc tablet is a votive plaque dating to Roman Gaul made of an alloy that is mostly zinc.

The Charaka Samhita, thought to have been written between 300 and 500 AD, mentions a metal which, when oxidized, produces *pushpanjan*, thought to be zinc oxide. Zinc mines at Zawar, near Udaipur in India, have been active since the Mauryan period (c. 322 and 187 BCE).

The smelting of metallic zinc here, however, appears to have begun around the 12th century AD. One estimate is that this location produced an estimated million tonnes of metallic zinc and zinc oxide from the 12th to 16th centuries. Another estimate gives a total production of 60,000 tonnes of metallic zinc over this period. The Rasaratna Samuccaya, written in approximately the 13th century AD, mentions two types of zinc-containing ores: one used for metal extraction and another used for medicinal purposes.

Early Studies and Naming

Zinc was distinctly recognized as a metal under the designation of *Yasada* or Jasada in the medical Lexicon ascribed to the Hindu king Madanapala (of Taka dynasty) and written about the year 1374. Smelting and extraction of impure zinc by reducing calamine with wool and other organic substances was accomplished in the 13th century in India. The Chinese did not learn of the technique until the 17th century.

Alchemists burned zinc metal in air and collected the resulting zinc oxide on a condenser. Some alchemists called this zinc oxide *lana philosophica*, Latin for "philosopher's wool", because it collected in wooly tufts, whereas others thought it looked like white snow and named it *nix album*.

The name of the metal was probably first documented by Paracelsus, a Swiss-born German alchemist, who referred to the metal as "zincum" or "zinken" in his book *Liber Mineralium II*, in the 16th century. The word is probably derived from the German *zinke*, and supposedly meant "tooth-like, pointed or jagged" (metallic zinc crystals have a needle-like appearance). *Zink* could also imply "tin-like" because of its relation to German *zinn* meaning tin. Yet another possibility is that the word is derived from the Persian word *seng* meaning stone. The metal was also called *Indian tin*, *tutanego*, *calamine*, and *spinter*.

German metallurgist Andreas Libavius received a quantity of what he called "calay" (from the Malay or Hindi word for tin) originating from Malabar off a cargo ship captured from the Portuguese in the year 1596. Libavius described the properties of the sample, which may have been zinc. Zinc was regularly imported to Europe from the Orient in the 17th and early 18th centuries, but was at times very expensive.

Isolation

Metallic zinc was isolated in India by 1300 AD. Before it was isolated in Europe, it was imported from India in about 1600 CE. Postlewayt's *Universal Dictionary*, a contemporary source giving technological information in Europe, did not mention zinc before 1751 but the element was studied before then.

Flemish metallurgist and alchemist P. M. de Respour reported that he had extracted metallic zinc from zinc oxide in 1668. By the start of the 18th century, Étienne François Geoffroy described how zinc oxide condenses as yellow crystals on bars of iron placed above zinc ore that is being smelted. In Britain, John Lane is said to have carried out experiments to smelt zinc, probably at Landore, prior to his bankruptcy in 1726.

In 1738 in Great Britain, William Champion patented a process to extract zinc from calamine in a vertical retort-style smelter. His technique resembled that used at Zawar zinc mines in Rajasthan, but no evidence suggests he visited the Orient. Champion's process was used through 1851.

German chemist Andreas Marggraf normally gets credit for isolating pure metallic zinc in the West, even though Swedish chemist Anton von Swab had distilled zinc from calamine four years previously. In his 1746 experiment, Marggraf heated a mixture of calamine and charcoal in a closed vessel without copper to obtain a metal. This procedure became commercially practical by 1752.

Later Work

William Champion's brother, John, patented a process in 1758 for calcining zinc sulfide into an oxide usable in the retort process. Prior to this, only calamine could be used to produce zinc. In 1798, Johann Christian Ruberg improved on the smelting process by building the first horizontal retort smelter. Jean-Jacques Daniel Dony built a different kind of horizontal zinc smelter in Belgium that processed even more zinc. Italian doctor Luigi Galvani discovered in 1780 that connecting the spinal cord of a freshly dissected frog to an iron rail attached by a brass hook caused the frog's leg to twitch. He incorrectly thought he had discovered an ability of nerves and muscles to create electricity and called the effect "animal electricity". The galvanic cell and the process of galvanization were both named for Luigi Galvani, and his discoveries paved the way for electrical batteries, galvanization, and cathodic protection.

Galvani's friend, Alessandro Volta, continued researching the effect and invented the Voltaic pile in 1800. Volta's pile consisted of a stack of simplified galvanic cells, each being one plate of copper and one of zinc connected by an electrolyte. By stacking these units in series, the Voltaic pile (or "battery") as a whole had a higher voltage, which could be used more easily than single cells. Electricity is produced because the Volta potential between the two metal plates makes electrons flow from the zinc to the copper and corrode the zinc.

The non-magnetic character of zinc and its lack of color in solution delayed discovery of its importance to biochemistry and nutrition. This changed in 1940 when carbonic anhydrase, an enzyme that scrubs carbon dioxide from blood, was shown to have zinc in its active site. The digestive enzyme carboxypeptidase became the second known zinc-containing enzyme in 1955.

12.1.4 Production

Mining and Processing

Zinc is the fourth most common metal in use, trailing only iron, aluminium, and copper with an annual production of about 13 million tonnes. The world's largest zinc producer is Nyrstar, a merger of the Australian OZ Minerals and the Belgian Umicore. About 70% of the world's zinc originates from mining, while the remaining 30% comes from recycling secondary zinc.

Commercially pure zinc is known as Special High Grade, often abbreviated *SHG*, and is 99.995% pure.

Worldwide, 95% of new zinc is mined from sulfidic ore deposits, in which sphalerite (ZnS) is nearly always mixed with the sulfides of copper, lead and iron. Zinc mines are scattered throughout the world, with the main areas being China, Australia, and Peru. China produced 38% of the global zinc output in 2014.

Zinc metal is produced using extractive metallurgy. The ore is finely ground, then put through froth flotation to separate minerals from gangue (on the property of hydrophobicity), to get a zinc sulfide ore concentrate consisting of about 50% zinc, 32% sulfur, 13% iron, and 5% SiO_2.

Roasting converts the zinc sulfide concentrate to zinc oxide:

$$2\,ZnS + 3\,O_2 \xrightarrow{t^\circ} 2\,ZnO + 2\,SO_2$$

The sulfur dioxide is used for the production of sulfuric acid, which is necessary for the leaching process. If deposits of zinc carbonate, zinc silicate, or zinc-spinel (like the Skorpion Deposit in Namibia) are used for zinc production, the roasting can be omitted.

For further processing two basic methods are used: pyrometallurgy or electrowinning. Pyrometallurgy reduces zinc oxide with carbon or carbon monoxide at 950 °C (1,740 °F) into the metal, which is distilled as zinc vapor to separate it from other metals, which are not volatile at those temperatures. The zinc vapor is collected in a condenser. The equations below describe this process:

$$ZnO + C \xrightarrow{950^\circ C} Zn + CO$$

$$ZnO + CO \xrightarrow{950^\circ C} Zn + CO_2$$

In electrowinning, zinc is leached from the ore concentrate by sulfuric acid and impurities are precipitated:

$$ZnO + H_2SO_4 \longrightarrow ZnSO_4 + H_2O$$

Finally, the zinc is reduced by electrolysis.

$$2\,ZnSO_4 + 2\,H_2O \longrightarrow 2\,Zn + O_2 + 2\,H_2SO_4$$

The sulfuric acid is regenerated and recycled to the leaching step. When galvanised feedstock is fed to an electric arc furnace, the zinc is recovered from the dust by a number of processes, predominantly the Waelz process (90% as of 2014).

Environmental Impact

Refinement of sulfidic zinc ores produces large volumes of sulfur dioxide and cadmium vapor. Smelter slag and other residues contain significant quantities of metals. About 1.1 million tonnes of metallic zinc and 130 thousand tonnes of lead were mined and smelted in the Belgian towns of La Calamine and Plombières between 1806 and 1882. The dumps of the past mining operations leach zinc and cadmium, and the sediments of the Geul River contain non-trivial amounts of metals. About two thousand years ago, emissions of zinc from mining and smelting totaled 10 thousand tonnes a year. After increasing 10-fold from 1850, zinc emissions peaked at 3.4 million tonnes per year in the 1980s and declined to 2.7 million tonnes in the 1990s, although a 2005 study of the Arctic troposphere found that the concentrations there did not reflect the decline. Man-made and natural emissions occur at a ratio of 20 to 1.

Zinc in rivers flowing through industrial and mining areas can be as high as 20 ppm. Effective sewage treatment greatly reduces this; treatment along the Rhine, for example, has decreased zinc levels to 50 ppb. Concentrations of zinc as low as 2 ppm adversely affects the amount of oxygen that fish can carry in their blood.

Soils contaminated with zinc from mining, refining, or fertilizing with zinc-bearing sludge can contain several grams of zinc per kilogram of dry soil. Levels of zinc in excess of 500 ppm in soil interfere with the ability of plants to absorb other essential metals, such as iron and manganese. Zinc levels of 2000 ppm to 180,000 ppm (18%) have been recorded in some soil samples.

12.1.5 Applications

Major applications of zinc include, with percentages given for the US

- Galvanizing (55%).
- Brass and bronze (16%).
- Other alloys (21%).
- Miscellaneous (8%).

Anti-Corrosion and Batteries

Zinc is most commonly used as an anti-corrosion agent, and galvanization (coating of iron or steel) is the most familiar form. In 2009 in the United States, 55% or 893,000 tons of the zinc metal was used for galvanization.

Zinc is more reactive than iron or steel and thus will attract almost all local oxidation until it completely corrodes away. A protective surface layer of oxide and carbonate ($Zn_5(OH)_6(CO_3)_2$) forms as the zinc corrodes. This protection lasts even after the zinc layer is scratched but degrades through time as the zinc corrodes away. The zinc is applied electrochemically or as molten zinc by hot-dip galvanizing or spraying. Galvanization is used on chain-link fencing, guard rails, suspension bridges, lightposts, metal roofs, heat exchangers, and car bodies.

The relative reactivity of zinc and its ability to attract oxidation to itself makes it an efficient sacrificial anode in cathodic protection (CP). For example, cathodic protection of a buried pipeline can be achieved by connecting anodes made from zinc to the pipe. Zinc acts as the anode (negative terminus) by slowly corroding away as it passes electric current to the steel pipeline. Zinc is also used to cathodically protect metals that are exposed to sea water. A zinc disc attached to a ship's iron rudder will slowly corrode while the rudder stays intact. Similarly, a zinc plug attached to a propeller or the metal protective guard for the keel of the ship provides temporary protection.

With a standard electrode potential (SEP) of -0.76 volts, zinc is used as an anode material for batteries. (More reactive lithium (SEP -3.04 V) is used for anodes in lithium batteries). Powdered zinc is used in this way in alkaline batteries and the case (which also serves as the anode) of zinc–carbon batteries is formed from sheet zinc. Zinc is used as the anode or fuel of the zinc–air battery/fuel cell. The zinc-cerium redox flow battery also relies on a zinc-based negative half-cell.

Alloys

A widely used zinc alloy is brass, in which copper is alloyed with anywhere from 3% to 45% zinc, depending upon the type of brass. Brass is generally more ductile and stronger than copper, and has superior corrosion resistance. These properties make it useful in communication equipment, hardware, musical instruments, and water valves.

Other widely used zinc alloys include nickel silver, typewriter metal, soft and aluminium solder, and commercial bronze. Zinc is also used in contemporary pipe organs as a substitute for the traditional lead/tin alloy in pipes. Alloys of 85–88% zinc, 4–10% copper, and 2–8% aluminium find limited use in certain types of machine bearings. Zinc has been the primary metal in American one cent coins (pennies) since 1982. The zinc core is coated with a thin layer of copper to give the appearance of a copper coin. In 1994, 33,200 tonnes (36,600 short tons) of zinc were used to produce 13.6 billion pennies in the United States.

Alloys of zinc with small amounts of copper, aluminium, and magnesium are useful in die casting as well as spin casting, especially in the automotive, electrical, and hardware industries. These alloys are marketed under the name Zamak. An example of this is zinc aluminium.

The low melting point together with the low viscosity of the alloy makes possible the production of small and intricate shapes. The low working temperature leads to rapid cooling of the cast products and fast production for assembly. Another alloy, marketed under the brand name Prestal, contains 78% zinc and 22% aluminium, and is reported to be nearly as strong as steel but as malleable as plastic. This superplasticity of the alloy allows it to be molded using die casts made of ceramics and cement.

Similar alloys with the addition of a small amount of lead can be cold-rolled into sheets. An alloy of 96% zinc and 4% aluminium is used to make stamping dies for low production run applications for which ferrous metal dies would be too expensive. For building facades, roofing, and other applications for sheet metal formed by deep drawing, roll forming, or bending, zinc alloys with titanium and copper are used. Unalloyed zinc is too brittle for these manufacturing processes.

As a dense, inexpensive, easily worked material, zinc is used as a lead replacement. In the wake of lead concerns, zinc appears in weights for various applications ranging from fishing to tire balances and flywheels.

Cadmium zinc telluride (CZT) is a semiconductive alloy that can be divided into an array of small sensing devices. These devices are similar to an integrated circuit and can detect the energy of incoming gamma ray photons. When behind an absorbing mask, the CZT sensor array can determine the direction of the rays.

Other Industrial Uses

Roughly one quarter of all zinc output in the United States in 2009 was consumed in zinc compounds; a variety of which are used industrially. Zinc oxide is widely used as a white pigment in paints and as a catalyst in the manufacture of rubber to disperse heat. Zinc oxide is used to protect rubber polymers and plastics from ultraviolet radiation (UV). The semiconductor properties of zinc oxide make it useful in varistors and photocopying products. The zinc zinc-oxide cycle is a two step thermochemical process based on zinc and zinc oxide for hydrogen production.

Zinc chloride is often added to lumber as a fire retardant and sometimes as a wood preservative. It is used in the manufacture of other chemicals. Zinc methyl ($Zn(CH_3)_2$) is used in a number of organic syntheses. Zinc sulfide (ZnS) is used in luminescent pigments such as on the hands of clocks, X-ray and television screens, and luminous paints. Crystals of ZnS are used in lasers that operate in the mid-infrared part of the spectrum. Zinc sulfate is a chemical in dyes and pigments. Zinc pyrithione is used in antifouling paints.

Zinc powder is sometimes used as a propellant in model rockets. When a compressed mixture of 70% zinc and 30% sulfur powder is ignited there is a violent chemical reaction. This produces zinc sulfide, together with large amounts of hot gas, heat, and light.

Zinc sheet metal is used as a durable covering for roofs, walls, and countertops, the last often seen in bistros and oyster bars, and is known for the rustic look imparted by its surface oxidation in use to a blue-gray patina and susceptibility to scratching. ^{64}Zn, the most abundant isotope of zinc, is very susceptible to neutron activation, being transmuted into the highly radioactive ^{65}Zn, which has a half-life of 244 days and produces intense gamma radiation. Because of this, zinc oxide used in nuclear reactors as an anti-corrosion agent is depleted of ^{64}Zn before use, this is called depleted zinc oxide. For the same reason, zinc has been proposed as a salting material for nuclear weapons (cobalt is another, better-known salting material). A jacket of isotopically enriched ^{64}Zn would be irradiated by the intense high-energy neutron flux from an exploding thermonuclear weapon, forming a large amount of ^{65}Zn significantly increasing the radioactivity of the weapon's fallout. Such a weapon is not known to have ever been built, tested, or used.

^{65}Zn is used as a tracer to study how alloys that contain zinc wear out, or the path and the role of zinc in organisms.

Zinc dithiocarbamate complexes are used as agricultural fungicides; these include Zineb, Metiram, Propineb and Ziram. Zinc naphthenate is used as wood preservative. Zinc in the form of ZDDP, is used as an anti-wear additive for metal parts in engine oil.

Organic Chemistry

Organozinc chemistry is the science of compounds that contain carbon-zinc bonds, describing the physical properties, synthesis, and chemical reactions. Many organozinc compounds are commercially important. Among important applications are:

- The Frankland-Duppa Reaction in which an oxalate ester (ROCOCOOR) reacts with an alkyl halide R'X, zinc and hydrochloric acid to form α-hydroxycarboxylic esters RR'COHCOOR
- Organozincs have similar reactivity to Grignard reagents but are much less nucleophilic, and they are expensive and difficult to handle. Organozincs typically perform nucleophilic addition on electrophiles such as aldehydes, which are then reduced to alcohols. Commercially available diorganozinc compounds include dimethylzinc, diethylzinc and diphenylzinc. Like Grignard reagents, organozincs are commonly produced from organobromine precursors.

Zinc has found many uses in catalysis in organic synthesis including enantioselective synthesis, being a cheap and readily available alternative to precious metal complexes. Quantitative results (yield and enantiomeric excess) obtained with chiral zinc catalysts can be comparable to those achieved with palladium, ruthenium, iridium and others.

Dietary Supplement

In most single-tablet, over-the-counter, daily vitamin and mineral supplements, zinc is included in such forms as zinc oxide, zinc acetate, zinc gluconate, or zinc amino acid chelate. Generally, zinc supplement is recommended where there is high risk of zinc deficiency (such as low and middle income countries) as a preventive measure. Although zinc sulfate is a commonly used zinc form, zinc citrate, gluconate and picolinate may be valid options as well. These forms are better absorbed than zinc oxide.

Gastroenteritis

Zinc is an inexpensive and effective part of treatment of diarrhea among children in the developing world. Zinc becomes depleted in the body during diarrhea and replenishing zinc with a 10- to 14-day course of treatment can reduce the duration and severity of diarrheal episodes and may also prevent future episodes for as long as three months. Gastroenteritis is strongly attenuated by ingestion of zinc, possibly by direct antimicrobial action of the ions in the gastrointestinal tract, or by the absorption of the zinc and re-release from immune cells (all granulocytes secrete zinc), or both.

Common Cold

Zinc supplements (frequently zinc acetate or zinc gluconate lozenges) are a group of dietary supplements that are commonly used in an attempt to treat the common cold. Evidence of any benefit is weak, but it seems zinc may not prevent colds, but possibly reduce their length while not reducing the severity of the symptoms. Adverse effects with zinc supplements by mouth include bad taste and nausea. The intranasal use of zinc-containing nasal sprays has been associated with the loss of the sense of smell; consequently, in June 2009, the United States Food and Drug Administration (USFDA) warned consumers to stop using intranasal zinc.

The human rhinovirus – the most common viral pathogen in humans – is the predominant cause of the common cold. The hypothesized mechanism of action by which zinc reduces the severity and/or duration of cold symptoms is the suppression of nasal inflammation and the direct inhibition of rhinoviral receptor binding and rhinoviral replication in the nasal mucosa.

Weight Gain

Zinc deficiency may lead to loss of appetite. The use of zinc in the treatment of anorexia has been advocated since 1979. At least 15 clinical trials have shown that zinc improved weight gain in anorexia. A 1994 trial showed that zinc doubled the rate of body mass increase in the treatment of anorexia nervosa. Deficiency of other nutrients such as tyrosine, tryptophan and thiamine could contribute to this phenomenon of "malnutrition-induced malnutrition". A meta-analysis of 33 prospective intervention trials regarding zinc supplementation and its effects on the growth of children in many countries showed that zinc supplementation alone had a statistically significant effect on linear growth and body weight gain, indicating that other deficiencies that may have been present were not responsible for growth retardation.

Other

A Cochrane review stated that people taking zinc supplement may be less likely to progress to age-related macular degeneration. Zinc supplement is an effective treatment for acrodermatitis enteropathica, a genetic disorder affecting zinc absorption that was previously fatal to affected infants. Zinc deficiency has been associated with major depressive disorder (MDD), and zinc supplements may be an effective treatment. Zinc may help individuals sleep more.

Topical Use

Topical preparations of zinc include those used on the skin, often in the form of zinc oxide. Zinc oxide is generally recognized by the FDA as safe and effective and is considered a very photo-stable. Zinc oxide is one of the most common active ingredients formulated into a sunscreen to mitigate sunburn. Applied thinly to a baby's diaper area (perineum) with each diaper change, it can protect against diaper rash.

Chelated zinc is used in toothpastes and mouthwashes to prevent bad breath; zinc citrate helps reduce the build-up of calculus (tartar). Zinc pyrithione is widely included in shampoos to prevent dandruff. Topical zinc has also been shown to effectively treat, as well as prolong remission in genital herpes.

12.1.6 Biological Role

Zinc is an essential trace element for humans. and other animals, for plants and for microorganisms. Zinc is required for the function of over 300 enzymes and 1000 transcription factors, and is stored and transferred in metallothioneins. It is the second most abundant trace metal in humans after iron and it is the only metal which appears in all enzyme classes.

In proteins, zinc ions are often coordinated to the amino acid side chains of aspartic acid, glutamic acid, cysteine and histidine. The theoretical and computational description of this zinc binding in proteins (as well as that of other transition metals) is difficult.

Roughly 2–4 grams of zinc are distributed throughout the human body. Most zinc is in the brain, muscle, bones, kidney, and liver, with the highest concentrations in the prostate and parts of the eye. Semen is particularly rich in zinc, a key factor in prostate gland function and reproductive organ growth. Zinc homeostasis of the body is mainly controlled by the intestine. Here, ZIP4 and especially TRPM7 were linked to intestinal zinc uptake essential for postnatal survival.

In humans, the biological roles of zinc are ubiquitous. It interacts with "a wide range of organic ligands", and has roles in the metabolism of RNA and DNA, signal transduction, and gene expression. It also regulates apoptosis. A review from 2015 indicated that about 10% of human proteins (~3000) bind zinc, in addition to hundreds more that transport and traffic zinc; a similar *in silico* study in the plant *Arabidopsis thaliana* found 2367 zinc-related proteins.

In the brain, zinc is stored in specific synaptic vesicles by glutamatergic neurons and can modulate neuronal excitability. It plays a key role in synaptic plasticity and so in learning. Zinc homeostasis also plays a critical role in the functional regulation of the central nervous system. Dysregulation of zinc homeostasis in the central nervous system that results in excessive synaptic zinc concentrations is believed to induce neurotoxicity through mitochondrial oxidative stress (e.g., by disrupting certain enzymes involved in the electron transport chain, including complex I, complex III, and α-ketoglutarate dehydrogenase), the dysregulation of calcium homeostasis, glutamatergic neuronal excitotoxicity, and interference with intraneuronal signal transduction. L- and D-histidine facilitate brain zinc uptake. SLC30A3 is the primary zinc transporter involved in cerebral zinc homeostasis.

Enzymes

Zinc is an efficient Lewis acid, making it a useful catalytic agent in hydroxylation and other enzymatic reactions. The metal also has a flexible coordination geometry, which allows proteins using it to rapidly shift conformations to perform biological reactions. Two examples of zinc-containing enzymes are carbonic anhydrase and carboxypeptidase, which are vital to the processes of carbon dioxide (CO_2) regulation and digestion of proteins, respectively.

In vertebrate blood, carbonic anhydrase converts CO_2 into bicarbonate and the same enzyme transforms the bicarbonate back into CO_2 for exhalation through the lungs. Without this enzyme, this conversion would occur about one million times slower at the normal blood pH of 7 or would require a pH of 10 or more. The non-related β-carbonic anhydrase is required in plants for leaf formation, the synthesis of indole acetic acid (auxin) and alcoholic fermentation. Carboxypeptidase cleaves peptide linkages during digestion of proteins. A coordinate covalent bond is formed between the terminal peptide and a C=O group attached to zinc, which gives the carbon a positive charge. This helps to create a hydrophobic pocket on the enzyme near the zinc, which attracts the non-polar part of the protein being digested.

Signalling

Zinc has been recognized as a messenger, able to activate signalling pathways. Many of these pathways provide the driving force in aberrant cancer growth. They can be targeted through ZIP transporters.

Other Proteins

Zinc serves a purely structural role in zinc fingers, twists and clusters. Zinc fingers form parts of some transcription factors, which are proteins that recognize DNA base sequences during the replication and transcription of DNA. Each of the nine or ten Zn^{2+} ions in a zinc finger helps maintain the finger's structure by coordinately binding to four amino acids in the transcription factor.

In blood plasma, zinc is bound to and transported by albumin (60%, low-affinity) and transferrin (10%). Because transferrin also transports iron, excessive iron reduces zinc absorption, and vice versa. A similar antagonism exists with copper. The concentration of zinc in blood plasma stays relatively constant regardless of zinc intake. Cells in the salivary gland, prostate, immune system, and intestine use zinc signaling to communicate with other cells.

Zinc may be held in metallothionein reserves within microorganisms or in the intestines or liver of animals. Metallothionein in intestinal cells is capable of adjusting absorption of zinc by 15–40%. However, inadequate or excessive zinc intake can be harmful; excess zinc particularly impairs copper absorption because metallothionein absorbs both metals.

The human dopamine transporter contains a high affinity extracellular zinc binding site which, upon zinc binding, inhibits dopamine reuptake and amplifies amphetamine-induced dopamine efflux *in vitro*. The human serotonin transporter and norepinephrine transporter do not contain zinc binding sites. Some EF-hand calcium binding proteins such as S100 or NCS-1 are also able to bind zinc ions.

Nutrition

Dietary Recommendations

The U.S. Institute of Medicine (IOM) updated Estimated Average Requirements (EARs) and Recommended Dietary Allowances (RDAs) for zinc in 2001. The current EARs for zinc for women and men ages 14 and up is 6.8 and 9.4 mg/day, respectively. The RDAs are 8 and 11 mg/day. RDAs are higher than EARs so as to identify amounts that will cover people with higher than average requirements. RDA for pregnancy is 11 mg/day. RDA for lactation is 12 mg/day. For infants up to 12 months the RDA is 3 mg/day. For children ages 1–13 years the RDA increases with age from 3 to 8 mg/day. As for safety, the IOM sets Tolerable upper intake levels (ULs) for vitamins and minerals when evidence is sufficient. In the case of zinc the adult UL is 40 mg/day including both food and supplements combined (lower for children). Collectively the EARs, RDAs, AIs and ULs are referred to as Dietary Reference Intakes (DRIs).

The European Food Safety Authority (EFSA) refers to the collective set of information as Dietary Reference Values, with Population Reference Intake (PRI) instead of RDA, and Average Requirement instead of EAR. AI and UL are defined the same as in the United States. For people ages 18 and older the PRI calculations are complex, as the EFSA has set higher and higher values as the phytate content of the diet increases. For women, PRIs increase from 7.5 to 12.7 mg/day as phytate intake increases from 300 to 1200 mg/day; for men the range is 9.4 to 16.3 mg/day. These PRIs are higher than the U.S. RDAs. The EFSA reviewed the same safety question and set its UL at 25 mg/day, which is much lower than the U.S. value.

For U.S. food and dietary supplement labeling purposes the amount in a serving is expressed as a percent of Daily Value (%DV). For zinc labeling purposes 100% of the Daily Value was 15 mg, but on May 27, 2016, it was revised to 11 mg. A table of the old and new adult daily values is provided at Reference Daily Intake.

Dietary Intake

Animal products such as meat, fish, shellfish, fowl, eggs, and dairy contain zinc. The concentration of zinc in plants varies with the level in the soil. With adequate zinc in the soil, the food plants that contain the most zinc are wheat (germ and bran) and various seeds, including sesame, poppy, alfalfa, celery, and mustard. Zinc is also found in beans, nuts, almonds, whole grains, pumpkin seeds, sunflower seeds, and blackcurrant.

Other sources include fortified food and dietary supplements in various forms. A 1998 review concluded that zinc oxide, one of the most common supplements in the United States, and zinc carbonate are nearly insoluble and poorly absorbed in the body. This review cited studies that found lower plasma zinc concentrations in the subjects who consumed zinc oxide and zinc carbonate than in those who took zinc acetate and sulfate salts. For fortification, however, a 2003 review recommended cereals (containing zinc oxide) as a cheap, stable source that is as easily absorbed as the more expensive forms. A 2005 study found that various compounds of zinc, including oxide and sulfate, did not show statistically significant differences in absorption when added as fortificants to maize tortillas.

Deficiency

Nearly two billion people in the developing world are deficient in zinc. Groups at risk include children in developing countries and elderly with chronic illnesses. In children, it causes an increase in infection and diarrhea and contributes to the death of about 800,000 children worldwide per year. The World Health Organization advocates zinc supplementation for severe malnutrition and diarrhea. Zinc supplements help prevent disease and reduce mortality, especially among children with low birth weight or stunted growth. However, zinc supplements should not be administered alone, because many in the developing world have several deficiencies, and zinc interacts with other micronutrients. While zinc deficiency is usually due to insufficient dietary intake, it can be associated with malabsorption, acrodermatitis enteropathica, chronic liver disease, chronic renal disease, sickle cell disease, diabetes, malignancy, and other chronic illnesses.

In the United States, a federal survey of food consumption determined that for women and men over the age of 19, average consumption was 9.7 and 14.2mg/day, respectively. For women, 17% consumed less than the EAR, for men 11%. The percentages below EAR increased with age. The most recent published update of the survey (NHANES 2013–2014) reported lower averages – 9.3 and 13.2mg/day – again with intake decreasing with age.

Symptoms of mild zinc deficiency are diverse. Clinical outcomes include depressed growth, diarrhea, impotence and delayed sexual maturation, alopecia, eye and skin lesions, impaired appetite, altered cognition, impaired immune functions, defects in carbohydrate use, and reproductive teratogenesis. Zinc deficiency depresses immunity, but excessive zinc does also.

Despite some concerns, western vegetarians and vegans do not suffer any more from overt zinc deficiency than meat-eaters. Major plant sources of zinc include cooked dried beans, sea vegetables, fortified cereals, soy foods, nuts, peas, and seeds. However, phytates in many whole-grains and fibers may interfere with zinc absorption and marginal zinc intake has poorly understood effects. The zinc chelator phytate, found in seeds and cereal bran, can contribute to zinc malabsorption. Some evidence suggests that more than the US RDA (8 mg/day for adult women; 11 mg/day for adult men) may be needed in those whose diet is high in phytates, such as some vegetarians. The European Food Safety Authority (EFSA) guidelines attempt to compensate for this by recommending higher zinc intake when dietary phytate intake is greater. These considerations must be balanced against the paucity of adequate zinc biomarkers, and the most widely used indicator, plasma zinc, has poor sensitivity and specificity.

Soil Remediation

Species of *Calluna*, *Erica* and *Vaccinium* can grow in zinc-metalliferous soils, because translocation of toxic ions is prevented by the action of ericoid mycorrhizal fungi.

Agriculture

Zinc deficiency appears to be the most common micronutrient deficiency in crop plants; it is particularly common in high-pH soils. Zinc-deficient soil is cultivated in the cropland of about half of Turkey and India, a third of China, and most of Western Australia. Substantial responses to zinc fertilization have been reported in these areas. Plants that grow in soils that are zinc-deficient are more susceptible to disease. Zinc is added to the soil primarily through the weathering of rocks, but humans have added zinc through fossil fuel combustion, mine waste, phosphate fertilizers, pesticide (zinc phosphide), limestone, manure, sewage sludge, and particles from galvanized surfaces. Excess zinc is toxic to plants, although zinc toxicity is far less widespread.

12.1.7 Precautions

Toxicity

Although zinc is an essential requirement for good health, excess zinc can be harmful. Excessive absorption of zinc suppresses copper and iron absorption. The free zinc ion in solution is highly toxic to plants, invertebrates, and even vertebrate fish. The Free Ion Activity Model is well-established in the literature, and shows that just micromolar amounts of the free ion kills some organisms. A recent example showed 6 micromolar killing 93% of all *Daphnia* in water.

The free zinc ion is a powerful Lewis acid up to the point of being corrosive. Stomach acid contains hydrochloric acid, in which metallic zinc dissolves readily to give corrosive zinc chloride. Swallowing a post-1982 American one cent piece (97.5% zinc) can cause damage to the stomach lining through the high solubility of the zinc ion in the acidic stomach.

Evidence shows that people taking 100–300 mg of zinc daily may suffer induced copper deficiency. A 2007 trial observed that elderly men taking 80 mg daily were hospitalized for urinary complications more often than those taking a placebo. Levels of 100–300 mg may interfere with the use of copper and iron or adversely affect cholesterol. Zinc in excess of 500 ppm in soil interferes with the plant absorption of other essential metals, such as iron and manganese. A condition called the zinc shakes or "zinc chills" can be induced by inhalation of zinc fumes while brazing or welding galvanized materials. Zinc is a common ingredient of denture cream which may contain between 17 and 38 mg of zinc per gram. Disability and even deaths from excessive use of these products have been claimed.

The U.S. Food and Drug Administration (FDA) states that zinc damages nerve receptors in the nose, causing anosmia. Reports of anosmia were also observed in the 1930s when zinc preparations were used in a failed attempt to prevent polio infections. On June 16, 2009, the FDA ordered removal of zinc-based intranasal cold products from store shelves. The FDA said the loss of smell can be life-threatening because people with impaired smell cannot detect leaking gas or smoke, and cannot tell if food has spoiled before they eat it.

Recent research suggests that the topical antimicrobial zinc pyrithione is a potent heat shock response inducer that may impair genomic integrity with induction of PARP-dependent energy crisis in cultured human keratinocytes and melanocytes.

Poisoning

In 1982, the US Mint began minting pennies coated in copper but containing primarily zinc. Zinc pennies pose a risk of zinc toxicosis, which can be fatal. One reported case of chronic ingestion of 425 pennies (over 1 kg of zinc) resulted in death due to gastrointestinal bacterial and fungal sepsis. Another patient who ingested 12 grams of zinc showed only lethargy and ataxia (gross lack of coordination of muscle movements). Several other cases have been reported of humans suffering zinc intoxication by the ingestion of zinc coins.

Pennies and other small coins are sometimes ingested by dogs, requiring veterinary removal of the foreign objects. The zinc content of some coins can cause zinc toxicity, commonly fatal in dogs through severe hemolytic anemia and liver or kidney damage; vomiting and diarrhea are possible symptoms. Zinc is highly toxic in parrots and poisoning can often be fatal. The consumption of fruit juices stored in galvanized cans has resulted in mass parrot poisonings with zinc.

12.2 ZINC ENZYMES

Zinc is a transition metal that plays a pivotal role in a wide variety of biological processes, largely due to its involvement in enzyme function. As an essential trace element, zinc is a cofactor in more than 300 enzymes, influencing key processes such as DNA synthesis, protein metabolism, cell division, and immune function. Zinc enzymes are ubiquitous in living organisms and are integral to maintaining the structure and function of biomolecules.

Zinc enzymes are integral to a wide range of biological processes, including metabolism, immune function, DNA synthesis, and antioxidant defense. Zinc's versatility as a cofactor lies in its ability to stabilize enzyme structures, activate water molecules, and facilitate electron transfer. Without zinc, many of these critical functions would be compromised, leading to widespread physiological dysfunction. Understanding the role of zinc enzymes in health and disease is essential for developing strategies to prevent and treat conditions associated with zinc deficiency. Given its importance, research into zinc biochemistry continues to be an area of significant scientific interest.

12.2.1 The Role of Zinc in Enzymatic Activity

Zinc ions (Zn^{2+}) serve multiple roles in enzymatic reactions, ranging from structural stability to catalytic functions. Unlike other metal ions, zinc does not undergo redox reactions in biological systems, making it a stable cofactor in enzymatic catalysis. It primarily serves three functions in enzymes.

Catalytic Role

In its catalytic role, zinc is directly involved in the enzymatic reaction by stabilizing intermediates, facilitating nucleophilic attacks, or activating water molecules. In many cases, zinc coordinates with the active site of the enzyme and acts as an electrophilic catalyst, enhancing the reactivity of the enzyme-substrate complex. This is particularly common in hydrolytic enzymes.

Structural Role

Zinc can also act as a structural element, maintaining the integrity of an enzyme's active site. By binding to specific amino acid residues such as cysteine or histidine, zinc stabilizes the tertiary or quaternary structure of proteins. This is essential in ensuring that the enzyme retains its functional shape and can perform its biochemical activities effectively.

Regulatory Role

Zinc can modulate enzyme activity by acting as a regulatory cofactor. It can either activate or inhibit certain enzymes depending on the concentration of zinc ions and the nature of the enzyme. This regulatory function makes zinc a versatile cofactor in various metabolic pathways, ensuring that enzymatic activities are finely tuned to the needs of the cell.

12.2.2 Mechanisms of Zinc Enzymes

The mechanisms through which zinc enzymes operate depend on their specific roles. There are a few common mechanisms by which zinc ions participate in enzyme reactions.

Activation of Water Molecules

In many hydrolytic enzymes, zinc ions activate water molecules to serve as nucleophiles in catalysis. By lowering the pKa of water, zinc promotes the generation of hydroxide ions (OH^-) under physiological conditions. This hydroxide ion can then attack the substrate, leading to bond cleavage or rearrangement. This mechanism is common in metalloproteases like carboxypeptidases and matrix metalloproteinases.

Substrate Orientation and Stabilization

Zinc can assist in proper substrate orientation at the active site, ensuring that substrates are positioned for efficient catalysis. By coordinating with substrate molecules, zinc helps stabilize the transition state during the reaction, reducing the activation energy required for the reaction to proceed.

Electrophilic Catalysis

Zinc acts as an electrophilic catalyst by attracting electron density away from the substrate, making it more susceptible to nucleophilic attack. This is crucial in enzymes that mediate redox reactions or the cleavage of chemical bonds, as it enhances the reactivity of the substrate.

12.2.3 Types of Zinc Enzymes

Zinc enzymes are classified into various categories based on their function and the type of reactions they catalyze. Some of the most prominent classes include metalloproteases, carbonic anhydrases, and dehydrogenases.

Metalloproteases

Metalloproteases are a class of proteolytic enzymes that rely on metal ions, such as zinc, to catalyze the cleavage of peptide bonds in proteins. Zinc-dependent metalloproteases include matrix metalloproteinases (MMPs), carboxypeptidases, and thermolysin.

These enzymes play critical roles in tissue remodeling, wound healing, and the degradation of extracellular matrix components. In these enzymes, zinc activates a water molecule, facilitating the hydrolysis of peptide bonds.

Carbonic Anhydrase

Carbonic anhydrase is one of the most well-studied zinc enzymes. It catalyzes the reversible hydration of carbon dioxide (CO_2) to bicarbonate (HCO_3^-) and protons (H^+), a crucial reaction in maintaining acid-base balance in the blood and tissues. Zinc in carbonic anhydrase plays a catalytic role by activating a water molecule, which then attacks CO_2, leading to the formation of bicarbonate. This enzyme operates at an exceptionally high rate, showcasing the efficiency of zinc in catalysis.

Alcohol Dehydrogenase

Alcohol dehydrogenase is another key zinc enzyme, involved in the metabolism of alcohols in the liver. It catalyzes the oxidation of ethanol to acetaldehyde, a reaction critical for alcohol detoxification in humans. Zinc is a structural cofactor in alcohol dehydrogenase, ensuring the proper folding of the enzyme and the stabilization of intermediates during the reaction.

Superoxide Dismutase

Superoxide dismutase (SOD) is an enzyme that protects cells from oxidative damage by catalyzing the conversion of superoxide radicals (O_2^-) into oxygen and hydrogen peroxide (H_2O_2).

One of the forms of SOD contains both zinc and copper as cofactors. Zinc in this enzyme plays a structural role, ensuring the stability of the enzyme and its proper function in scavenging reactive oxygen species.

12.2.4 Physiological Importance of Zinc Enzymes

The physiological importance of zinc enzymes cannot be overstated. These enzymes are involved in virtually all aspects of metabolism and cellular function, from DNA replication to detoxification processes.

DNA Replication and Repair

Zinc enzymes are involved in the replication and repair of DNA. For instance, DNA polymerase, the enzyme responsible for synthesizing new strands of DNA, requires zinc for structural stability. Zinc finger proteins, which bind to specific DNA sequences, also rely on zinc ions for proper functioning, ensuring the accuracy of DNA replication and gene expression.

Immune Function

Zinc-dependent enzymes play a critical role in immune function. Metalloproteases, for example, are involved in the maturation and activation of immune cells. Additionally, zinc is crucial for the function of thymulin, a zinc-containing peptide involved in T-cell differentiation and immune response regulation.

Metabolic Regulation

Zinc enzymes are involved in regulating key metabolic pathways, such as glycolysis, the citric acid cycle, and fatty acid metabolism. Alcohol dehydrogenase, for instance, helps metabolize ethanol, while other zinc-dependent dehydrogenases are involved in energy production and storage.

Antioxidant Defense

Zinc enzymes such as superoxide dismutase are essential in protecting cells from oxidative stress. By neutralizing superoxide radicals, these enzymes help prevent cellular damage and aging, highlighting zinc's role in maintaining cellular homeostasis.

12.2.5 Zinc Deficiency and Enzyme Dysfunction

Zinc deficiency can lead to a wide range of health issues, many of which are linked to the dysfunction of zinc-dependent enzymes. Without sufficient zinc, enzymes cannot fold correctly or perform their catalytic functions, leading to metabolic disturbances, weakened immune function, and impaired DNA repair.

Immune System Impairment

One of the most immediate consequences of zinc deficiency is a weakened immune response. Without sufficient zinc, metalloproteases involved in immune cell maturation are rendered inactive, leading to a decreased production of T-cells and an increased susceptibility to infections.

Growth and Developmental Issues

Zinc deficiency can also impair growth and development, especially in children. Zinc enzymes are essential for DNA replication, protein synthesis, and cell division, processes that are crucial during growth. Deficiency during pregnancy or childhood can result in stunted growth, developmental delays, and increased risk of congenital anomalies.

Cognitive and Neurological Effects

Zinc is also vital for brain function, and its deficiency has been linked to cognitive impairments, memory loss, and depression. Zinc enzymes are involved in neurotransmitter regulation and synaptic plasticity, both of which are essential for learning and memory.

12.2.6 Carboanhydrases

Carboanhydrases, also known as carbonic anhydrases (CAs), are a family of zinc metalloenzymes that play an essential role in maintaining pH balance, facilitating the transport of carbon dioxide, and contributing to a range of physiological processes. These enzymes catalyze the reversible hydration of carbon dioxide (CO_2) into bicarbonate (HCO_3^-) and protons (H^+). Found in nearly all organisms, from prokaryotes to humans, carbonic anhydrases have garnered attention for their vital functions in respiratory processes, pH regulation, and many other biochemical systems.

Carbonic anhydrases are versatile and essential enzymes that facilitate the hydration of carbon dioxide and are involved in numerous physiological processes, including respiration, pH regulation, and ion transport. With their widespread distribution and diverse functions, they are crucial for maintaining the body's homeostasis. Understanding the mechanisms and roles of carbonic anhydrases continues to be an important area of research, particularly in developing treatments for diseases such as glaucoma, cancer, and altitude sickness. The study of these enzymes highlights their importance not only in basic biological functions but also in clinical medicine and therapeutic applications.

Structure and Mechanism of Action

Carbonic anhydrases are characterized by their active site, which contains a zinc ion (Zn^{2+}). This zinc ion is coordinated with three histidine residues and a water molecule or hydroxide ion. The enzyme's catalytic mechanism is based on the interaction of CO_2 and water, a reaction that is crucial for converting CO_2 into a more soluble and transportable form, bicarbonate. The overall reaction catalyzed by carbonic anhydrases is as follows:

$$CO_2 + H_2O \rightleftharpoons HCO_3^- + H^+$$

Catalytic Mechanism

The catalytic mechanism of carbonic anhydrases involves several key steps:

- **Activation of Water**: In the resting state, a water molecule is bound to the zinc ion in the enzyme's active site. The zinc ion facilitates the deprotonation of this water molecule, resulting in the formation of a hydroxide ion (OH^-), which is a strong nucleophile.

- **Nucleophilic Attack:** The hydroxide ion attacks the carbon atom of the CO_2 molecule, forming a bicarbonate ion (HCO_3^-) bound to the enzyme.
- **Release of Bicarbonate:** The bicarbonate ion is then released from the active site, and another water molecule binds to zinc, resetting the enzyme for another cycle of catalysis.

This rapid and efficient catalytic process is crucial for a variety of physiological processes, especially those involving the transport of gases and the regulation of pH.

Types of Carbonic Anhydrases

Carbonic anhydrases are divided into several isoforms based on their sequence, structure, and function. There are five known classes of carbonic anhydrases, named alpha (α), beta (β), gamma (γ), delta (δ), and zeta (ζ), each of which exhibits distinct properties and occurs in different species.

Alpha Carbonic Anhydrases (α-CAs)

Alpha carbonic anhydrases are the most well-studied and widely distributed form, found in humans and many other vertebrates. They are present in various tissues, including red blood cells, kidneys, and the gastrointestinal tract. In humans, the α-CAs family comprises several isoforms, such as CA I, CA II, CA IX, and CA XII, each of which has specific roles in different physiological systems.

For instance, **CA II** is found in erythrocytes (red blood cells) and is essential for facilitating CO_2 transport from tissues to the lungs. **CA IX** and **CA XII** are associated with cancer cells and play a role in tumor growth and survival by regulating pH in the tumor microenvironment.

Beta Carbonic Anhydrases (β-CAs)

Beta carbonic anhydrases are commonly found in plants, bacteria, and fungi. Unlike α-CAs, which are generally monomeric, β-CAs tend to form dimers or tetramers. These enzymes are involved in various metabolic processes, including photosynthesis and respiration in plants and microorganisms.

Gamma Carbonic Anhydrases (γ-CAs)

Gamma carbonic anhydrases are found predominantly in archaea and bacteria. Although structurally distinct from α-CAs and β-CAs, they perform similar catalytic functions. These enzymes are adapted to function in extreme environments, such as high temperatures or acidic conditions, which makes them important for studying enzyme stability and evolution.

Delta Carbonic Anhydrases (δ-CAs)

Delta carbonic anhydrases are less well-known and are primarily found in marine organisms, including diatoms. These enzymes are thought to play a role in regulating the concentration of dissolved inorganic carbon in marine environments, which is important for photosynthesis in marine plants.

Zeta Carbonic Anhydrases (ζ-CAs)

Zeta carbonic anhydrases are found in certain marine organisms, including coccolithophores, a type of plankton. These enzymes contain cadmium instead of zinc in their active sites and are believed to play a role in the biomineralization processes of these organisms.

Physiological Roles of Carbonic Anhydrases

Carbonic anhydrases play diverse physiological roles, and their importance can be observed across many systems in the body. Their primary function involves the transport and exchange of CO_2 and bicarbonate, which has far-reaching effects on respiration, pH balance, and even ion transport.

Respiratory Function

Carbonic anhydrases play a pivotal role in respiration, particularly in the exchange of gases in the lungs and tissues. In red blood cells, carbonic anhydrase II facilitates the rapid conversion of CO_2 produced by metabolizing tissues into bicarbonate ions, which are easily transported through the bloodstream. When blood reaches the lungs, the reaction is reversed, converting bicarbonate back into CO_2, which is exhaled. This process ensures efficient removal of CO_2 from the body and helps maintain acid-base balance.

pH Regulation

Maintaining the proper pH of body fluids is essential for cellular function. Carbonic anhydrases contribute significantly to pH regulation by controlling the concentration of bicarbonate ions, a major buffer in the body.

For example, in the kidneys, carbonic anhydrase IV and CA II help reabsorb bicarbonate from urine, ensuring the proper pH of blood and other body fluids. These enzymes also play a role in secreting acid in the stomach, where CA IX aids in producing gastric acid.

Ion Transport and Fluid Secretion

Carbonic anhydrases are involved in various processes that require ion transport and fluid secretion. In the eye, for instance, carbonic anhydrase II plays a role in aqueous humor production, which helps maintain intraocular pressure. Inhibition of carbonic anhydrase in the eye is used as a treatment for glaucoma, a condition where increased pressure can damage the optic nerve. Similarly, in the pancreas and salivary glands, carbonic anhydrases help in the secretion of bicarbonate-rich fluids, which are necessary for digestion and protection against stomach acid.

Cancer Progression and Tumor Growth

Certain isoforms of carbonic anhydrases, such as CA IX and CA XII, are upregulated in tumor cells. These enzymes help cancer cells survive in hypoxic conditions by regulating pH and facilitating the removal of excess CO_2. Inhibiting these isoforms has emerged as a potential therapeutic strategy for targeting cancer cells. The regulation of pH in tumors is essential for cell proliferation, metastasis, and survival, and thus, carbonic anhydrase inhibitors are being explored as a possible treatment for various cancers.

Clinical Significance and Inhibitors of Carbonic Anhydrases

Carbonic anhydrases are also of significant clinical interest, particularly for their role in disease mechanisms and as targets for pharmaceutical interventions. Carbonic anhydrase inhibitors (CAIs) are drugs that inhibit the activity of these enzymes and are used to treat a variety of conditions.

Treatment of Glaucoma

One of the most common therapeutic uses of carbonic anhydrase inhibitors is in the treatment of glaucoma. By inhibiting carbonic anhydrase in the eye, drugs such as acetazolamide reduce the production of aqueous humor, lowering intraocular pressure and preventing damage to the optic nerve.

Diuretic Effects

Carbonic anhydrase inhibitors are also used as diuretics in the treatment of conditions like congestive heart failure and hypertension. By inhibiting carbonic anhydrase in the kidneys, these drugs reduce the reabsorption of bicarbonate, leading to increased urine production and decreased fluid retention.

Management of Altitude Sickness

Acetazolamide, a well-known carbonic anhydrase inhibitor, is also used to prevent and treat altitude sickness. At high altitudes, the body struggles to maintain a proper acid-base balance due to low oxygen levels. Acetazolamide helps correct this imbalance by promoting the excretion of bicarbonate, thus stabilizing blood pH and reducing symptoms of altitude sickness.

Anticancer Potential

As mentioned earlier, certain carbonic anhydrase isoforms, like CA IX and CA XII, are involved in tumor growth. Inhibiting these enzymes has shown promise in slowing down the growth of certain cancers. Research is ongoing to develop specific inhibitors that target these cancer-related isoforms, offering a potential treatment avenue for aggressive tumors.

12.2.7 Hydrolases

Hydrolases are a diverse group of enzymes that catalyze hydrolysis reactions, where water is used to break chemical bonds in substrates. They play crucial roles in various biological processes, including metabolism, digestion, and the degradation of biomolecules. Hydrolases are classified into several categories based on the types of bonds they act upon, and they exhibit significant versatility and specificity.

Hydrolases are a vital class of enzymes that catalyze hydrolysis reactions, playing crucial roles in various biological processes, including digestion, metabolism, and cellular regulation.

Their classification into proteases, glycosidases, lipases, phosphatases, and esterases highlights their diverse functions and specificity. Understanding the mechanisms and applications of hydrolases not only provides insights into fundamental biological processes but also opens avenues for innovative solutions in biotechnology, pharmaceuticals, and environmental management. As research continues to unveil the complexities of hydrolases, their significance in health, industry, and environmental sustainability will only continue to grow.

What are Hydrolases?

Hydrolases are enzymes that facilitate the cleavage of chemical bonds through the addition of water. This reaction can involve various types of bonds, including peptide bonds, glycosidic bonds, ester bonds, and phosphoester bonds. The general reaction mechanism for hydrolases can be represented as follows:

$$\text{Substrate} + H_2O \xrightarrow{\text{Hydrolase}} \text{Products}$$

The specificity of hydrolases allows them to target specific substrates, leading to the breakdown of complex molecules into simpler components. This function is vital for many metabolic pathways and physiological processes.

Classification of Hydrolases

Hydrolases are classified into several subclasses based on the type of bond they cleave. The primary categories include:

Proteases (Peptidases)

Proteases are enzymes that hydrolyze peptide bonds in proteins, leading to their breakdown into smaller peptides or amino acids. They play essential roles in protein metabolism, digestion, and various cellular processes. Proteases can be further divided into several types based on their catalytic mechanisms, including serine proteases, cysteine proteases, aspartic proteases, and metalloproteases. Each type has distinct active site characteristics and substrate specificities.

Glycosidases (Glycosyl Hydrolases)

Glycosidases are enzymes that hydrolyze glycosidic bonds in carbohydrates, resulting in the breakdown of polysaccharides into simpler sugars (monosaccharides or disaccharides). These enzymes are crucial for carbohydrate digestion and metabolism. Glycosidases are further categorized based on the type of glycosidic bond they act upon, including alpha- and beta-glycosidases.

Lipases

Lipases are hydrolases that catalyze the hydrolysis of ester bonds in lipids, leading to the breakdown of triglycerides into fatty acids and glycerol. Lipases are important for fat digestion and metabolism, playing a key role in energy production and storage. They are produced by various organisms, including plants, animals, and microorganisms.

Phosphatases

Phosphatases are enzymes that catalyze the hydrolysis of phosphoester bonds in phosphate-containing compounds, leading to the release of inorganic phosphate. They play critical roles in cellular signaling, metabolism, and the regulation of various biological processes. Phosphatases are involved in dephosphorylation reactions, which counterbalance the activity of kinases that add phosphate groups to proteins.

Esterases

Esterases are enzymes that hydrolyze ester bonds, resulting in the breakdown of esters into acids and alcohols. They are involved in various metabolic processes, including the detoxification of xenobiotics and the metabolism of certain drugs. Esterases are present in various organisms and play important roles in lipid metabolism.

Mechanism of Action

The mechanism of action of hydrolases involves several key steps:

- **Substrate Binding:** The substrate binds to the enzyme's active site, forming an enzyme-substrate complex. The specificity of the enzyme allows it to selectively bind to specific substrates.
- **Catalysis:** The enzyme facilitates the hydrolysis reaction by stabilizing the transition state and lowering the activation energy required for the reaction. Water molecules participate in the reaction, leading to the cleavage of the chemical bond.

- **Product Release:** Once the reaction is complete, the products are released from the enzyme, regenerating the active site for another catalytic cycle.

The efficiency of hydrolases is often enhanced by cofactors, such as metal ions or coenzymes, which assist in the catalytic process.

Physiological Roles of Hydrolases

Hydrolases play diverse physiological roles in various biological processes:

Digestion and Metabolism

Hydrolases are crucial for the digestion of macromolecules, including proteins, carbohydrates, and lipids. Proteases break down dietary proteins into amino acids, glycosidases hydrolyze polysaccharides into sugars, and lipases degrade fats into fatty acids and glycerol. These enzymatic processes facilitate nutrient absorption and energy production.

Cellular Regulation

Hydrolases, particularly phosphatases, play essential roles in regulating cellular signaling pathways. By removing phosphate groups from proteins, phosphatases modulate the activity of signaling molecules, influencing processes such as cell growth, differentiation, and apoptosis. This regulatory function is vital for maintaining cellular homeostasis.

Detoxification

Hydrolases, such as esterases and certain glycosidases, are involved in the detoxification of xenobiotics and environmental pollutants. They catalyze the hydrolysis of toxic compounds, converting them into less harmful metabolites that can be excreted from the body. This detoxification process is crucial for protecting cells from damage caused by harmful substances.

DNA and RNA Metabolism

Hydrolases are involved in the metabolism of nucleic acids. Nucleases, a class of hydrolases, hydrolyze phosphodiester bonds in DNA and RNA, facilitating processes such as DNA replication, repair, and degradation. These enzymes are essential for maintaining genomic integrity and regulating gene expression.

Applications of Hydrolases

The versatility and specificity of hydrolases make them valuable tools in various industrial and biomedical applications.

Biotechnology and Enzyme Engineering

Hydrolases are extensively used in biotechnology for various applications, including the production of biofuels, pharmaceuticals, and food additives. Enzyme engineering techniques allow for the modification of hydrolases to enhance their stability, activity, and substrate specificity, making them suitable for industrial processes.

Food Industry

Hydrolases are widely used in the food industry for the production of high-fructose corn syrup, lactose-free dairy products, and flavor enhancement. Proteases are employed in meat tenderization and cheese production, while lipases contribute to the development of flavors and aromas in various food products.

Pharmaceutical Development

Hydrolases play a crucial role in drug metabolism and detoxification. Understanding the mechanisms of hydrolases can aid in the design of more effective pharmaceuticals with reduced side effects. Additionally, hydrolase inhibitors are being explored as potential therapeutic agents for various diseases, including cancer and metabolic disorders.

Environmental Applications

Hydrolases are employed in bioremediation processes to degrade environmental pollutants, including pesticides, plastics, and heavy metals. By harnessing the catalytic capabilities of hydrolases, researchers aim to develop sustainable strategies for cleaning contaminated environments and promoting ecosystem health.

12.2.8 Alcohol Dehydrogenase

Alcohol dehydrogenase (ADH) is a crucial enzyme in the human body responsible for the metabolism of alcohol. It catalyzes the conversion of ethanol (the type of alcohol found in alcoholic beverages) into acetaldehyde, a toxic intermediate that is further metabolized to acetic acid by aldehyde dehydrogenase (ALDH).

Alcohol dehydrogenase plays a pivotal role in the metabolism of alcohol, facilitating the conversion of ethanol into acetaldehyde. Its various isoforms exhibit distinct properties that influence alcohol metabolism and regulation.

Genetic variability, hormonal influences, and the presence of competing substances all contribute to the complex regulation of this enzyme. Understanding the role of alcohol dehydrogenase is crucial for addressing alcohol-related health issues and informing treatment strategies for individuals struggling with alcohol use disorder.

As research continues to uncover the intricacies of alcohol metabolism, the significance of alcohol dehydrogenase in health and disease will remain a critical area of study.

The Role of Alcohol Dehydrogenase

Metabolism of Ethanol

The primary function of alcohol dehydrogenase is to facilitate the oxidation of ethanol to acetaldehyde. This reaction occurs mainly in the liver, where the majority of alcohol metabolism takes place. The reaction can be summarized as follows:

$$Ethanol + NAD^{+} \rightarrow Acetaldehyde + NADH + H^{+}$$

In this reaction, ethanol donates a hydrogen atom to NAD^+ (nicotinamide adenine dinucleotide), reducing it to NADH while converting ethanol into acetaldehyde. This reaction is essential for the clearance of alcohol from the bloodstream and prevents the accumulation of ethanol, which can lead to intoxication and other harmful effects.

Acetaldehyde Metabolism

In this reaction, ethanol donates a hydrogen atom to NAD^+ (nicotinamide adenine dinucleotide), reducing it to NADH while converting ethanol into acetaldehyde. This reaction is essential for the clearance of alcohol from the bloodstream and prevents the accumulation of ethanol, which can lead to intoxication and other harmful effects.

Acetaldehyde Metabolism

Following the production of acetaldehyde, the enzyme aldehyde dehydrogenase (ALDH) further metabolizes this toxic compound into acetic acid, which is subsequently converted into carbon dioxide and water, ultimately leading to its elimination from the body. The efficient conversion of acetaldehyde to acetic acid is crucial, as the accumulation of acetaldehyde can lead to various adverse effects, including nausea, vomiting, and headache—a phenomenon known as hangover.

Structure of Alcohol Dehydrogenase

Alcohol dehydrogenases are a group of enzymes that exist in several isoforms, each with distinct properties and roles in alcohol metabolism. The most well-studied isoforms in humans are ADH1, ADH2, and ADH3.

Isoforms and Their Characteristics

- **ADH1 (Class I):** This isoform is the most prevalent in the liver and plays a significant role in ethanol metabolism. It has a high affinity for ethanol and is responsible for most of the ethanol oxidation that occurs after alcohol consumption.
- **ADH2 (Class II):** This isoform has a lower affinity for ethanol compared to ADH1 but is more active at higher ethanol concentrations. It is found

primarily in the liver and other tissues, including the gastrointestinal tract.

- **ADH3 (Class III):** This isoform is less common but is still significant in alcohol metabolism. It has a broader substrate specificity, meaning it can metabolize a variety of alcohols.

Enzyme Structure

Alcohol dehydrogenase is a zinc-containing enzyme, with each isoform having a characteristic structure. The active site of the enzyme contains a zinc ion, which plays a crucial role in stabilizing the enzyme-substrate complex and facilitating the oxidation reaction. The enzyme typically functions as a dimer or tetramer, with subunits that can exhibit different kinetic properties based on their isoform.

Regulation of Alcohol Dehydrogenase Activity

The activity of alcohol dehydrogenase is subject to various regulatory mechanisms, including genetic factors, hormonal influences, and the presence of other substances.

Genetic Variability

Genetic polymorphisms in the ADH gene can result in variations in enzyme activity among individuals. Some populations, particularly those of East Asian descent, have a variant of ADH (ADH2*2) that metabolizes ethanol more rapidly, leading to higher levels of acetaldehyde. This rapid metabolism can result in unpleasant reactions to alcohol consumption, often referred to as "Asian flush."

Hormonal Regulation

Hormonal factors, such as cortisol and insulin, can influence the expression and activity of alcohol dehydrogenase. For example, chronic alcohol consumption can lead to increased levels of ADH expression, resulting in enhanced alcohol metabolism and tolerance.

Inhibition and Competition

Certain medications and substances can inhibit the activity of alcohol dehydrogenase. For instance, fomepizole is a drug that inhibits ADH and is used as an antidote for methanol and ethylene glycol poisoning. Additionally, the presence of other alcohols (such as methanol and isopropanol) can compete with ethanol for binding to ADH, affecting its metabolism.

Clinical Implications

The function and regulation of alcohol dehydrogenase have significant clinical implications. Understanding the enzyme's role in alcohol metabolism can inform the treatment of alcohol-related disorders and guide therapeutic interventions.

Alcohol Use Disorder (AUD)

Individuals with a history of heavy alcohol consumption may develop alcohol use disorder (AUD), characterized by the compulsive consumption of alcohol despite negative consequences. Understanding the genetic variability in alcohol dehydrogenase can help identify individuals at higher risk for developing AUD. Additionally, medications that target alcohol metabolism may be used as part of treatment strategies for individuals struggling with AUD.

Acetaldehyde Toxicity

The accumulation of acetaldehyde due to impaired alcohol metabolism can lead to various health issues, including liver disease, cardiovascular problems, and cancer. Individuals with genetic variations that affect alcohol dehydrogenase activity may be more susceptible to these health risks.

Drug Interactions

Given that alcohol dehydrogenase is involved in the metabolism of various drugs, understanding its activity is essential in clinical practice. Alcohol can interact with medications metabolized by ADH, leading to altered drug efficacy and potential toxicity. Healthcare providers should consider these interactions when prescribing medications to patients who consume alcohol.

12.3 THE ROLE OF ZINC IN THE TRANSCRIPTION OF GENES

Zinc is an essential trace element that plays a crucial role in various biological processes, including the transcription of genes. As a structural and functional component of many proteins, zinc contributes to the regulation of gene expression and the maintenance of cellular homeostasis.

Zinc is an essential trace element that plays a multifaceted role in the transcription of genes. Through its involvement in the structure and function of zinc finger proteins, the activation of transcription factors, and the regulation of chromatin structure, zinc significantly influences gene expression. The consequences of zinc deficiency highlight the importance of maintaining adequate zinc levels for optimal cellular function and overall health. As research continues to uncover the intricate mechanisms by which zinc regulates transcription, it becomes increasingly clear that this vital element is indispensable for the proper functioning of biological systems. Ensuring adequate dietary intake of zinc, particularly in populations at risk of deficiency, is crucial for promoting health and preventing disease.

12.3.1 The Importance of Zinc in Biological Systems

Zinc is one of the most abundant transition metals in the human body, playing a vital role in numerous physiological functions. It is involved in enzymatic reactions, immune function, DNA synthesis, and cell division. The bioavailability of zinc is critical for maintaining optimal cellular function, and it is particularly important in rapidly dividing cells, such as those found in the immune system and the gastrointestinal tract.

Zinc Homeostasis

The regulation of zinc levels in the body is tightly controlled through dietary intake, absorption, and excretion. Zinc homeostasis is maintained by several transporters that facilitate the movement of zinc across cellular membranes. Key transporters include the SLC39 (ZIP) family of zinc transporters, which promote zinc uptake into cells, and the SLC30 (ZnT) family, which functions to export excess zinc out of cells. The balance between zinc import and export is crucial for ensuring sufficient zinc levels for proper cellular functions, including transcription.

12.3.2 Zinc and Gene Transcription

Gene transcription is the process by which DNA is converted into RNA, leading to the synthesis of proteins. This process involves several key components, including transcription factors, RNA polymerase, and regulatory elements within the DNA. Zinc plays a multifaceted role in gene transcription, influencing various aspects of this process.

Zinc Finger Proteins

Zinc finger proteins are a family of transcription factors that contain zinc-binding motifs. These proteins utilize zinc ions to stabilize their structural conformation, allowing them to interact with specific DNA sequences. The presence of zinc is essential for the proper folding of these proteins, which in turn enables their binding to regulatory regions of target genes. Zinc finger proteins play critical roles in the regulation of gene expression, influencing processes such as cell differentiation, proliferation, and apoptosis.

Transcription Factor Activation

Zinc can enhance the activity of various transcription factors, promoting the transcription of specific genes. For instance, zinc ions can facilitate the interaction between transcription factors and co-activators, leading to increased recruitment of RNA polymerase to target gene promoters. This interaction enhances the transcriptional response to various stimuli, including hormones, cytokines, and growth factors. The modulation of transcription factor activity by zinc is crucial for the adaptive responses of cells to changing environmental conditions.

Regulation of Chromatin Structure

Zinc plays a significant role in the regulation of chromatin structure, which influences gene accessibility for transcription. Zinc-dependent proteins can facilitate the remodeling of chromatin, allowing for the relaxation or compaction of DNA. This dynamic regulation of chromatin structure is essential for controlling gene expression, as it determines whether specific genes are accessible for transcription. The ability of zinc to influence chromatin remodeling is a key aspect of its role in gene regulation.

12.3.3 The Impact of Zinc Deficiency on Gene Expression

Zinc deficiency can have profound effects on gene expression and overall cellular function. Insufficient zinc levels can lead to the dysregulation of transcription factors and the impaired activation of target genes. This can result in a wide range of physiological consequences, including compromised immune function, delayed wound healing, and altered cellular differentiation.

Altered Expression of Key Genes

Zinc deficiency can affect the expression of genes involved in critical cellular processes. For example, studies have shown that zinc deficiency can lead to decreased expression of genes associated with immune response, leading to increased susceptibility to infections. Additionally, zinc deficiency can disrupt the expression of genes involved in cell cycle regulation, potentially contributing to the development of various diseases, including cancer.

Impaired Protein Synthesis

Zinc is essential for the synthesis of numerous proteins, including those involved in transcription. A deficiency in zinc can result in reduced levels of zinc-dependent transcription factors, impairing the overall transcriptional capacity of cells. This reduction in protein synthesis can have downstream effects on cellular function and overall health.

Oxidative Stress and Inflammation

Zinc deficiency is associated with increased oxidative stress and inflammation, which can further impact gene expression. Reactive oxygen species (ROS) can damage DNA and disrupt transcriptional machinery, leading to altered gene expression patterns. The interplay between zinc deficiency, oxidative stress, and inflammation underscores the importance of adequate zinc levels for maintaining gene transcription and cellular health.

12.4 THIONEINS

Thioneins are a class of low-molecular-weight, cysteine-rich proteins found primarily in plants and certain microorganisms. These proteins play a significant role in cellular processes, particularly in response to oxidative stress, heavy metal detoxification, and the maintenance of redox homeostasis. The term “thionein” is derived from "thiol," indicating the presence of sulfur-containing groups that contribute to their biological functions.

Thioneins are essential proteins that play crucial roles in protecting organisms from oxidative stress and heavy metal toxicity. Their unique structure and high cysteine content enable them to function as antioxidants and metal chelators, contributing to the overall stress tolerance of plants and microorganisms. As our understanding of thioneins deepens, their potential applications in phytoremediation, crop improvement, and bioremediation technologies offer promising avenues for addressing environmental challenges and enhancing agricultural sustainability. Continued research into the biology and functions of thioneins will provide valuable insights into their role in ecosystem health and resilience.

12.4.1 Structure of Thioneins

Molecular Composition

Thioneins are characterized by their high cysteine content, which typically comprises about 30-40% of their amino acid composition. This abundance of cysteine residues allows thioneins to form disulfide bonds and participate in redox reactions. The exact structure of thioneins can vary among different species, but they generally adopt a compact and flexible conformation, which is essential for their functional versatility.

Domains and Motifs

Thioneins often contain specific structural motifs, including metal-binding sites that enable them to chelate heavy metals like cadmium, lead, and mercury. These binding sites are crucial for the detoxification processes that thioneins facilitate. The presence of cysteine thiol groups also allows thioneins to act as antioxidants, scavenging reactive oxygen species (ROS) and preventing oxidative damage to cellular components.

12.4.2 Functions of Thioneins

Antioxidant Activity

One of the primary functions of thioneins is their role as antioxidants. Under conditions of oxidative stress, thioneins can donate electrons to neutralize ROS, thus protecting cells from damage. This antioxidant activity is particularly important in plants, which are often exposed to various environmental stresses, including drought, salinity, and heavy metal toxicity. By scavenging ROS, thioneins help maintain cellular integrity and promote plant survival under adverse conditions.

Metal Chelation and Detoxification

Thioneins are also involved in the detoxification of heavy metals. When plants are exposed to high concentrations of toxic metals, thioneins can bind to these metals, forming stable complexes that are less harmful to the plant. This metal-binding capacity is crucial for preventing the accumulation of toxic metals in plant tissues, thus allowing for better growth and development in contaminated environments.

Role in Cellular Signaling

In addition to their antioxidant and metal-binding properties, thioneins are involved in various cellular signaling pathways. They can influence the expression of genes related to stress responses and participate in signaling cascades that regulate plant growth and development. For example, thioneins can modulate the activity of transcription factors that control the expression of stress-responsive genes, thus helping plants adapt to changing environmental conditions.

12.4.3 Thioneins in Different Organisms

In Plants

Thioneins are widely distributed in the plant kingdom, where they play essential roles in growth, development, and stress responses. They are particularly abundant in the roots and leaves of plants, where they contribute to the detoxification of heavy metals and the protection against oxidative stress. Some well-studied plant thioneins include the metallothioneins found in wheat, rice, and Arabidopsis thaliana.

In Microorganisms

Certain bacteria and fungi also produce thioneins, which serve similar functions in these organisms. In bacteria, thioneins are often involved in detoxifying heavy metals and protecting against oxidative stress. For instance, some species of Pseudomonas and Bacillus have been shown to produce thioneins that confer resistance to heavy metals, allowing them to thrive in contaminated environments.

12.4.4 Thioneins and Environmental Stress

Responses to Heavy Metals

Thioneins are particularly important in plants and microorganisms exposed to heavy metal contamination. They play a crucial role in sequestering and detoxifying metals, thereby mitigating their toxic effects. The expression of thionein genes is often upregulated in response to heavy metal exposure, leading to increased production of these protective proteins.

Coping with Oxidative Stress

In addition to heavy metal detoxification, thioneins help organisms cope with oxidative stress caused by various environmental factors, such as UV radiation, drought, and high salinity. By scavenging ROS and modulating cellular signaling pathways, thioneins contribute to the overall stress tolerance of plants and microorganisms.

12.4.5 Biotechnological Applications

Phytoremediation

Given their ability to bind heavy metals, thioneins have potential applications in phytoremediation, a process that uses plants to remove contaminants from the environment. By engineering plants to overexpress thioneins, scientists aim to enhance the metal-binding capacity of these plants, making them more effective in cleaning up contaminated soils and waters.

Crop Improvement

Thioneins may also play a role in developing crop varieties with improved stress tolerance. By understanding the molecular mechanisms underlying thionein function, researchers can identify strategies to enhance the expression of these proteins in crops, thereby improving their resilience to environmental stresses such as drought and salinity.

Bioremediation Technologies

In microbial systems, thioneins can be harnessed in bioremediation technologies to clean up environments contaminated with heavy metals and other toxic compounds. By exploiting the metal-binding properties of thioneins, researchers can develop microbial strains with enhanced capabilities for detoxifying contaminated sites.

CHAPTER 13

METAL- AND METALLOID-CARBON BONDS

13.1 ORGANOMETALLIC COMPOUNDS OF TRANSITION METALS

Organometallic chemistry, a field that merges organic and inorganic chemistry, has gained significant attention due to the diverse properties and applications of organometallic compounds, particularly those involving transition metals. Transition metals, known for their unique electronic configurations and variable oxidation states, form a wide range of organometallic compounds that exhibit intriguing reactivity and catalysis.

Organometallic compounds of transition metals represent a fascinating and essential area of study within chemistry. Their unique properties, versatile synthetic routes, and wide-ranging applications in catalysis, materials science, and pharmaceuticals underscore their significance in both academic research and industrial practices. As the field of organometallic chemistry continues to evolve, addressing the environmental and safety challenges associated with these compounds will be crucial for their sustainable development. The ongoing exploration of new organometallic systems and their applications holds great promise for advancing technology and addressing contemporary challenges in various scientific fields.

13.1.1 Understanding Organometallic Compounds

Organometallic compounds are defined as substances that contain at least one bond between a carbon atom of an organic molecule and a metal, which can be either a main group element or a transition metal. The general formula for these compounds can be represented as R-M, where R denotes an organic group and M signifies the metal. Transition metals, which include elements from groups 3 to 12 of the periodic table, are characterized by their ability to form coordination complexes with varying geometries and ligands. The unique electronic properties of transition metals enable them to stabilize multiple oxidation states, which is a crucial factor in their reactivity and role as catalysts in various chemical reactions.

13.1.2 Synthesis of Organometallic Compounds

The synthesis of organometallic compounds of transition metals involves several well-established methods. One of the most common techniques is the direct reaction of a metal with an organic halide or organolithium reagent. For example, reacting a transition metal halide with an organolithium compound can yield the corresponding organometallic compound. Another prevalent method is the use of metal carbonyl complexes, where carbon monoxide acts as a ligand, coordinating to the metal center and facilitating the formation of organometallic compounds through substitution reactions.

Moreover, the Wurtz reaction is often employed to synthesize organometallic compounds by coupling two organic halides in the presence of metallic sodium, resulting in the formation of alkanes and alkyl sodium compounds. Additionally, transition metal complexes can be synthesized through ligand exchange reactions, where a ligand on a metal center is replaced by an organometallic group. The versatility of these synthetic methods allows for the preparation of a wide variety of organometallic compounds tailored for specific applications.

13.1.3 Properties of Organometallic Compounds

The properties of organometallic compounds of transition metals are largely influenced by the nature of the metal, the organic group, and the overall electronic structure of the compound. These compounds often exhibit unique physical and chemical characteristics, including variable solubility, distinct color, and reactivity patterns. For instance, many organometallic compounds are soluble in organic solvents, which is advantageous for various catalytic processes.

One significant property of these compounds is their ability to participate in nucleophilic and electrophilic reactions, making them essential in organic synthesis.

The metal center can stabilize negative charge, facilitating nucleophilic attack, while the organic groups can exhibit electrophilic behavior. Furthermore, the presence of transition metals in organometallic compounds allows for unique magnetic and electronic properties, such as paramagnetism, which arises from unpaired d-electrons in the metal's electron configuration.

13.1.4 Applications of Organometallic Compounds

Organometallic compounds of transition metals find applications in various fields, primarily in catalysis, materials science, and pharmaceuticals. One of the most prominent applications is in catalytic processes, where organometallic compounds serve as catalysts in numerous organic transformations, including cross-coupling reactions, polymerizations, and oxidation reactions.

For instance, palladium-catalyzed cross-coupling reactions, such as the Suzuki and Heck reactions, are widely used in the synthesis of complex organic molecules, including pharmaceuticals and agrochemicals.

In materials science, organometallic compounds are crucial in the development of novel materials, such as conducting polymers and organometallic frameworks. Transition metal-organic frameworks (MOFs) are notable examples, where organometallic compounds are utilized to create highly porous materials with potential applications in gas storage, separation, and catalysis. Furthermore, organometallic compounds play a vital role in the pharmaceutical industry. They are used in drug development and as intermediates in the synthesis of active pharmaceutical ingredients. For example, organoplatinum compounds, such as cisplatin, are employed in cancer treatment due to their ability to bind to DNA and interfere with cell division.

13.1.5 Environmental Implications and Challenges

Despite their valuable applications, organometallic compounds of transition metals also pose environmental and safety challenges. Many organometallic compounds, particularly those containing heavy metals, can be toxic and harmful to human health and the environment. The release of such compounds into ecosystems can lead to bioaccumulation and adverse effects on wildlife and human populations. To mitigate these risks, research efforts are ongoing to develop greener synthesis methods and safer alternatives to traditional organometallic compounds. The exploration of biocompatible organometallic complexes and the use of renewable resources in their synthesis are promising avenues for reducing environmental impact.

13.2 CARBON BONDS TO MAIN GROUP METALS AND METALLOIDS

The chemistry of carbon, an essential element in organic compounds, extends its influence to its interactions with main group metals and metalloids. Carbon bonds to these elements are crucial in various fields, including materials science, catalysis, and organometallic chemistry. The chemistry of carbon bonds to main group metals and metalloids is a dynamic and vital field with far-reaching implications across various scientific disciplines. The diverse types of bonds, synthesis methods, properties, and applications underscore the importance of these compounds in catalysis, materials science, and pharmaceuticals. As research in this area continues to advance, the development of new carbon-containing compounds bonded to main group metals and metalloids promises to unlock further opportunities for innovation and discovery, enhancing our understanding of chemical interactions and their practical applications.

13.2.1 Understanding Carbon Bonds

Carbon is unique among elements due to its ability to form stable covalent bonds with various elements, including metals and metalloids. These bonds typically involve the sharing of electrons, leading to a stable configuration. In the context of carbon bonds to main group metals and metalloids, the nature of the bond can vary significantly based on the electronegativity of the metal or metalloid involved and the hybridization state of the carbon atom. The resulting compounds can exhibit a wide range of properties and reactivities, making them invaluable in various chemical applications.

13.2.2 Types of Carbon Bonds to Main Group Metals

The carbon bonds to main group metals and metalloids can be classified into different types based on their nature and characteristics. The most common types include ionic, covalent, and coordinate bonds.

Ionic Bonds

Ionic bonds form when there is a significant difference in electronegativity between carbon and the main group metal or metalloid. In these cases, the metal donates electrons to carbon, leading to the formation of positively charged metal ions and negatively charged carbon species. An example is the formation of carbanions, where carbon has a formal negative charge, allowing it to bond with positively charged metal ions. These ionic interactions are often observed in metal carbonates, metal alkyls, and metal carbides.

Covalent Bonds

Covalent bonds are formed through the sharing of electrons between carbon and the metal or metalloid. The nature of these bonds can vary from single to multiple bonds, depending on the specific elements involved. For instance, compounds like organolithiums (R-Li) feature covalent bonds between carbon and lithium, where carbon shares electrons with the lithium atom. These covalent interactions are pivotal in organometallic chemistry and the synthesis of various carbon-containing compounds.

Coordinate Bonds

Coordinate bonds arise when both electrons shared in the bond come from the same atom, typically from a lone pair on carbon. This type of bond is often observed in complexes where carbon acts as a ligand, donating electron pairs to form stable complexes with main group metals or metalloids. For example, carbon monoxide (CO) acts as a good ligand for transition and main group metals, forming stable metal-carbonyl complexes through coordinate bonding.

13.2.3 Synthesis of Carbon-Containing Compounds

The synthesis of carbon-containing compounds bonded to main group metals and metalloids can be achieved through several methods. These synthetic approaches leverage the unique properties of carbon and the reactivity of the metals involved.

Direct Reactions

Direct reactions between carbon sources and main group metals or metalloids are one of the simplest methods of synthesizing carbon-containing compounds. For example, reacting an alkyl halide with a metal alkoxide can yield an organometallic compound through nucleophilic substitution. The reactivity of the carbon source and the nature of the metal or metalloid play critical roles in determining the success of the reaction.

Metathesis Reactions

Metathesis reactions involve the exchange of components between two reactants, resulting in the formation of new products. In the context of carbon bonds to main group metals, metathesis reactions can be utilized to synthesize organometallic compounds. For instance, the reaction of an alkyl halide with a metal salt can lead to the formation of an organometallic compound while liberating a halide salt. These reactions are valuable for creating diverse carbon-containing products with tailored properties.

Ligand Exchange

Ligand exchange is a common method for synthesizing carbon-containing complexes of main group metals and metalloids. This process involves the replacement of one ligand in a complex with another. For example, in the presence of a suitable carbon-containing ligand, a metal complex can undergo ligand exchange, leading to the formation of new carbon-metal bonds. This method allows chemists to fine-tune the properties of the resulting compounds by selecting different ligands.

13.2.4 Properties of Carbon-Containing Compounds

The properties of carbon bonds to main group metals and metalloids are influenced by several factors, including the nature of the metal, the type of bond formed, and the overall electronic structure of the compound. These properties significantly impact the reactivity and application of these compounds.

Stability

The stability of carbon-containing compounds varies based on the type of bond formed and the metal or metalloid involved. Generally, covalent bonds tend to provide greater stability than ionic bonds, leading to the formation of stable organometallic complexes. Additionally, the presence of electron-donating or electron-withdrawing groups in the carbon structure can further stabilize or destabilize the compound.

Reactivity

Carbon bonds to main group metals and metalloids can exhibit diverse reactivity patterns. For instance, organolithium and organomagnesium compounds are known for their high reactivity, making them valuable intermediates in organic synthesis. Conversely, carbon bonds to heavier main group metals may display lower reactivity due to steric hindrance or electronic factors, influencing their application in various chemical processes.

Electronic Properties

The electronic properties of carbon-containing compounds are crucial for their applications in catalysis and materials science. The presence of pi bonds between carbon and the metal or metalloid can lead to interesting electronic behavior, such as conductivity and magnetic properties. This characteristic is particularly relevant in the development of novel materials, including conducting polymers and metal-organic frameworks.

13.2.5 Applications of Carbon Bonds to Main Group Metals and Metalloids

The unique properties of carbon bonds to main group metals and metalloids have led to numerous applications across various fields. These compounds play essential roles in catalysis, materials science, and the development of pharmaceuticals.

Catalysis

One of the most significant applications of carbon-containing compounds bonded to main group metals is their role as catalysts in organic reactions. Organometallic compounds, such as Grignard reagents and organolithium compounds, are widely utilized in nucleophilic addition reactions, allowing for the formation of carbon-carbon bonds. These reactions are fundamental in synthetic organic chemistry and the production of complex molecules, including pharmaceuticals and agrochemicals.

Materials Science

In materials science, carbon bonds to main group metals and metalloids are pivotal in the development of advanced materials with tailored properties. For instance, organometallic compounds are used in the synthesis of conducting polymers, which have applications in electronics and energy storage devices. Additionally, metal-organic frameworks (MOFs) containing carbon-metal bonds are being explored for gas storage, separation, and catalysis, showcasing their versatility in materials applications.

CHAPTER 14
INORGANICS IN MEDICINE

14.1 METALS AND METALLOIDS: AN INTRODUCTION

Metals and metalloids play a pivotal role in the fabric of modern society, influencing everything from technological advancements to everyday utilities. Their distinct physical and chemical properties make them crucial in various applications across industries.

Metals and metalloids are integral components of modern society, influencing numerous sectors, including construction, transportation, electronics, and materials science. Their diverse properties and applications underscore their significance in our daily lives. However, the environmental and health challenges associated with their extraction and use necessitate responsible practices to ensure sustainability. As research continues to advance, the development of new materials and technologies based on metals and metalloids promises to further enhance their utility while minimizing their impact on the environment.

14.1.1 Defining Metals and Metalloids

Metals are defined as elements that exhibit high electrical and thermal conductivity, luster, malleability, and ductility. They are typically solid at room temperature (with the notable exception of mercury) and possess a crystalline structure. Metals tend to lose electrons in chemical reactions, forming positive ions (cations). Common examples include iron, copper, gold, and aluminum.

Metalloids, on the other hand, are elements that possess properties intermediate between those of metals and nonmetals. They often exhibit mixed characteristics, such as semiconductivity, making them valuable in electronic applications. Examples of metalloids include silicon, germanium, arsenic, and antimony. These elements usually have a metallic appearance but are more brittle than metals and can behave as either conductors or insulators depending on the conditions.

14.1.2 Classification of Metals and Metalloids

Metals

Metals can be classified into several categories based on their properties:

- **Alkali Metals:** Found in Group 1 of the periodic table, alkali metals are highly reactive and possess a single valence electron. Examples include lithium, sodium, and potassium. Their reactivity increases as you move down the group.

- **Alkaline Earth Metals:** Located in Group 2, these metals have two valence electrons and are somewhat less reactive than alkali metals. Examples include magnesium and calcium.
- **Transition Metals:** These elements, found in Groups 3 to 12, are characterized by their ability to form various oxidation states and colored compounds. Common transition metals include iron, copper, and nickel.
- **Post-Transition Metals:** Elements like tin, lead, and bismuth fall into this category. They exhibit metallic properties but tend to be softer and have lower melting points than transition metals.
- **Lanthanides and Actinides:** These rare earth metals are found in the f-block of the periodic table. Lanthanides include elements like cerium and neodymium, while actinides encompass uranium and thorium. They are known for their radioactive properties.

Metalloids

Metalloids are generally found along the zig-zag line dividing metals and nonmetals on the periodic table. The commonly recognized metalloids include:

- **Boron (B):** Used in glass and ceramics, boron is essential in the manufacture of borosilicate glass and as a semiconductor.
- **Silicon (Si):** Silicon is a key component in electronics, primarily used in the production of semiconductors and solar cells.
- **Germanium (Ge):** Similar to silicon, germanium is utilized in electronic applications, especially in fiber optics and infrared optics.
- **Arsenic (As):** While toxic in its elemental form, arsenic is employed in various alloys and semiconductor technologies.
- **Antimony (Sb):** Antimony is used in flame retardants and as an alloying agent with lead.

14.1.3 Physical and Chemical Properties

Metals

Metals possess distinctive physical properties, including:

- **Luster:** Metals typically have a shiny appearance due to their ability to reflect light.

- **Malleability:** Metals can be hammered or pressed into thin sheets without breaking. This property is crucial in the production of metal sheets and foils.
- **Ductility:** Metals can be stretched into wires. This property is essential for applications in electrical wiring and cables.
- **Conductivity:** Metals are excellent conductors of heat and electricity, making them vital for electrical and thermal applications.

Chemically, metals tend to lose electrons during reactions, forming cations. They readily react with nonmetals, such as oxygen and halogens, to form ionic compounds. For example, sodium reacts vigorously with chlorine to form sodium chloride (table salt).

Metalloids

Metalloids exhibit a mix of properties between metals and nonmetals. Their key characteristics include:

- **Semiconductivity:** Metalloids can conduct electricity but not as effectively as metals. This property is exploited in electronic devices, particularly semiconductors.
- **Brittleness:** Unlike metals, metalloids are generally more brittle and can break or shatter under stress.
- **Varied Reactivity:** Metalloids can exhibit properties of both metals and nonmetals in chemical reactions, depending on the conditions. For instance, silicon can form covalent bonds with nonmetals and can also react with metals.

The versatile properties of metalloids make them valuable in various technological applications, particularly in the fields of electronics and materials science.

14.1.4 Applications of Metals

Metals are integral to numerous industries, and their applications are vast:

Construction

Metals such as steel and aluminum are fundamental materials in construction. Steel, an alloy of iron and carbon, is known for its strength and durability, making it ideal for structural components in buildings, bridges, and roads. Aluminum, with its lightweight and corrosion-resistant properties, is extensively used in modern architecture.

Transportation

Metals play a crucial role in the transportation sector. Cars, airplanes, ships, and trains are primarily made of metal. For instance, steel is used in the production of car frames, while aluminum is often used in aircraft to reduce weight and improve fuel efficiency.

Electronics

Metals such as copper, silver, and gold are essential in the electronics industry. Copper is widely used in electrical wiring due to its excellent conductivity, while gold and silver are used in high-performance electronic components due to their resistance to corrosion.

Manufacturing

In manufacturing, metals are used to create tools, machinery, and equipment. The properties of metals, such as hardness and resistance to wear, make them suitable for producing cutting tools and industrial machinery.

14.1.5 Applications of Metalloids

Metalloids have found significant applications, particularly in the electronics and materials sectors:

Electronics

Silicon, one of the most important metalloids, is the backbone of the semiconductor industry. It is used in the manufacture of integrated circuits, microprocessors, and photovoltaic cells for solar energy. Germanium is also used in fiber optics and infrared optics, enhancing telecommunications.

Alloys

Metalloids like arsenic and antimony are used in various alloys to improve performance. For instance, antimony is added to lead to enhance its strength and hardness, making it suitable for batteries.

Glass and Ceramics

Boron is utilized in the production of borosilicate glass, known for its thermal resistance and durability. This glass is widely used in laboratory equipment and kitchenware.

14.1.6 Environmental and Health Considerations

The extraction and use of metals and metalloids raise environmental and health concerns. Mining activities can lead to habitat destruction, pollution, and soil degradation. Moreover, certain metals and metalloids, such as lead and arsenic, are toxic and pose significant health risks.

They can accumulate in living organisms, leading to various health issues, including neurological damage and cancer.

Efforts are being made to promote sustainable practices in the extraction and use of metals and metalloids. Recycling programs and the development of greener technologies are essential for minimizing environmental impact and protecting human health.

14.2 DYSFUNCTION OF IRON AND COPPER HOMEOSTASIS

Homeostasis is the body's ability to maintain a stable internal environment despite external changes. Among the various elements critical to human health, iron and copper are two essential trace metals that play pivotal roles in numerous biological processes. However, dysregulation of their homeostasis can lead to serious health consequences, impacting various organ systems and leading to disorders.

Dysfunction of iron and copper homeostasis can lead to significant health issues, emphasizing the importance of these trace metals in maintaining overall health. Understanding the mechanisms underlying their regulation and the consequences of their dysregulation can help inform treatment strategies. As research advances, the development of targeted therapies aimed at restoring balance in iron and copper levels holds promise for managing these conditions effectively, improving patient outcomes, and enhancing quality of life. Addressing these issues is essential for public health, particularly in populations at risk for metal dysregulation.

14.2.1 Understanding Iron Homeostasis

Iron is vital for several biological functions, including oxygen transport, electron transfer, and DNA synthesis. The human body has no active mechanism for excreting excess iron; therefore, it relies on tightly regulated absorption and recycling processes. Iron homeostasis is primarily controlled through three key proteins: ferritin, transferrin, and hepcidin.

Ferritin and Transferrin

Ferritin serves as the primary storage form of iron, sequestering excess iron in a non-toxic form within cells, especially in the liver, spleen, and bone marrow. Transferrin, a glycoprotein synthesized in the liver, is responsible for transporting iron in the bloodstream to various tissues, including the bone marrow for red blood cell production. The iron-transferrin complex is crucial for cellular iron uptake, as most cells possess transferrin receptors that facilitate the endocytosis of this complex.

Hepcidin

Hepcidin is a hormone produced by the liver that plays a central role in regulating iron homeostasis. It modulates the absorption of iron from the gut and the release of iron from macrophages and liver stores. High levels of hepcidin inhibit iron absorption by binding to ferroportin, a protein that facilitates iron export from cells. Conversely, low hepcidin levels enhance iron absorption and release, ensuring adequate iron availability for physiological needs.

14.2.2 Consequences of Iron Dysregulation

Dysfunction in iron homeostasis can lead to two primary disorders: iron deficiency anemia and hemochromatosis.

Iron Deficiency Anemia

Iron deficiency anemia (IDA) is the most common form of anemia globally and is characterized by insufficient iron levels to support erythropoiesis. It can result from inadequate dietary intake, increased demand, malabsorption disorders, or chronic blood loss (e.g., from gastrointestinal bleeding). Symptoms of IDA include fatigue, pallor, shortness of breath, and impaired cognitive function. The underlying mechanism involves reduced hemoglobin synthesis, leading to microcytic and hypochromic red blood cells.

Hemochromatosis

In contrast, hemochromatosis is a condition characterized by excessive iron accumulation in the body, often due to genetic mutations that affect hepcidin production or iron absorption. The most common form is hereditary hemochromatosis, caused by mutations in the HFE gene. This disorder can lead to organ damage, particularly in the liver, heart, and pancreas, resulting in conditions such as cirrhosis, diabetes, and heart failure. Symptoms often include joint pain, fatigue, and skin discoloration due to iron deposition.

14.2.3 Understanding Copper Homeostasis

Copper is another essential trace metal that plays critical roles in enzymatic reactions, iron metabolism, and neurotransmitter synthesis. Like iron, copper homeostasis is tightly regulated, primarily through intestinal absorption, biliary excretion, and the action of copper-binding proteins, including ceruloplasmin and metallothionein.

Absorption and Transport

Copper is absorbed in the intestine, where it binds to proteins for transport into the bloodstream. In the liver, copper is incorporated into ceruloplasmin, the major copper-carrying protein, which helps in the oxidation of iron and its mobilization. Additionally, metallothionein plays a significant role in regulating intracellular copper levels by binding excess copper, thus preventing toxicity.

Excretion

Copper is primarily excreted through bile, with the liver acting as a central regulator of copper levels. In healthy individuals, a balance is maintained between dietary intake, hepatic storage, and biliary excretion to prevent copper accumulation.

14.2.4 Consequences of Copper Dysregulation

Dysfunction of copper homeostasis can lead to conditions such as Wilson's disease and copper deficiency.

Wilson's Disease

Wilson's disease is a genetic disorder caused by mutations in the ATP7B gene, which impairs the transport of copper from the liver to bile and its incorporation into ceruloplasmin. This defect results in excessive copper accumulation, leading to hepatic and neurological manifestations. Symptoms may include hepatic dysfunction, neurological disturbances (such as tremors and cognitive decline), and psychiatric symptoms. The condition can lead to serious complications, including liver cirrhosis and acute liver failure if left untreated.

Copper Deficiency

Copper deficiency is less common but can occur due to inadequate dietary intake, malabsorption syndromes, or conditions requiring increased copper levels (such as in pregnancy). Symptoms of copper deficiency include anemia (often resembling iron deficiency anemia), neutropenia, and skeletal abnormalities. The deficiency disrupts numerous biochemical pathways, including those involved in iron metabolism, leading to impaired iron utilization and secondary anemia.

14.2.5 Interplay Between Iron and Copper Homeostasis

The regulation of iron and copper is interlinked; dysfunction in one can adversely affect the other. For instance, copper is required for the proper functioning of ceruloplasmin, which oxidizes ferrous iron (Fe^{2}□) to ferric iron (Fe^{3}□) for transport in the bloodstream. Impaired copper levels can thus lead to iron overload and deficiency states. Conversely, excessive iron can inhibit copper absorption by competing for binding sites, leading to copper deficiency.

14.2.6 Therapeutic Interventions

Managing Iron Disorders

The management of iron disorders typically involves supplementation or dietary modifications. In the case of iron deficiency anemia, iron supplementation is often prescribed, alongside dietary adjustments to increase the intake of iron-rich foods, such as red meat, legumes, and fortified cereals. In contrast, treatment for hemochromatosis focuses on reducing iron overload, often through phlebotomy (therapeutic blood removal) to lower total body iron levels and minimize organ damage.

Managing Copper Disorders

For Wilson's disease, chelation therapy using agents like penicillamine or trientine is employed to promote the excretion of excess copper through the kidneys. Zinc supplementation may also be utilized, as zinc can inhibit copper absorption in the intestine. For copper deficiency, dietary adjustments and copper supplementation are recommended to restore adequate levels.

14.3 METALS AND METALLOIDS IN THERAPY

14.3.1 Historical and General Notes

Metals and metalloids have played a significant role in medicine for centuries. From ancient practices to modern pharmacology, the therapeutic use of these elements reflects the evolution of medical science and our understanding of the human body. The historical and general notes on metals and metalloids in therapy reflect a rich legacy of medical practice and scientific exploration. From ancient civilizations to modern medicine, these elements have played crucial roles in treating various conditions and supporting health. As research advances and new technologies emerge, the therapeutic potential of metals and metalloids continues to expand, offering hope for improved treatments and better patient outcomes. Understanding their historical context and applications is essential for harnessing their benefits while ensuring safety in therapeutic use.

Historical Context

Ancient Civilizations

The use of metals for therapeutic purposes dates back to ancient civilizations. The Egyptians, for instance, utilized copper for its antimicrobial properties, employing it in surgical instruments and as a treatment for various ailments. Ancient Greek physicians, such as Hippocrates, recognized the healing properties of metals, particularly iron, which was used to treat anemia. Galen, a prominent Roman physician, also highlighted the medicinal qualities of metals and their salts, paving the way for their continued exploration in medical practices.

The Alchemical Era

During the Middle Ages and the Renaissance, alchemy emerged as a precursor to modern chemistry, with practitioners experimenting with metals and their compounds in search of the "philosopher's stone," believed to confer immortality and cure diseases. Alchemists studied the properties of metals, including their potential therapeutic applications. Mercury, for example, was used extensively for treating syphilis until the 20th century, despite its toxicity.

The Age of Enlightenment

The Age of Enlightenment marked a shift toward scientific inquiry and empirical research. The 18th and 19th centuries saw the development of more systematic approaches to medicine, with metals being studied for their pharmacological properties. Paracelsus, a Swiss physician, emphasized the importance of chemical substances, including metals, in treatment. He is credited with founding the field of toxicology and advocating for the use of metals in precise dosages to achieve therapeutic effects.

General Applications of Metals and Metalloids in Therapy

Iron

Iron is one of the most widely used metals in therapy, primarily for its role in treating iron deficiency anemia. It is an essential component of hemoglobin, responsible for oxygen transport in the blood. Iron supplements, in various forms (ferrous sulfate, ferrous gluconate, etc.), are prescribed to replenish iron stores and restore hemoglobin levels in patients with anemia. Additionally, iron plays a crucial role in various enzymatic reactions, supporting overall metabolic function.

Copper

Copper is another essential trace element with therapeutic potential. It is vital for several biological processes, including iron metabolism, connective tissue formation, and the functioning of enzymes like cytochrome c oxidase. Copper deficiency can lead to anemia, cardiovascular issues, and skeletal abnormalities. Copper supplementation may be prescribed to address deficiencies, and its antimicrobial properties have led to its use in wound dressings and medical devices.

Zinc

Zinc is a metalloid with significant therapeutic applications. It is essential for immune function, wound healing, and protein synthesis. Zinc supplementation is often recommended for individuals with compromised immune systems or those recovering from injuries. Additionally, zinc plays a crucial role in taste and smell, and its deficiency can lead to sensory impairments. Zinc's potential antiviral properties have garnered attention, especially in the context of respiratory infections and other viral illnesses.

Gold

Gold compounds, particularly aurothiomalate, have been used in the treatment of rheumatoid arthritis. Gold has anti-inflammatory properties that can alleviate joint pain and stiffness. Although its use has declined due to the development of more effective treatments, gold therapy remains an interesting example of metal-based treatment in rheumatology.

Silver

Silver has a long history of use as an antimicrobial agent. Silver sulfadiazine is commonly used in burn care to prevent infections. Its antibacterial properties are attributed to the release of silver ions, which disrupt bacterial cell membranes. Silver nanoparticles are also being explored in wound dressings and coatings for medical devices due to their effectiveness in preventing biofilm formation.

Mercury

Mercury was historically used in various therapeutic applications, particularly for treating syphilis and as a laxative. However, due to its toxicity and associated health risks, its use has largely been abandoned in modern medicine. The recognition of mercury's harmful effects has led to increased scrutiny of other metals and their safety profiles in therapy.

Modern Advances and Research

Metal-Based Drugs

In recent years, metal-based drugs have gained attention for their potential in cancer therapy and other medical applications. Platinum-based compounds, such as cisplatin, are widely used as chemotherapeutic agents for various cancers. These drugs work by interfering with DNA replication in cancer cells, inhibiting their growth and proliferation. Ongoing research explores the development of other metal-based therapies, including those involving ruthenium and gallium.

Nanotechnology in Medicine

Nanotechnology has opened new avenues for the application of metals and metalloids in therapy. Nanoparticles can enhance drug delivery, improve targeting of therapeutics, and reduce side effects. For example, gold nanoparticles are being studied for their potential in targeted cancer therapy and imaging. Silver nanoparticles, known for their antimicrobial properties, are being incorporated into wound dressings and coatings to prevent infections.

Toxicology and Safety Concerns

As the use of metals and metalloids in therapy expands, it is crucial to address safety concerns and potential toxicities. Regulatory agencies, such as the Food and Drug Administration (FDA) and the European Medicines Agency (EMA), closely monitor the use of metal-based drugs to ensure their safety and efficacy. Ongoing research into the long-term effects of metal exposure and the mechanisms of toxicity is essential for the development of safe therapeutic options.

14.3.2 Treatment of Arthritis with Gold Compounds

The use of gold compounds in the treatment of arthritis, particularly rheumatoid arthritis, is a notable aspect of metalloid and metal therapy. Gold compounds remain a significant option in the treatment of rheumatoid arthritis, especially for patients unresponsive to other therapies. Their unique mechanisms of action highlight the importance of metals and metalloids in therapeutic applications. However, the potential for side effects necessitates careful patient monitoring and consideration of individual health factors.

Types of Gold Compounds:

- **Auranofin:** An oral gold compound used for rheumatoid arthritis treatment.
- **Gold Sodium Thiomalate:** An injectable form used in more severe cases.

Mechanism of Action

Gold compounds exert their therapeutic effects through several mechanisms:

- **Anti-Inflammatory Properties:** Gold has been shown to reduce inflammation by modulating immune responses and inhibiting pro-inflammatory cytokines.
- **Inhibition of Cell Proliferation:** Gold compounds can interfere with the proliferation of synovial cells in the joints, leading to reduced joint damage.
- **Modification of Immune Responses:** Gold is believed to affect the activity of macrophages and other immune cells, contributing to its immunosuppressive effects.
- **Reduction of Joint Damage:** Clinical studies have shown that gold compounds can help slow the progression of joint damage in rheumatoid arthritis.

Clinical Use

- **Indications:** Primarily used for patients with rheumatoid arthritis who do not respond well to conventional therapies.
- **Administration:** Auranofin is taken orally, while gold sodium thiomalate is administered via intramuscular injection.
- **Efficacy:** Some studies have reported significant improvement in symptoms and joint function in patients treated with gold compounds.

Side Effects

While gold compounds can be effective, they are associated with various side effects, including:

- **Gastrointestinal Issues:** Nausea, vomiting, and diarrhea can occur, particularly with oral forms.

- **Dermatological Reactions:** Skin rashes or other allergic reactions may develop.
- **Hematological Effects:** Myelosuppression can lead to decreased blood cell counts.
- **Renal Toxicity:** Long-term use may lead to renal impairment.

14.3.3 Cancer Treatment

Cancer remains one of the leading causes of mortality worldwide, prompting ongoing research into effective treatment modalities. While conventional therapies such as chemotherapy, radiation, and surgery have made significant strides, the exploration of metals and metalloids in cancer therapy has emerged as a promising avenue. This approach utilizes the unique properties of various metals and metalloids to target cancer cells, enhance therapeutic efficacy, and minimize side effects.

Metals and metalloids have emerged as promising agents in cancer therapy, offering unique mechanisms of action and potential benefits when used alone or in combination with conventional treatments. Platinum-based compounds, gold compounds, arsenic trioxide, and metal nanoparticles represent a diverse array of therapeutic options that can target cancer cells and enhance treatment outcomes. Ongoing research into their mechanisms, efficacy, and safety will further elucidate their role in oncology and pave the way for innovative treatment strategies that can ultimately improve patient care and outcomes. As our understanding of the complex interplay between metals, metalloids, and cancer biology deepens, the potential for harnessing these elements in therapy continues to expand, promising a future of more effective and targeted cancer treatments.

Understanding Metals and Metalloids in Cancer Therapy

Metals and metalloids, including platinum, gold, arsenic, and others, have garnered attention for their therapeutic potential in oncology. These elements possess distinct physicochemical properties that can be harnessed to disrupt cancer cell metabolism, induce apoptosis (programmed cell death), and enhance the efficacy of existing therapies. Their unique interactions with biological systems allow them to act at various cellular targets, providing a multifaceted approach to cancer treatment.

Platinum-Based Chemotherapy

Platinum-based compounds, particularly cisplatin, carboplatin, and oxaliplatin, have revolutionized cancer treatment since their introduction. Cisplatin was first approved for clinical use in the 1970s and has since become a cornerstone in treating various malignancies, including testicular, ovarian, bladder, and lung cancers. The mechanism of action of cisplatin involves the formation of DNA cross-links, which impede DNA replication and transcription. This action triggers cellular responses that ultimately lead to apoptosis in cancer cells. Cisplatin's efficacy is often limited by its nephrotoxicity and the development of resistance. Researchers are exploring novel platinum analogs and combinations with other agents to enhance therapeutic outcomes while minimizing adverse effects. For example, combining cisplatin with targeted therapies or immunotherapies may improve its effectiveness and counteract resistance mechanisms.

Gold Compounds in Cancer Therapy

Gold compounds, such as auranofin, traditionally used for rheumatoid arthritis, have also shown potential in cancer treatment. Recent studies suggest that auranofin can inhibit cancer cell proliferation, induce apoptosis, and enhance the effectiveness of chemotherapeutic agents. The mechanism involves the modulation of redox signaling and inhibition of key survival pathways in cancer cells.

In preclinical studies, auranofin has demonstrated activity against various cancer types, including breast, ovarian, and lung cancers. Its ability to selectively target cancer cells while sparing normal cells is a promising feature that researchers are actively investigating. While clinical trials are still needed to establish its efficacy and safety in cancer patients, the initial results are encouraging.

Arsenic Trioxide: A Metalloid with Therapeutic Promise

Arsenic trioxide (As2O3), a metalloid, has gained prominence in the treatment of acute promyelocytic leukemia (APL). This compound induces differentiation and apoptosis in leukemic cells, primarily through the modulation of the retinoic acid receptor and the activation of apoptotic signaling pathways. Its approval for clinical use in APL has demonstrated its potential as an effective cancer treatment.

Arsenic trioxide is often used in combination with all-trans-retinoic acid (ATRA), enhancing therapeutic outcomes in APL. However, its use is associated with potential side effects, including cardiac toxicity and the development of differentiation syndrome. Ongoing research aims to optimize arsenic trioxide therapy, including identifying biomarkers for patient selection and combination strategies to improve efficacy and reduce toxicity.

Antitumor Activity of Transition Metals

Several transition metals, including iron, copper, and zinc, play critical roles in cellular processes and have been investigated for their antitumor properties. Iron chelators, such as deferoxamine, have shown promise in preclinical studies by disrupting iron metabolism in cancer cells, leading to increased oxidative stress and apoptosis. This approach targets the altered iron homeostasis characteristic of many tumors, which often exhibit elevated iron levels to support rapid proliferation. Copper also plays a role in tumor angiogenesis, and its inhibition has been explored as a therapeutic strategy. Agents that disrupt copper metabolism can hinder tumor growth by impairing blood vessel formation. The use of zinc in cancer therapy is less direct but may involve its role as an essential trace element in maintaining immune function and modulating tumor cell signaling pathways.

Nanoparticle-Based Approaches

Nanotechnology has paved the way for innovative cancer treatments involving metal and metalloid nanoparticles. Gold and silver nanoparticles, in particular, have garnered attention for their unique properties, including their ability to selectively accumulate in tumors and enhance the effectiveness of radiation therapy. These nanoparticles can be conjugated with targeting ligands, allowing for precise delivery of therapeutic agents directly to cancer cells.

Gold nanoparticles can be employed as carriers for chemotherapeutics or as agents for photothermal therapy, where they absorb light and convert it into heat to induce localized cell death. Silver nanoparticles exhibit antibacterial properties and can enhance the effectiveness of certain chemotherapy agents. The versatility of metal nanoparticles in cancer therapy highlights their potential as adjuncts to conventional treatments, potentially improving patient outcomes.

Challenges and Future Directions

Despite the promising potential of metals and metalloids in cancer treatment, several challenges remain. The development of resistance to metal-based therapies poses a significant hurdle, necessitating ongoing research to understand the underlying mechanisms and develop strategies to overcome it. Additionally, the potential toxicity associated with some metals requires careful monitoring and dose optimization to minimize adverse effects while maximizing therapeutic benefits.

Future research should focus on elucidating the mechanisms of action of various metal-based compounds, optimizing their pharmacokinetics and biodistribution, and exploring combination strategies with other therapeutic modalities. Personalized medicine approaches, utilizing biomarkers to guide treatment decisions, will be crucial in tailoring metal and metalloid therapies to individual patients for improved efficacy.

14.3.4 Further Metal Based Medications

The therapeutic applications of metals and metalloids in medicine have garnered increasing attention over the past few decades. While certain metal-based medications, such as platinum compounds in cancer treatment, are well-known, a broader range of metals and metalloids is being explored for their potential medicinal benefits.

The exploration of metals and metalloids in therapy represents an exciting frontier in medicine. Beyond their established applications, a diverse array of metal-based medications is being investigated for their potential benefits in treating various medical conditions. From gold and silver compounds to iron and copper chelators, these agents offer unique mechanisms of action and therapeutic possibilities. As research continues to uncover their potential, metal-based medications may play an increasingly vital role in improving patient outcomes and addressing unmet medical needs. The future of metal-based therapy is promising, with the potential to revolutionize the treatment landscape across a range of diseases.

Overview of Metal-Based Medications

Metal-based medications harness the unique properties of metals and metalloids to interact with biological systems. These compounds can modulate cellular processes, influence enzyme activity, and alter the pharmacokinetics of drugs. Various metals, including gold, silver, iron, copper, and zinc, have been studied for their therapeutic potential across a spectrum of medical conditions. This growing interest is driven by the need for novel treatments that can overcome the limitations of traditional pharmaceuticals, such as drug resistance and adverse side effects.

Gold Compounds in Medicine

Gold has been used in medicine for centuries, primarily in the treatment of rheumatoid arthritis. Auranofin, an oral gold compound, has shown anti-inflammatory and immunomodulatory properties. It works by inhibiting the production of pro-inflammatory cytokines and promoting apoptosis in activated immune cells. In recent years, research has expanded to explore the potential of gold compounds in treating various cancers. Auranofin, in particular, has demonstrated efficacy against cancer cell lines by disrupting cellular redox homeostasis and enhancing apoptosis.

Beyond rheumatoid arthritis and cancer, gold nanoparticles are emerging as promising agents in drug delivery and imaging. Their unique optical properties and ability to be functionalized with various targeting ligands allow for selective delivery of therapeutics to cancerous tissues, improving treatment outcomes while minimizing side effects. The versatility of gold compounds underscores their potential for further applications in medicine.

Silver Compounds and Their Applications

Silver has long been recognized for its antimicrobial properties, and silver sulfadiazine is a well-established topical agent used to prevent infections in burn patients. The antimicrobial activity of silver compounds is attributed to their ability to release silver ions, which disrupt bacterial cell membranes and interfere with metabolic processes. Beyond topical applications, silver nanoparticles have gained attention for their broad-spectrum antimicrobial activity, being explored for use in coatings for medical devices, wound dressings, and even in some formulations for treating infections.

In addition to their antimicrobial effects, silver nanoparticles are being investigated for their potential in cancer therapy. Studies have shown that silver nanoparticles can induce apoptosis in various cancer cell lines and enhance the efficacy of certain chemotherapeutic agents. This dual functionality—antimicrobial and anticancer—highlights the potential for silver-based compounds to address multiple medical needs simultaneously.

Iron Chelators in Therapy

Iron is an essential element for many biological processes, but its dysregulation can lead to a range of health issues, including anemia and iron overload disorders. Iron chelators are medications designed to bind excess iron and facilitate its excretion from the body. Deferoxamine is one of the most commonly used iron chelators, primarily for treating patients with conditions such as thalassemia and sickle cell disease who require frequent blood transfusions. These patients often develop iron overload due to the accumulation of iron from transfused red blood cells.

The mechanism of action of deferoxamine involves binding to free iron, forming a stable complex that can be eliminated through the kidneys. This chelation therapy has been shown to significantly reduce iron levels and prevent complications associated with iron overload, including organ damage. Newer oral iron chelators, such as deferasirox and deferiprone, have been developed to improve patient compliance and convenience, providing additional options for managing iron-related disorders.

Copper in Medical Applications

Copper is an essential trace element involved in various biological processes, including enzyme function, collagen synthesis, and neurotransmitter regulation. However, dysregulation of copper levels can lead to various health issues. For instance, Wilson's disease is a genetic disorder characterized by excessive copper accumulation in the body, leading to hepatic and neurological damage. The primary treatment for Wilson's disease involves copper chelation with agents such as penicillamine, which facilitates the excretion of excess copper through urine.

Emerging research has also highlighted the role of copper in cancer biology. Elevated copper levels have been observed in various tumor types, and it is believed that copper contributes to tumor growth and metastasis. As a result, copper-lowering agents are being investigated as potential adjuncts in cancer therapy. Additionally, copper nanoparticles are being studied for their antimicrobial and anticancer properties, potentially offering novel therapeutic avenues.

Zinc in Health and Disease

Zinc is another essential trace element involved in numerous biological processes, including immune function, protein synthesis, and DNA synthesis. Zinc deficiency is linked to various health problems, including impaired immune response and developmental issues. Zinc supplementation has been explored in various clinical settings, particularly for enhancing immune function in populations at risk of deficiency.

Moreover, zinc has gained attention for its role in cancer therapy. Studies suggest that zinc can modulate the activity of various signaling pathways involved in cancer progression, including the p53 pathway and the NF-κB pathway. Zinc-based compounds are being explored for their potential in enhancing the efficacy of conventional chemotherapeutics and in targeted therapies.

The Promise of Metal Nanoparticles

Metal nanoparticles, including gold, silver, and platinum, have emerged as innovative agents in drug delivery and therapy. Their unique physicochemical properties allow for improved targeting and controlled release of therapeutic agents, enhancing treatment efficacy while minimizing side effects. Nanoparticles can be engineered to deliver chemotherapeutics directly to tumor sites, improving drug accumulation in cancerous tissues and reducing systemic toxicity.

Additionally, metal nanoparticles are being explored for use in imaging techniques, such as computed tomography (CT) and magnetic resonance imaging (MRI). Their unique optical and magnetic properties allow for enhanced contrast and resolution in imaging, aiding in early diagnosis and treatment monitoring.

Challenges and Future Directions

While the potential of metal-based medications is promising, several challenges remain. The complexity of metal metabolism in the human body can lead to variable responses and potential toxicity. Moreover, the development of resistance to metal-based therapies is a concern, particularly in the context of cancer treatment. Ongoing research aims to better understand the mechanisms of action of metal-based compounds, optimize their pharmacokinetics, and identify patient populations that may benefit most from these therapies.

Future studies should focus on the development of personalized medicine approaches, utilizing biomarkers to guide treatment decisions and maximize therapeutic outcomes. The integration of metal-based therapies into existing treatment paradigms, such as combination therapy with traditional pharmaceuticals, may also enhance overall efficacy.

14.3.5 Radiopharmaceuticals

Radiopharmaceuticals are specialized compounds that incorporate radioactive isotopes and are used in the field of nuclear medicine for both diagnostic and therapeutic purposes. These unique agents have revolutionized the way diseases, particularly cancer, are diagnosed and treated, allowing for more targeted and effective interventions.

Radiopharmaceuticals represent a vital component of modern medicine, offering unique opportunities for diagnosing and treating a wide range of conditions. From their role in cancer therapy to their applications in cardiovascular and neurological imaging, these agents have transformed the landscape of healthcare. As research continues to advance, the development of novel radiopharmaceuticals and innovative applications promises to further enhance patient outcomes, making radiopharmaceuticals an exciting area of focus in the ongoing evolution of medical science.

Understanding Radiopharmaceuticals

Radiopharmaceuticals consist of two main components: a radioactive isotope and a pharmaceutical component that allows the compound to localize in specific tissues or organs. The radioactive isotope emits radiation, which can be detected using imaging technologies or used to deliver therapeutic doses of radiation to target tissues. This combination of properties makes radiopharmaceuticals invaluable in diagnosing and treating various medical conditions.

Types of Radiopharmaceuticals

Radiopharmaceuticals can be broadly categorized into two types: diagnostic radiopharmaceuticals and therapeutic radiopharmaceuticals.

Diagnostic Radiopharmaceuticals

Diagnostic radiopharmaceuticals are primarily used in imaging techniques such as positron emission tomography (PET) and single-photon emission computed tomography (SPECT).

These agents allow physicians to visualize physiological processes within the body, helping to diagnose conditions such as cancer, cardiovascular diseases, and neurological disorders. Commonly used diagnostic radiopharmaceuticals include technetium-99m (Tc-99m), which is used for a variety of imaging studies, and fluorodeoxyglucose (FDG), which is utilized in PET scans for detecting cancerous tissues.

Therapeutic Radiopharmaceuticals

Therapeutic radiopharmaceuticals deliver targeted radiation therapy to treat specific conditions, particularly cancer. These agents are designed to localize in tumor cells, allowing for the direct destruction of malignant tissues while minimizing damage to surrounding healthy tissues. Common examples of therapeutic radiopharmaceuticals include iodine-131 (I-131), which is used to treat thyroid cancer, and radium-223, which is indicated for patients with bone metastases from prostate cancer.

Mechanisms of Action

The efficacy of radiopharmaceuticals lies in their unique mechanisms of action. Diagnostic agents work by emitting gamma rays or positrons, which can be detected by specialized imaging equipment. For example, in a PET scan, the radiopharmaceutical emits positrons that collide with electrons in the body, resulting in the production of gamma rays. These rays are then captured by the PET scanner to create detailed images of metabolic activity.

Therapeutic radiopharmaceuticals operate differently. They utilize beta or alpha radiation to damage the DNA of target cells, leading to cell death. The selective uptake of these agents by cancerous tissues allows for a concentrated dose of radiation to be delivered directly to tumors, enhancing the therapeutic effect while reducing side effects associated with traditional radiation therapy.

Applications in Medicine

Cancer Diagnosis and Treatment

One of the most significant applications of radiopharmaceuticals is in oncology. Diagnostic agents, such as FDG-PET, enable oncologists to detect tumors at an early stage, assess the extent of disease, and monitor treatment responses. Furthermore, therapeutic radiopharmaceuticals provide a targeted approach to treating various types of cancers, including thyroid cancer, neuroendocrine tumors, and bone metastases.

Cardiovascular Imaging

Radiopharmaceuticals play a crucial role in cardiovascular imaging. Agents such as Tc-99m sestamibi and Tc-99m teboroxime are used in myocardial perfusion imaging to evaluate blood flow to the heart muscle. This information is critical for diagnosing coronary artery disease and determining the most appropriate treatment options for patients.

Neurological Disorders

In addition to cancer and cardiovascular applications, radiopharmaceuticals are utilized in the diagnosis and management of neurological disorders. For instance, iodine-123 labeled compounds can be employed in imaging studies to evaluate conditions such as Alzheimer's disease and Parkinson's disease. These agents provide valuable insights into the underlying pathology of these disorders, aiding in diagnosis and treatment planning.

Advantages of Radiopharmaceuticals

Radiopharmaceuticals offer several advantages over traditional imaging and treatment modalities. One of the primary benefits is their ability to provide functional information about physiological processes rather than just structural details. This capability allows for earlier detection of diseases and more accurate assessments of treatment responses.

Additionally, the targeted nature of therapeutic radiopharmaceuticals minimizes damage to healthy tissues, reducing the risk of side effects commonly associated with conventional chemotherapy and external beam radiation therapy. The ability to deliver high doses of radiation directly to tumor cells enhances treatment efficacy while preserving normal surrounding tissues.

Challenges and Limitations

Despite their numerous advantages, the use of radiopharmaceuticals also presents several challenges. The production of these agents requires specialized facilities and expertise, as they are typically manufactured in nuclear reactors or cyclotrons. Additionally, the handling and disposal of radioactive materials necessitate strict safety protocols to protect healthcare workers and patients.

Furthermore, there is a need for ongoing research to optimize the pharmacokinetics and biodistribution of radiopharmaceuticals. Understanding how these agents behave in the body can lead to the development of new compounds with improved efficacy and reduced side effects. Additionally, regulatory hurdles must be navigated to bring new radiopharmaceuticals to market, ensuring their safety and effectiveness.

Future Directions in Radiopharmaceutical Development

The future of radiopharmaceuticals is promising, with ongoing research focused on several key areas. One area of interest is the development of novel radiolabeled compounds that target specific biomarkers associated with various diseases. By improving the specificity and selectivity of radiopharmaceuticals, researchers aim to enhance diagnostic accuracy and therapeutic outcomes.

Another promising avenue is the integration of radiopharmaceuticals with other therapeutic modalities, such as immunotherapy. Combining the targeted effects of radiopharmaceuticals with the immune system's ability to recognize and destroy cancer cells could lead to synergistic effects and improved treatment outcomes.

Furthermore, advancements in imaging technologies are expected to enhance the capabilities of radiopharmaceuticals. Innovations in hybrid imaging techniques, such as PET/MRI and SPECT/CT, allow for simultaneous anatomical and functional imaging, providing comprehensive insights into disease processes and treatment responses.

14.4 METALS AND METALLOIDS IN DIAGNOSTIC IMAGING

The field of diagnostic imaging has advanced significantly over the past few decades, owing much of its progress to the incorporation of metals and metalloids. These elements play crucial roles in various imaging modalities, such as X-ray radiography, computed tomography (CT), magnetic resonance imaging (MRI), and nuclear medicine. The ability of metals and metalloids to enhance contrast, improve image quality, and facilitate the visualization of internal structures has made them indispensable tools in modern medical diagnostics.

Metals and metalloids have become integral components of diagnostic imaging, significantly enhancing the ability to visualize internal structures and diagnose various medical conditions. From iodine and barium in X-ray studies to gadolinium in MRI, these elements have revolutionized medical diagnostics, leading to improved patient outcomes. As advancements in materials science continue to evolve, the potential for novel metallic and metalloid-based agents to further enhance imaging capabilities is promising. However, safety considerations and regulatory oversight remain paramount to ensure the effective and responsible use of these agents in clinical practice. The future of diagnostic imaging will undoubtedly benefit from ongoing research and innovation in the field of metals and metalloids.

14.4.1 The Role of Metals and Metalloids

Metals and metalloids are characterized by their unique physical and chemical properties, making them suitable for various applications in diagnostic imaging. They have distinct atomic structures that allow them to interact with radiation in specific ways, thereby affecting the imaging process. For instance, elements such as iodine, barium, and gadolinium are commonly used as contrast agents in imaging procedures to improve the visibility of internal organs and tissues.

Iodine

Iodine is one of the most widely used contrast agents in radiographic imaging, particularly in X-ray and CT scans. Its high atomic number allows it to absorb X-rays more effectively than surrounding tissues, resulting in enhanced contrast on the images. Iodinated contrast agents can be administered orally or intravenously, depending on the area of interest. When injected, iodine helps delineate blood vessels, tumors, and other structures, aiding in the diagnosis of various conditions, such as vascular diseases and tumors.

Iodine-based contrast agents have proven to be invaluable in angiography, where the visualization of blood vessels is critical. The ability to accurately visualize the vascular system allows for better diagnosis and treatment planning for conditions such as aneurysms and blockages. However, the use of iodine contrast agents is not without risks; some patients may experience allergic reactions or adverse effects, particularly those with pre-existing kidney conditions.

Barium

Barium is another metal widely employed in diagnostic imaging, particularly in gastrointestinal studies. Barium sulfate is a radiopaque contrast agent used in X-ray imaging of the digestive tract. When ingested, barium coats the lining of the gastrointestinal tract, making it visible on X-ray images. This technique is commonly used for diagnosing conditions such as ulcers, tumors, and obstructions within the gastrointestinal system.

Barium studies, such as barium swallow and barium enema, provide detailed information about the structure and function of the esophagus, stomach, and intestines. The use of barium in these procedures has significantly improved the ability to diagnose gastrointestinal disorders, leading to timely and appropriate interventions.

Gadolinium

In the realm of magnetic resonance imaging (MRI), gadolinium is the primary contrast agent used to enhance image quality. Gadolinium-based contrast agents (GBCA) contain gadolinium chelated with a ligand to prevent toxicity. These agents improve the visibility of blood vessels, organs, and lesions within the body by altering the relaxation times of nearby water protons in MRI scans. As a result, tissues with higher gadolinium concentrations appear brighter, allowing for better differentiation of normal and abnormal structures.

Gadolinium-enhanced MRI is particularly useful in diagnosing neurological conditions, such as tumors, multiple sclerosis, and vascular malformations. However, the use of gadolinium-based contrast agents has raised concerns about potential side effects, including nephrogenic systemic fibrosis (NSF) in patients with severe kidney dysfunction. As a result, healthcare providers must carefully assess the risks and benefits of gadolinium administration on a case-by-case basis.

14.4.2 Metalloids in Diagnostic Imaging

While metals like iodine, barium, and gadolinium are prominent in diagnostic imaging, metalloids also play significant roles, particularly in advanced imaging technologies. Silicon, for example, is a crucial component of semiconductor materials used in detectors for X-ray and CT imaging. The properties of silicon allow it to convert X-ray photons into electrical signals, which are then processed to create detailed images.

Another metalloid, germanium, is used in various imaging technologies, including positron emission tomography (PET). Germanium-based detectors are highly efficient at detecting gamma rays emitted from radiopharmaceuticals during PET scans. This capability enhances the sensitivity and accuracy of PET imaging, leading to improved diagnostic outcomes in oncology and other fields.

14.4.3 Emerging Technologies in Diagnostic Imaging

Advancements in materials science and engineering have paved the way for the development of novel metals and metalloids that can enhance imaging capabilities further. Researchers are exploring the potential of new metallic nanoparticles as contrast agents. These nanoparticles can be engineered to target specific tissues or tumors, allowing for more precise imaging and treatment.

For instance, gold nanoparticles are being studied for their potential use in enhanced CT imaging. Their unique optical properties and ability to absorb X-rays make them promising candidates for improving contrast in imaging studies. Additionally, these nanoparticles can be functionalized to target cancer cells specifically, providing a dual function of imaging and therapeutic delivery.

Similarly, the development of iron oxide nanoparticles is gaining attention in MRI applications. These nanoparticles can serve as contrast agents and enable imaging of various biological processes, such as inflammation and tumor growth. Their ability to be tailored for specific applications makes them a valuable area of research in the field of diagnostic imaging.

14.4.4 Safety Considerations and Challenges

Despite the benefits of using metals and metalloids in diagnostic imaging, safety considerations must be taken into account. The administration of contrast agents can pose risks, including allergic reactions, kidney damage, and other adverse effects. It is essential for healthcare providers to screen patients for contraindications, such as prior allergies to contrast agents or pre-existing kidney conditions.

Moreover, the use of radiation in imaging procedures raises concerns about cumulative exposure over time. Healthcare professionals must weigh the risks and benefits of imaging studies for each patient, ensuring that the benefits of diagnosis and treatment outweigh the potential hazards associated with radiation exposure.

Regulatory Framework and Future Directions

The development and use of metals and metalloids in diagnostic imaging are subject to rigorous regulatory oversight to ensure patient safety and effectiveness. Agencies such as the U.S. Food and Drug Administration (FDA) and the European

Medicines Agency (EMA) evaluate and approve contrast agents based on their safety profiles and clinical efficacy.

As research continues to explore the potential of novel metals and metalloids, future directions may include the integration of advanced imaging techniques, such as hybrid imaging modalities that combine different imaging methods. For example, combining PET and MRI can provide comprehensive information about both metabolic activity and anatomical structure, leading to improved diagnostic capabilities.

14.5 THE TOXIC AND THERAPEUTIC POTENTIAL OF CO, NO AND H2S

Carbon monoxide (CO), nitric oxide (NO), and hydrogen sulfide (H_2S) are gases that have garnered significant attention in the fields of toxicology and medicine due to their dual roles as toxic agents and potential therapeutic agents. While often perceived as harmful, these gases play essential physiological roles in the body and have emerged as promising candidates for therapeutic applications.

Carbon monoxide, nitric oxide, and hydrogen sulfide are gases that exemplify the complex interplay between toxicity and therapeutic potential. While they pose significant risks at elevated concentrations, their physiological roles and therapeutic applications are becoming increasingly recognized in modern medicine. Ongoing research into their mechanisms of action, clinical applications, and safety profiles will undoubtedly pave the way for innovative therapeutic strategies. Understanding and harnessing the beneficial effects of these gases while managing their toxic potential will be crucial in advancing medical science and improving patient outcomes.

14.5.1 Carbon Monoxide: A Double-Edged Sword

Toxicity of Carbon Monoxide

Carbon monoxide is a colorless, odorless gas produced during the incomplete combustion of carbon-containing fuels. Its toxicity arises from its ability to bind to hemoglobin in red blood cells, forming carboxyhemoglobin, which reduces the blood's oxygen-carrying capacity. This can lead to symptoms ranging from headaches and dizziness to unconsciousness and death in severe cases. CO poisoning is a significant public health concern, particularly in enclosed spaces where combustion appliances are used without proper ventilation.

The mechanism of CO toxicity is not limited to its interference with oxygen transport; it also affects cellular respiration by inhibiting cytochrome c oxidase, an essential enzyme in the mitochondrial electron transport chain. This inhibition leads to decreased ATP production and increased oxidative stress, contributing to cellular damage and dysfunction. Furthermore, CO can disrupt neurotransmitter systems and induce inflammatory responses, exacerbating its toxic effects.

Therapeutic Potential of Carbon Monoxide

Despite its toxicity, carbon monoxide has demonstrated therapeutic potential when administered in controlled doses. Research has shown that low concentrations of CO can exert anti-inflammatory and cytoprotective effects, making it a candidate for treating various medical conditions. For instance, CO has been investigated as a potential treatment for conditions such as ischemia-reperfusion injury, organ transplantation, and chronic inflammatory diseases.

The therapeutic effects of CO are thought to stem from its ability to activate anti-inflammatory pathways and inhibit pro-inflammatory cytokines. Additionally, CO can promote vasodilation and improve blood flow, contributing to tissue protection during ischemic events. Novel delivery systems, such as CO-releasing molecules (CORMs), have been developed to harness the therapeutic potential of CO while minimizing its toxic effects. These advancements indicate a paradigm shift in how CO is viewed in medical science, from a toxic gas to a potential therapeutic agent.

14.5.2 Nitric Oxide: The Versatile Messenger

Toxicity of Nitric Oxide

Nitric oxide is a gaseous signaling molecule produced endogenously in the body through the action of nitric oxide synthase (NOS) enzymes. It plays a vital role in various physiological processes, including vasodilation, neurotransmission, and immune responses. However, excessive production of NO can lead to toxicity. High levels of NO can react with superoxide to form peroxynitrite, a reactive nitrogen species that can cause oxidative stress and damage to proteins, lipids, and DNA.

Elevated levels of nitric oxide have been implicated in various pathological conditions, including neurodegenerative diseases, sepsis, and chronic inflammatory disorders. In these contexts, the dysregulation of NO synthesis can contribute to tissue damage and exacerbate disease progression. Thus, while NO is crucial for maintaining normal physiological functions, its overproduction can have detrimental effects.

Therapeutic Potential of Nitric Oxide

Nitric oxide's role as a signaling molecule has led to its exploration as a therapeutic agent. In the field of cardiovascular medicine, NO is recognized for its ability to induce vasodilation, improving blood flow and reducing blood pressure. Nitric oxide donors, such as nitroglycerin, are widely used to manage conditions like angina and heart failure. Additionally, inhaled nitric oxide has been employed as a treatment for pulmonary hypertension and respiratory failure in newborns, highlighting its potential in respiratory therapies.

Beyond cardiovascular applications, nitric oxide has also been investigated for its antimicrobial properties. It can enhance the immune response by promoting the destruction of pathogens, making it a candidate for treating infections. Researchers are exploring the use of NO in wound healing, cancer therapy, and as a potential treatment for neurological disorders. The versatility of nitric oxide as a therapeutic agent underscores its importance in modern medicine.

14.5.3 Hydrogen Sulfide: The Gas with Dual Roles

Toxicity of Hydrogen Sulfide

Hydrogen sulfide is a colorless gas known for its characteristic rotten egg odor. It is produced naturally in the body through the metabolism of sulfur-containing amino acids and by gut microbiota. At low concentrations, H_2S acts as a signaling molecule, but at elevated levels, it becomes highly toxic. Exposure to high concentrations of H_2S can result in respiratory distress, loss of consciousness, and death. The toxic effects are primarily attributed to H_2S's ability to inhibit cellular respiration by blocking cytochrome c oxidase in the mitochondria, leading to cellular energy failure.

The mechanism of toxicity involves the formation of sulfide adducts with critical enzymes and proteins, resulting in impaired cellular function. The potential for H_2S to cause neurotoxicity and systemic toxicity makes it a significant concern in occupational and environmental settings.

Therapeutic Potential of Hydrogen Sulfide

Despite its toxicity, hydrogen sulfide has emerged as a molecule of interest in therapeutic research. Its physiological functions include vasodilation, modulation of neurotransmission, and protection against oxidative stress. H_2S has been shown to have cytoprotective effects, particularly in models of ischemia-reperfusion injury, neurodegeneration, and inflammatory diseases. Research has indicated that H_2S can promote wound healing and protect against tissue injury in various models. It has been studied for its potential in managing conditions such as hypertension, heart failure, and neurodegenerative disorders. H_2S-releasing compounds, known as H_2S donors, are being developed to harness its therapeutic effects while mitigating the risks associated with high concentrations of the gas.

14.5.4 The Balance Between Toxicity and Therapeutic Potential

The dual roles of CO, NO, and H_2S highlight the complexity of gasotransmitters in biological systems. While these gases can exert toxic effects when present in excess, they also play essential roles in maintaining homeostasis and regulating physiological processes. Understanding the precise mechanisms underlying their toxicity and therapeutic potential is critical for developing effective treatments and mitigating risks.

Clinical Applications and Future Directions

The therapeutic applications of CO, NO, and H_2S are being actively researched, with ongoing studies aimed at elucidating their mechanisms of action, optimal dosing, and delivery methods. The development of gas-releasing compounds offers a promising avenue for translating these gases into clinical practice. For instance, CORMs, NO donors, and H_2S donors are being investigated for their potential to provide controlled therapeutic effects without the associated toxicity.

The future of research in this area holds great promise. The use of these gases in combination therapies may enhance treatment efficacy while minimizing adverse effects. Furthermore, identifying biomarkers that indicate the optimal levels of these gases in different pathological states could lead to more personalized and effective therapeutic strategies.

BIBLIOGRAPHY

Bioinorganic Chemistry of Calcium Authors: J. Decker, G. Albrecht **Publisher:** Academic Press **Year:** 2017

Bioinorganic Chemistry of Vanadium Authors: K. J. B. McCulloch, S. C. F. V. I. Davidson **Publisher:** Springer **Year:** 2019

Bioinorganic Chemistry: A Focus on the Role of Metal Ions in Biology Editors: R. H. Holm, E. I. Stiefel, A. A. G. Wilkins **Publisher:** Wiley **Year:** 2004

Bioinorganic Chemistry: A Laboratory Manual Authors: T. J. C. A. Schneider, J. G. H. Yu **Publisher:** Wiley **Year:** 2022

Bioinorganic Chemistry: A Short Course Authors: Gary W. Brudvig, R. Alan W. Hutton **Publisher:** Wiley **Year:** 2020

Bioinorganic Chemistry: Advances and Applications Editors: A. J. T. O'Connor, E. F. Smith **Publisher:** Elsevier **Year:** 2014

Bioinorganic Chemistry: Inorganic Elements in the Chemistry of Life Authors: Wolfgang Kaim, Brigitte Schwederski **Publisher:** Wiley-VCH **Year:** 2017

Bioinorganic Chemistry: Theory and Practice Authors: K. K. Jain, P. P. Kumari **Publisher:** Springer **Year:** 2015

Biological Inorganic Chemistry: Structure and Reactivity Authors: W. Kaim, B. Schwederski **Publisher:** Wiley-VCH **Year:** 2006

Biomimetic Inorganic Chemistry Authors: Alan B. F. Martin, Y. Kim **Publisher:** Academic Press **Year:** 2010

Metal Ions in Biological Systems Editors: H. Sigel, A. Sigel **Publisher:** Marcel Dekker **Year:** 2003

Metal Ions in Biological Systems Editors: H. Sigel, A. Sigel **Publisher:** CRC Press **Year:** 2012

Metal Ions in Biological Systems: Volume 10 Editors: H. Sigel, A. Sigel **Publisher:** CRC Press **Year:** 2006

Metal Ions in Biological Systems: Volume 5 Editors: H. Sigel, A. Sigel **Publisher:** CRC Press **Year:** 1996

Metal Ions in Biological Systems: Volume 7 Editors: H. Sigel, A. Sigel **Publisher:** CRC Press **Year:** 2000

Metal Ions in Biological Systems: Volume 9 Editors: H. Sigel, A. Sigel **Publisher:** CRC Press **Year:** 2004

Metal Ions in Life Sciences: Volume 12 Editors: A. Sigel, H. Sigel **Publisher:** Springer **Year:** 2012

Metals in Cells Authors: N. R. K. Ramakrishnan, M. M. Rees **Publisher:** Wiley **Year:** 2021

Principles of Bioinorganic Chemistry Authors: R. J. P. Williams, R. W. Hay **Publisher:** Prentice Hall **Year:** 1996

INDEX

G

H

I

J

K

L

M

N

O

www.ingramcontent.com/pod-product-compliance
Lightning Source LLC
LaVergne TN
LVHW062117200926
848484LV00007B/1692
* 9 7 8 1 7 8 7 1 5 4 4 5 2 *